METHODS IN MOLECULAR BIOLOGY

Series Editor
John M. Walker
School of Life and Medical Sciences,
University of Hertfordshire, Hatfield,
Hertfordshire AL10 9AB, UK

For further volumes:
http://www.springer.com/series/7651

Plant Germline Development

Methods and Protocols

Edited by

Anja Schmidt

Centre for Organismal Studies, University of Heidelberg
Heidelberg, Germany

Humana Press

Editor
Anja Schmidt
Centre for Organismal Studies
University of Heidelberg
Heidelberg, Germany

ISSN 1064-3745 ISSN 1940-6029 (electronic)
Methods in Molecular Biology
ISBN 978-1-4939-7285-2 ISBN 978-1-4939-7286-9 (eBook)
DOI 10.1007/978-1-4939-7286-9

Library of Congress Control Number: 2017950215

Cover Illustration Caption: Female gametophytes and a male gametophyte.

Printed on acid-free paper

This Humana Press imprint is published by Springer Nature
The registered company is Springer Science+Business Media LLC
The registered company address is: 233 Spring Street, New York, NY 10013, U.S.A.

Preface

The plant reproductive lineage or germline is one of the two heteromorphic generations constituting the life cycle of higher plants. The plant male and female germline lineages have long been recognized as powerful model systems to study developmental processes and cell type specification. This is due to their simple organization of a few highly specialized cell types at maturity that are forming from single sporophytic cells through only a few divisions and subsequent cellularization. The mature female gametophyte typically consists of the two female gametes (egg and central cell), the synergids important for pollen tube reception, and the accessory antipodal cells. The mature male gametophyte harbors the male gametes, the sperm cells, enclosed in the vegetative cell of the pollen. Upon double fertilization, the male and female gametes fuse and seed development initiates. Apart from the great potential of the plant reproductive lineages to serve as models for developmental studies, investigations of the processes related to germline and seed development are also of utmost importance for applications with respect to the improvement of crop seed production.

The aim of this volume on *Plant Germline Development* in the *Methods in Molecular Biology* series is to provide protocols that are of interest to a broad readership interested in plant reproduction and germline formation. It includes detailed descriptions of common and many specialized methods to study various aspects of plant germline formation and targeted manipulation, including imaging and hybridization techniques to study cell type specification, cell lineage, signaling and hormones, cell cycle, and the cytoskeleton. In addition, cell type-specific methods for targeted ablation or isolation are provided, protocols to apply "omics" technologies and to perform bioinformatic data analysis, as well as methods relevant for aspects of biotechnology or plant breeding. This includes protocols that are relevant for the targeted manipulation of pathways, for crop plant transformation, or for conditional induction of phenotypes. While most protocols are suitable for investigations on the widely used plant model species *Arabidopsis thaliana*, also methods suited for the crop species rice, maize, and sorghum are included, as well as a protocol to study aspects of germline development in the tree *Quercus suber*.

All methods chapters follow the established format of the series of *Methods in Molecular Biology* and provide detailed protocols to be used in the daily research. The chapters open with an introduction to the subject and method described, comprise a detailed summary of materials and chemicals required to conduct the protocol, and describe the procedure of the method step by step. Importantly, in the "Notes" section, attention is given to clarify important details, to give advice on difficult to conduct steps, and to describe possible pitfalls.

This book is targeted for a broad readership interested in plant germline development, developmental genetics, crop manipulation, but also advances in cell or tissue type-specific "omics" technologies. Specifically, the book is subdivided into six major parts: (1) introductory chapters on plant germline formation and its potential for applications in crop improvement, (2) developmental studies targeting investigations on cell type specification, cell lineage, cell cycle, hormonal activity, and signaling, (3) special methods including chromosome spreading, whole mount in situ, and transmission electron microscopy, (4) cell type-specific isolations, (5) "omics" technologies and bioinformatics, and (6) methods relevant for targeted manipulation of pathways, plant breeding, and crop improvement.

I hope this book will serve as a comprehensive tool not only to study basic questions related to different aspects of plant reproductive development, but also for state-of-the-art methods, in addition to being a source of inspiration for new approaches and research questions in many laboratories.

Heidelberg, Germany ***Anja Schmidt***

Contents

Contributors

Jack Aidley • *Department of Genetics, University of Leicester, Leicester, UK*
Maria Isabel Amorim • *Departamento de Biologia, Faculdade de Ciências, Universidade do Porto, Porto, Portugal; Biosystems and Integrative Sciences Institute (BioISI), Porto, Portugal*
Sarah Anderson • *Department of Plant Biology, University of California, Davis, CA, USA*
Célia Baroux • *Department of Plant and Microbial Biology, University of Zurich, Zurich, Switzerland; Zurich-Basel Plant Science Center, University of Zurich, Zurich, Switzerland*
Martin Bayer • *Department of Cell Biology, Max Planck Institute for Developmental Biology, Tuebingen, Germany*
Jörg D. Becker • *Instituto Gulbenkian de Ciência, Oeiras, Portugal*
Marco Di Berardino • *Amphasys AG, Lucerne, Switzerland*
Frederic Berger • *Gregor Mendel Institute (GMI), Austrian Academy of Sciences, Vienna Biocenter (VBC), Vienna, Austria*
Cláudia Bispo • *Instituto Gulbenkian de Ciência, Oeiras, Portugal; UCSF Flow Cytometry Core Facility, San Francisco, CA, USA*
Andrea Bleckmann • *Cell Biology and Plant Biochemistry, Biochemie-Zentrum Regensburg, University of Regensburg, Regensburg, Germany*
Aurélien Boisson-Dernier • *Biocenter, Botanical Institute, University of Cologne, Cologne, Germany*
Lynette Brownfield • *Department of Biochemistry, University of Otago, Dunedin, New Zealand*
Patrick Bürgel • *Department of Cell Biology, Max Planck Institute for Developmental Biology, Tuebingen, Germany*
Murray Cadzow • *Department of Biochemistry, University of Otago, Dunedin, New Zealand*
Mark A. Chamberlin • *DuPont Pioneer, Johnston, IA, USA*
Antony M. Chettoor • *Department of Plant Biology, Carnegie Institution for Science, Stanford, CA, USA*
Sílvia Coimbra • *Departamento de Biologia, Faculdade de Ciências, Universidade do Porto, Porto, Portugal; Biosystems and Integrative Sciences Institute (BioISI), Porto, Portugal*
Joann A. Conner • *Horticulture Department, University of Georgia, Tifton, GA, USA*
Mário Luís da Costa • *Departamento de Biologia, Faculdade de Ciências, Universidade do Porto, Porto, Portugal; Biosystems and Integrative Sciences Institute (BioISI), Porto, Portugal*
Nico Dissmeyer • *Independent Junior Research Group on Protein Recognition and Degradation, Leibniz Institute of Plant Biochemistry (IPB), Halle (Saale), Germany*
Thomas Dresselhaus • *Cell Biology and Plant Biochemistry, Biochemie-Zentrum Regensburg, University of Regensburg, Regensburg, Germany*
Maria Englhart • *Cell Biology and Plant Biochemistry, Biochemie-Zentrum Regensburg, University of Regensburg, Regensburg, Germany*
Matthew M. S. Evans • *Department of Plant Biology, Carnegie Institution for Science, Stanford, CA, USA*

JAN FÍLA • *Laboratory of Pollen Biology, Institute of Experimental Botany AS CR, Prague, Czech Republic*
CHRISTINA MARIA FRANCK • *Biocenter, Botanical Institute, University of Cologne, Cologne, Germany*
DANNY GEELEN • *In Vitro Biology and Horticulture, Department of Plant Production, Faculty of Bioscience Engineering, University of Ghent, Ghent, Belgium*
XIAOPING GOU • *School of Life Sciences, Lanzhou University, Lanzhou, Gansu, China*
UELI GROSSNIKLAUS • *Department of Plant and Microbial Biology, University of Zurich, Zurich, Switzerland; Zurich-Basel Plant Science Center, University of Zurich, Zurich, Switzerland*
AFIF HEDHLY • *Department of Plant and Microbial Biology, University of Zurich, Zurich, Switzerland; Zurich-Basel Plant Science Center, University of Zurich, Zurich, Switzerland*
IRIS HEIDMANN • *Enza Zaden Research and Development B.V., Enkhuizen, The Netherlands; Acepo Seyndersloot, Enkhuizen, The Netherlands*
DAVID HONYS • *Laboratory of Pollen Biology, Institute of Experimental Botany AS CR, Prague, Czech Republic*
PENG-FEI JIA • *State Key Laboratory of Molecular Developmental Biology, Institute of Genetics and Developmental Biology, Chinese Academy of Sciences, Beijing, China*
DANIEL S. JONES • *Department of Microbiology and Plant Biology, University of Oklahoma, Norman, OK, USA*
HANS DE JONG • *Laboratory of Genetics, Wageningen University & Research, Wageningen, The Netherlands; Centre for Agricultural Biotechnology, Kasetsart University, Nakhon Pathom, Thailand*
LAKSANA KANTAMA • *Faculty of Liberal Arts and Science, Kasetsart University, Nakhon Pathom, Thailand*
TOMOKAZU KAWASHIMA • *Gregor Mendel Institute (GMI), Austrian Academy of Sciences, Vienna Biocenter (VBC), Vienna, Austria; Department of Plant and Soil Sciences, University of Kentucky, Lexington, KY, USA*
BURCU NUR KEÇELI • *In Vitro Biology and Horticulture, Department of Plant Production, Faculty of Bioscience Engineering, University of Ghent, Ghent, Belgium*
SHAI J. LAWIT • *DuPont Pioneer, Johnston, IA, USA*
HONG-JU LI • *State Key Laboratory of Molecular Developmental Biology, Institute of Genetics and Developmental Biology, Chinese Academy of Sciences, Beijing, China*
ANA LÚCIA LOPES • *Departamento de Biologia, Faculdade de Ciências, Universidade do Porto, Porto, Portugal; Biosystems and Integrative Sciences Institute (BioISI), Porto, Portugal*
GERMAN MARTINEZ • *Department of Plant Biology, Swedish University of Agricultural Sciences, Uppsala, Sweden; Linnean Center of Plant Biology, Uppsala, Sweden*
DAISUKE MARUYAMA • *Kihara Institute for Biological Research, Yokohama City University, Yokohama, Kanagawa, Japan*
TAKASHI OKAMOTO • *Department of Biological Sciences, Tokyo Metropolitan University, Tokyo, Japan*
PEGGY OZIAS-AKINS • *Horticulture Department and Institute of Plant Breeding, Genetics & Genomics, University of Georgia, Tifton, GA, USA*
BENJAMIN PETERS • *Department of Biochemistry, University of Otago, Dunedin, New Zealand*
HOLGER PUCHTA • *Botanical Institute II, Karlsruhe Institute of Technology, Karlsruhe, Germany*

WEIHONG QI • *Functional Genomics Center Zurich, Zurich, Switzerland*
HUBERT REHRAUER • *Functional Genomics Center Zurich, Zurich, Switzerland*
PAVEL REICHMAN • *Independent Junior Research Group on Protein Recognition and Degradation, Leibniz Institute of Plant Biochemistry (IPB), Halle (Saale), Germany*
SCOTT D. RUSSELL • *Department of Microbiology and Plant Biology, University of Oklahoma, Norman, OK, USA*
MÁRIO R. SANTOS • *Instituto Gulbenkian de Ciência, Oeiras, Portugal*
ETHEL MENDOCILLA SATO • *Department of Plant and Microbial Biology, University of Zurich, Zurich, Switzerland; Zurich-Basel Plant Science Center, University of Zurich, Zurich, Switzerland*
RALPH SCHLAPBACH • *Functional Genomics Center Zurich, Zurich, Switzerland*
MARC W. SCHMID • *Department of Evolutionary Biology and Environmental Studies, University of Zurich, Zürich, Switzerland; Department of Plant and Microbial Biology, University of Zurich, Zürich, Switzerland; URPP Global Change and Biodiversity, University of Zurich, Zürich, Switzerland; S3IT, University of Zurich, Zürich, Switzerland*
CARLA SCHMIDT • *Botanical Institute II, Karlsruhe Institute of Technology, Karlsruhe, Germany*
WENJING SHE • *Department of Plant and Microbial Biology, University of Zurich, Zurich, Switzerland; Zurich-Basel Plant Science Center, University of Zurich, Zurich, Switzerland*
DANIEL SLANE • *Department of Cell Biology, Max Planck Institute for Developmental Biology, Tuebingen, Germany*
LUCIJA ŠOLJIĆ • *Cell Biology and Plant Biochemistry, Biochemie-Zentrum Regensburg, University of Regensburg, Regensburg, Germany*
STEFANIE SPRUNCK • *Cell Biology and Plant Biochemistry, Biochemie-Zentrum Regensburg, University of Regensburg, Regensburg, Germany*
JEANNETTE STEINERT • *Botanical Institute II, Karlsruhe Institute of Technology, Karlsruhe, Germany*
NICO DE STORME • *In Vitro Biology and Horticulture, Department of Plant Production, Faculty of Bioscience Engineering, University of Ghent, Ghent, Belgium*
VENKATESAN SUNDARESAN • *Department of Plant Biology, University of California, Davis, CA, USA*
DAICHI SUSAKI • *Kihara Institute for Biological Research, Yokohama City University, Yokohama, Kanagawa, Japan*
DAVID TWELL • *Department of Genetics, University of Leicester, Leicester, UK*
HANNES VOGLER • *Department of Plant and Microbial Biology, University of Zurich, Zurich, Switzerland; Zurich-Basel Plant Science Center, University of Zurich, Zurich, Switzerland*
XINKUN WANG • *NUSeq Core Research Facility, Northwestern University, Evanston, IL, USA*
JENS WESTERMANN • *Biocenter, Botanical Institute, University of Cologne, Cologne, Germany*
ERIK WIJNKER • *Laboratory of Genetics, Wageningen University & Research, Wageningen, The Netherlands*
EMILY WU • *DuPont Pioneer Ag Biotech R & D, Johnston, IA, USA*
WEI-CAI YANG • *State Key Laboratory of Molecular Developmental Biology, Institute of Genetics and Developmental Biology, Chinese Academy of Sciences, Beijing, China*
RAMESH YELAGANDULA • *Gregor Mendel Institute (GMI), Austrian Academy of Sciences, Vienna Biocenter (VBC), Vienna, Austria*
ZUO-YU ZHAO • *DuPont Pioneer Ag Biotech R & D, Honolulu, HI, USA*

Part I

Plant Germline Development: From Investigation to Application

Chapter 1

An Introduction to Male Germline Development

Hannes Vogler, Ueli Grossniklaus, and Afif Hedhly

Abstract

In this introductory chapter, we describe male germline development in plants taking *Arabidopsis thaliana* as a reference species. We first describe the transition from sporophytic to germline development, then microsporogenesis including meiosis, followed by male gametophyte development prior to pollination, and finally the progamic phase culminating in double fertilization, which leads to the formation of the embryo and the endosperm. For detailed information on some of these processes or on the molecular underpinning of certain fate transitions, we refer the reader to recent reviews. An important but often neglected aspect of male gametophyte development is the formation of the unique pollen cell wall. In contrast to that of other plant cells, the pollen cell wall is composed of two principal layers, the intine and exine. While the intine, the inner pecto-cellulosic cell wall layer, is biochemically and structurally similar to a "classical" plant cell wall, the exine is a unique composite with sporopollenin as its main component. Biosynthesis of the cell wall is remarkably similar between the spores of mosses and ferns, and pollen of seed plants, although slight differences exist, even between closely related species (reviewed in Wallace et al., AoB Plants 2011:plr027, 2011). In the latter sections of this chapter, we will present a brief overview of cell wall development in *Arabidopsis* pollen, where this aspect has been intensively studied.

Key words Male germline, Pollen, Microsporogenesis, Meiosis, Lineage model, Cluster model, Cell wall, Intine, Exine, Harmomegathy

1 Sporophyte-Germline Transition

Plant germline development, be it male or female, is initiated within the reproductive organs of the flower from one to few sporophytic cells denominated archesporial cells (collectively known as archesporia). This transition from a sporophytic to an archesprorial fate occurs late during the plant life cycle after many cell divisions of the undifferentiated cells in the shoot apical meristem to produce the above ground body of the plant. While anthers contain cells from all three floral meristems layers (L1, L2, and L3), sub-epidermal wall layers and sporogenous cells are only derived from L2. Based on the widely accepted lineage model for the transition from sporophytic to male germline development in *Arabidopsis* [1, 2], there is a single, prominent, sub-epidermal male archesporial cell in each of

Anja Schmidt (ed.), *Plant Germline Development: Methods and Protocols*, Methods in Molecular Biology, vol. 1669, DOI 10.1007/978-1-4939-7286-9_1, © Springer Science+Business Media LLC 2017

the anther microsporangia, from which all three sub-epidermal anther wall layers and the microspore mother cells (MiMCs) are derived (Fig. 1a). Thus, while the divisions in the epidermis are all anticlinal, a concerted sequence of periclinal followed by anticlinal divisions occur in the archespore and its daughter cells. The first periclinal division of the archesporial cell produces the primary parietal cell (outer cell) and the primary sporogenous cell (inner cell). This division is followed by anticlinal divisions of both the cells, leading to the formation of the primary parietal layer and the primary sporogenous layer (Fig. 1a). A new wave of periclinal divisions in both the layers then leads to the formation—from the outermost to the innermost layer—of the endothecium, the secondary parietal layer, and secondary sporogenous cells (Fig. 1a, b). Further periclinal divisions within the secondary parietal layer subsequently form the middle layer (outer) and the tapetum (inner) (Fig. 1a, c). After secondary sporogenous cells cease to divide, they

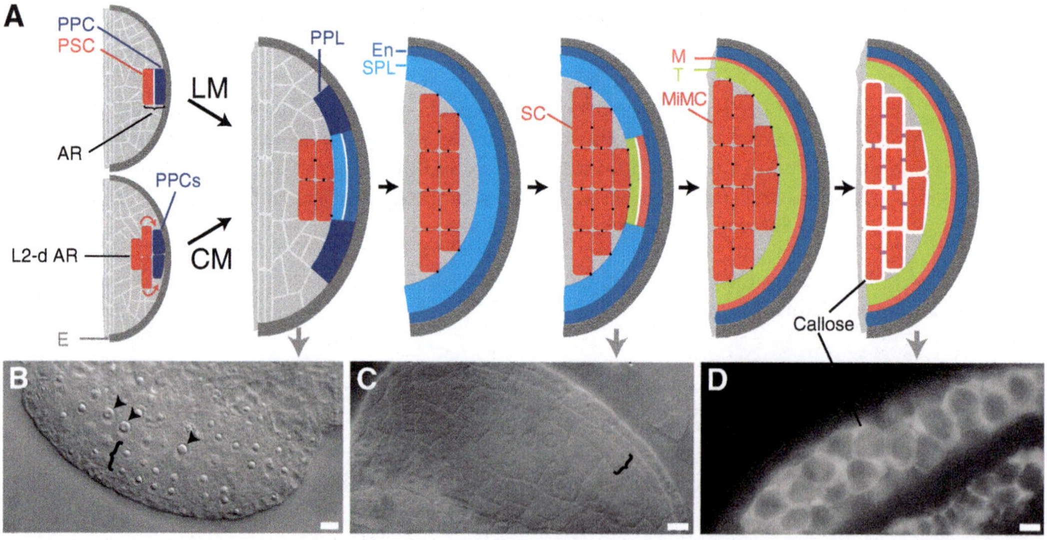

Fig. 1 Models of pre-meiotic anther development and formation of microsporocytes. (**a**) Schematic representation of the lineage (LM) and cluster (CM) models of early anther development. In the LM, division of the sub-epidermal archesporial cell (AR) gives rise to the primary parietal cell (PPC) and the primary sporogenous cell (PSC). In the CM, first, L2-derived ARs are specified deep beneath the epidermis (E). Subsequently, they possibly trigger undifferentiated, sub-epidermal L2 cells to become PPCs (*red, curved arrows*). The subsequent divisions are similar in both the models. Periclinal divisions in the primary parietal layer (PPL) give rise to the endothecium (En) and the secondary parietal layer (SPL). Periclinal divisions in the SPL give rise to the middle layer (M) and the tapetum (T). Before the onset or at very early stages of meiosis, the MiMCs are encased by a thick callose wall. Plasmodesmata (*black small dashes* connecting sporogenous cells [SCs]) and cytomictic channels (*purple dashes* within callose) are illustrated. (**b**) DIC image of a cleared anther representing periclinal divisions within the PPL (*brace*). Dividing SCs show prominent nucleoli (*arrowheads*). (**c**) DIC image of a cleared anther representing periclinal divisions within the SPL (*brace*). (**d**) Aniline blue staining shows the individual MiMCs surrounded by a thick callose wall. Size bars = 10 μm

differentiate into meiotically competent cells called MiMCs (also known as pollen mother cells, microsporocytes, or meiocytes).

It should be noted that in some publications archesporial cells and MiMCs are used interchangeably. Furthermore, because of the highly contrasting germline development in plants compared with most animals, which cell or group of cells should be considered the beginning of germline development in plants is under debate [3–5]. In most multicellular animals, diploid germline cells are set aside early during development to undergo meiosis at a later stage, with the meiotic products giving directly rise to gametes [6]. However, male archesporial cells and MiMCs in plants lead to the formation of accessory somatic cells besides gametic cells (sperm cells). Thus, some authors place germline initiation after the last accessory (or vegetative) cell was separated from the exclusively gamete-producing (or generative) cell within the male gametophyte [4]. However, a lot of evidence points to the earlier formation of the pre-meiotic archesporial cell as the initial specification of the cell fate for gamete production [3, 5, 6]. Although the above delineated lineage model is widely accepted for early anther development in *Arabidopsis*, there is only limited evidence for the existence of a single founder archesporial cell in this species [1]. Cellular markers are needed to provide a definitive answer to this question. Although the existence of a single archesporial cell has indeed been documented during the first half of the twentieth century in a variety of plant species, the difficulty to observe such a sub-epidermal cell in some of these species, or the existence of an L2-derived group of prominent archesporial cells deep inside the microsporangium in other plant species, was also reported ([7] and references therein). A recent comprehensive characterization of early anther development in maize, using 3D reconstruction from confocal microscopy combined to in vivo EdU-labeling (EdU is 5-ethynyl-2′-deoxyuridine), allowed the authors to refute most of the assumptive models for plant organ (intercalary meristem and exponential models) and anther (lineage or wave form models) development. The occurrence of local clustering of developmental events, where the middle layer and the tapetum seem to form a developmental unit, leads to the proposal of a cluster model for anther wall layer formation (Fig. 1a; for an in-depth discussion of these developmental models refer to [8]).

Mutant analyses in different plant species identified some key players during early germline development, such as *SPOROCYTELESS* and *EXCESS MICROSPOROCYTES1* (for further details *see* [2, 9–11]). Interestingly, redox-sensitive mechanisms known to mediate plant stress responses also appear to play a role (via glutaredoxins) in archesporial cell fate acquisition in different plant species [12–15]. In summary, this is the least well-characterized stage of male germline development and there are many unanswered questions that warrant further research: (1) contrasting the cluster model of the sporophyte-germline transition against the lineage model in different plant species; (2) characterizing the order and synchrony of cell divisions within the inner three anther wall layers; (3) characterizing the order and number of cell divisions within sporogenous cells; and (4) demonstrating the existence and the nature of possible morphogenetic signaling molecules and their interplay with the redox status.

Different classical microscopic techniques were used to characterize early anther development in diverse plant species (e.g., [16–19]). Recent advances in computer processing power and storage capacity, in fluorophore technology, and in using genetically engineered marker lines allowed wide-field and confocal microscopy to become standard techniques. Although classical ultrastructural and morphological microscopy techniques can be very useful, improvement in resolution of whole-mountWhole-mount staining specimen processing techniques makes such procedures much more efficient and informative, and should provide a definitive answer to how archesporial cell identity is determined and radial symmetry established within the anther. Thus, promising results can be expected from using state-of-the-art 3D microscopic analyses. Also, developing novel cell identity markers for archesporial cells, as well as for the different anther wall layers and sporogenous cells, will be very helpful in this endeavor.

2 Meiosis

The meiotic process in male floral organs has been well characterized in different plant species because of the high number of synchronously dividing microsporocytes that are easily accessible. Unlike in many animals, where meiosis can occupy a significant proportion of germline development, meiosis in most plant species is a relatively synchronous process that occurs in a comparably very short time. Exceptions are plant species that overwinter after flower bud initiation, in which germline development is arrested at different stages, often during meiosis, until bud dormancy is broken ([20] and references therein).

While dividing, sporogenous cells are connected by plasmodesmata across a thin cell wall at their contact zone [21]. MiMC differentiation is characterized by the deposition of callose surrounding each cell (Fig. 1a, d), the rupture of these plasmodesmata, and the isolation of the MiMC from the sporophytic maternal tissues [22]. Soon afterward, at the pre-leptotene stage, larger channels known as cytomictic channels develop between meiocytes and, in some species, between meiocytes and tapetal cells [21]. These cytomictic channels are reminiscent of protoplasmic bridges during animal spermiogenesis [23, 24] or the ring canals in the egg chambers of *Drosophila melanogaster* [25]. Their large diameter—up to

1750 nm as compared with 80 nm for plasmodesmata—allows not only the exchange of small molecules and ions but also of larger structures. Nuclei, nuclear material, chromosomes, plastids, and mitochondria were indeed observed to transit through these channels [21, 26, 27]. Since their discovery a century ago [28] interpretations of their possible function have changed. Cytomictic channels were frequently observed in genetically unbalanced plants such as hybrids, mutants, aneuploids, and polyploids, and were considered to be a pathological manifestation aimed at adjusting and balancing genomic defects. However, their occurrence during normal spermiogenesis and oogenesis in animals as well as in "balanced" plants reinforced the view of a non-pathological structure involved in a developmental process aimed at synchronizing meiosis [21, 24] or in a general recycling process during the transition from sporophytic to germline fate [27].

Meiosis in plants can be of the simultaneous type, such as in *Arabidopsis*, where there is no proper dyad formation, and where cytokinesis occurs after meiosis II. In other plant species, cytokinesis with cell wall deposition occurs after meiosis I (formation of dyad) and meiosis II (formation of tetrad). These different types of cytokinesis generate different types of tetrad arrangement. Callose deposition also appears to play an important role in tetrad formation (Fig. 2a, b). After the completion of telophase II in *Arabidopsis*, the four haploid nuclei of the coenocytic microsporocytes migrate to the periphery of the cell, where they are arranged in a nearly tetrahedral configuration. Protrusions from the thick callose wall surrounding the microsporocytes start to grow centripetally inward. They form constriction furrows between the nuclei, which meet in the center, thereby isolating the four microspores from each other to form the tetrad (Fig. 2b; [17]). At this stage, every microspore is completely enclosed by the freshly synthesized callose layer, and the tetrad is still encased by the original microsporocytic callose wall. At the same time tapetal cells, which are required for the normal development of the microspores, loose the cell walls facing the locule as well as their radial cell walls. Interestingly, callose deposition may also extend to these regions (e.g., [29]). The function of the callose wall is not completely clear; however, there is good evidence that it prevents cell cohesion between developing microspores. Species that disperse their pollen as tetrads often have reduced or missing callose cell walls [30–32]. Therefore, the callose wall may be involved in proper separation of individual microspores within the tetrad. The *quartet* mutant of *Arabidopsis*, in which pollen remains in the tetrad configuration, has also a reduced callose layer [33], further supporting this view. In addition, because microspores start to develop the primexine while still surrounded by callose [34], callose was shown to be necessary but not sufficient for the regular deposition of exine structures [35, 36]. A pattern for future exine deposition is thought to be

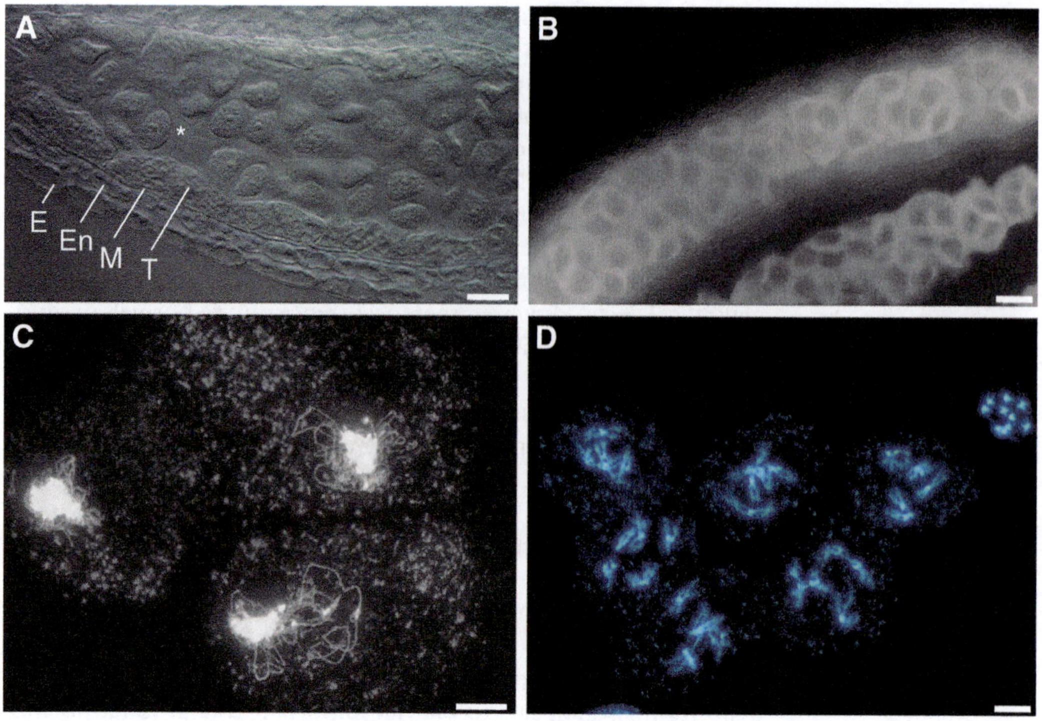

Fig. 2 Snapshots of microspore development in *Arabidopsis* during meiosis. (**a**) DIC image of a cleared anther at the tetrad stage. A single locule containing pollen tetrads surrounded with callose (*asterisk*) is shown. The four layers of sporophytic tissue are well visible in the lower part of the image. (**b**) A thick callose layer surrounds the tetrads as well as the individual microspores. (**c**) Microsporocytes at the zygotene stage of prophase I. (**d**) During diakinesis the chromosomes are further condensed and the pairing of the homologous chromosomes is clearly visible. Herr cleared (**a**) or aniline blue stained (**b**) anthers, and DAPI stained chromosome spreads (**c**, **d**). *E* epidermis, *En* endothecium, *M* middle layer, *T* tapetum. Size bars = 10 μm

imprinted on the still united microspores. The factors that are involved are not yet identified, but it has been suggested that subepidermal accumulations of endoplasmic reticulum form an "apertural shield," a region that is left blank from the later deposition of sporopollenin and forms the apertures [37]. Sporopollenin precursors are incorporated at specific positions of the primexine. The biopolymer sporopollenin is a mixture of long-chain, unbranched, polyhydroxylated fatty acids, oxygenated aromatic rings, and phenylpropanoids [38–40], forming a chemically inert structure. With respect to the multilayered cell wall and the production of sporopollenin, other durable cells, such as the spores of mosses and ferns, share remarkable similarities with the pollen cell wall [41].

Electron microscopy and resin or paraffin embedding-sectioning-staining techniques were pioneering techniques that allowed the uncovering of cellular, structural, and anatomical features of meiosis in plants (e.g., [17]). The development of chromosome spreading and staining techniques using hematoxylin, silver, DAPI, or propidium iodide (Fig. 2c, d; e.g., [42, 43]) led to a better understanding of chromosomal pairing, recombination, and chromosome segregation. These techniques are sometimes combined with immunolocalization of either the meiotic spindle and chromatin-associated proteins [43, 44], chromosomal fluorescent in situ hybridization [44, 45], or BrdU-labeling (BrDU is 5-bromo-2′-desoxyuridine; [46]). Although meiotic mutant collections were first generated in cultivated plant species decades ago (e.g., [47]), the tremendous progress in recent years is attributable to molecular genetic analyses in the model species *Arabidopsis*, maize, and rice in combination with the cytogenetic techniques described above (for a comprehensive list of known genes required for meiosis, *see* [48, 49]). As stated above, much more progress can be expected from taking advantage of "omics" studies and state-of-art 3D microscopy techniques. Characterizing the meiotic process is not only of interest for understanding the basis of heredity, but also to characterize developmental variations of reproduction, such as those leading to apomixis [50, 51] and polyploidization [52, 53]. Unraveling these developmental variations would not only provide insights into the developmental versatility of plant reproduction and its response to environmental fluctuations but is also of tremendous interest for seed production and plant breeding [54, 55].

3 Male Gametophyte Development and Function

The digestion of callose, and the subsequent release of microspores, marks the end of sporogenesis and the beginning of the male haploid phase of the plant life cycle. Plant microspores will give rise to a multicellular organism—the male gametophyte—and undergo microgametogenesis to produce the male gametes, i.e., the sperm cells. While in lower land plants, such as bryophytes and ferns, gametophytes are free-living, autonomous organisms, they are drastically reduced in flowering plants. The mature male gametophyte of all flowering plants consists of just three cells, a vegetative cell harboring two sperm cells, and its development is dependent on the sporophyte. The male gametophyte matures within the locules of the anther into pollen grains that are subsequently let free at anther dehiscence. Upon deposition on a receptive stigma, the pollen grain germinates and forms a tube that grows heterotrophically through the stigmatic, stylar, and ovarian tissues carrying two sperm cells as cargo. Growth through these tissues involves a multitude of signaling events between the pollen tube and female tissues (reviewed in [56]). After reaching the female gametophyte within the ovule, pollen tube reception by the receptive synergid occurs, followed by sperm discharge into this synergid [57, 58]. One of the sperm cells then fuses with the egg cell to form the zygote, while the second sperm cell fuses with

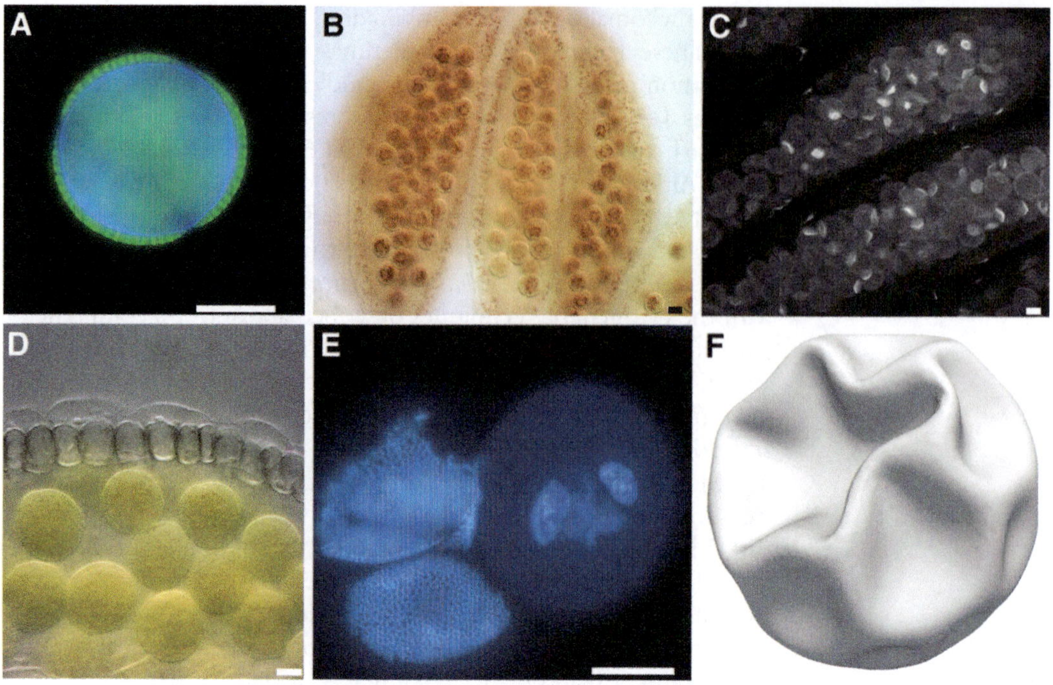

Fig. 3 Pollen cell wall formation. (**a**) Mature pollen grain showing intine (blue fluorescence, calcofluor staining) and exine (green fluorescence, auramine staining). (**b**) Bicellular pollen stage. Pollen grains and the endothecium show a conspicuous starch accumulation. (**c**) Callose deposition between the vegetative and the newly formed generative cell at the early bicellular stage. (**d**) DIC image of maturing tricellular pollen grains. (**e**) DAPI staining of a tricellular pollen grain. The vegetative nucleus is flanked by the two sperm cells. The exine has been stripped away by gently squeezing the pollen grain leaving the protoplast surrounded by the intine. (**f**) FEM model of a homogenous, thin spherical shell illustrating the process of harmomegathy. Image shown in (**f**) courtesy of Dr. G. Munglani. Size bars = 10 μm

the central cell to form the endosperm, an embryo-nourishing tissue in the developing seed.

In contrast to the stages of sporogenesis prior to meiosis, released microspores are physically separated from sporophytic tissues (except from early direct contact with tapetal cells) and are fed by the locular fluid. Major events accompanying microspore maturation into the pollen grain are the completion of exine development, the formation of the intine (Fig. 3a), one or two mitotic cycles, the accumulation of carbohydrate reserves (Fig. 3b; [59]), and the progressive dehydration in a species-specific manner (reviewed in [60]). The first pollen mitosis (PM I) is asymmetric (Fig. 3c), with the bigger cell (vegetative cell) subsequently engulfing the smaller one (the generative cell). Only the generative cell then undergoes a second, symmetric division (PM II), either shortly after PM I before pollen grain maturation and dehiscence (tricellular pollen grains occur in about one third of angiosperms; Fig. 3d, e) or shortly after migrating into the pollen tube following pollen grain shedding and germination (bicellular pollen grains in most angiosperms and all extant non-flowering seed plants [61]).

During maturation, the pollen cytoplasm dehydrates in a controlled manner, thereby reducing the weight and metabolic rate, leaving the pollen grain in a dormant state. This is necessary because, upon release from the anther, pollen leaves its protective environment and is exposed to air and sunlight. Even at moderate relative humidity, air has a strongly negative water potential, which could lead to rapid desiccation [62]. The exine is a marvelous structure, optimized to allow the dispersal of the pollen—sometimes over very long distances—and to protect it from desiccation. Dehydration leads to a process called harmomegathy [63], a characteristic infolding of the pollen grain (visualized in Fig. 3f), which usually happens at the apertures, where the exine is lacking and the intine is normally thickened [64], accompanied by a simultaneous reduction of cell volume. Shape and position of the apertures have a great influence on the folding pattern. Upon dehydration, elongated apertures (referred to as colpi) invaginate and the opposite margins approach each other to seal the aperture off, whereas pollen without elongated apertures tend to form depressions [65]. In heterocolpate pollen, the tasks of harmomegathy and germination are performed by different types of colpi. Compound apertures (colpori) possess endoapertures with a thickened cell wall correlating with the functions of the intine during germination. Pseudoapertures, on the other hand, lack endoapertures, which renders them useless for pollen tube germination but provide more intine flexibility for harmomegathy [66]. This functional diversification of the pollen exine can be seen as an evolutionary progression because heterocolpate pollen are considered to be both more efficient in harmomegathy and more successful in germination [67].

Once the pollen grain lands on a wet stigma or receives water from a stigmatic cell of a dry stigma, harmomegathy is reversed, and the pollen grain rehydrates and unfolds. Successful pollen tube germination depends on rapid water uptake once in contact with the stigma, with the apertures being the main structures for water uptake. Pollen tubes emerge from one of the apertures or directly through the exine and make their way through the extracellular space between papillar cells, or invade the papillar extracellular matrix before entering the style [68]. Within the style, pollen tubes grow through the transmitting tract in species with a solid style, or the stylar canal in species with a hollow style. Both the cases provide an environment that is enriched in extracellular nutrients, which fuel amazingly high pollen tube growth rates. Different guidance cues have been suggested to direct pollen tubes during their growth through female tissues toward the ovules. Finally, an individual pollen tube is directed to an ovule, enters through the micropyleMicropylar where it is received by the receptive synergid into which the sperm cells are discharged [56, 57, 69, 70]. To ensure successful fertilization, pollen tubes elongate extremely fast and growth is restricted to their tip. This implies high tensile stress on the cell wall induced by turgor pressure. Precise control of cell wall deposition and remodeling is crucial to resist this stress and to maintain the shape of the pollen tube during growth. Self-similar growth requires a non-expandable shank and a relatively soft apical region where cell expansion can occur. New cell wall material, mainly methyl-esterified pectins and crystalline cellulose, is constantly added to the tip of the growing pollen tube. A few micrometers away from the pollen tube tip, in the transition zone between the spherical tip and the cylindrical shank, pectins become de-esterified by pectin methylesterases and gelled by calcium crosslinking. In the same region callose is added to the cell wall, which leads to further reinforcement of the shank [71]. An *Arabidopsis* pollen tube model that simulates self-similar growth predicts a significant change in the mechanical properties of the cell wall in this transition region [72]. However, to maintain the diameter of the pollen tube in that model constant, i.e., to prevent expansion of the shank, it was necessary to set the elasticity (Young's modulus) of the shank to 4 GPa, which is an order of magnitude higher than the Young's modulus determined for the lily pollen tube cell wall (around 100 MPa) based on micro- and soft indentation experiments [73, 74]. Taking strain stiffening, a nonlinear increase in stress with increasing deformation into account would alter the Young's modulus to more accurate values. The recent development of specific microfluidic devices (e.g., [75–79]) will allow a more detailed investigation of the interplay between cytomechanics and cell physiology of pollen tubes.

Male gametophyte development is a field of vibrant research, facilitated by a combination of traditional, well-established methods and the continuous development of new approaches. Apart from refined microscopy techniques, the increasing availability of sensor lines and dyes for characterizing physiological parameters, and more and more sensitive "omics" methods, the involvement of engineering and mathematical modeling led to a rapid increase in rheological and mechanical studies in the recent past [80]. This book will cover some of the state-of-the-art methods and techniques in modern pollen research.

References

1. Canales C, Bhatt AM, Scott R et al (2002) EXS, a putative LRR receptor kinase, regulates male germline cell number and tapetal identity and promotes seed development in *Arabidopsis*. Curr Biol 12:1718–1727
2. Scott RJ, Spielman M, Dickinson HG (2004) Stamen structure and function. Plant Cell 16 (Suppl S46):560
3. Grossniklaus U (2011) Plant germline development: a tale of cross-talk, signaling, and cellular interactions. Sex Plant Reprod 24:91–95
4. Twell D (2011) Male gametogenesis and germline specification in flowering plants. Sex Plant Reprod 24:149–160
5. Schmidt A, Schmid MW, Grossniklaus U (2015) Plant germline formation: common

concepts and developmental flexibility in sexual and asexual reproduction. Development 142:229–241

6. Dickinson HG, Grant-Downton R (2009) Bridging the generation gap: flowering plant gametophytes and animal germlines reveal unexpected similarities. Biol Rev Camb Philos Soc 84:589–615
7. Maheshwari P (1950) An introduction to the embryology of angiosperms. McGraw-Hill, New York, NY
8. Kelliher T, Walbot V (2011) Emergence and patterning of the five cell types of the *Zea mays* anther locule. Dev Biol 350:32–49
9. Ma H (2005) Molecular genetic analyses of microsporogenesis and microgametogenesis in flowering plants. Annu Rev Plant Biol 56:393–434
10. Chang F, Wang Y, Wang S et al (2011) Molecular control of microsporogenesis in *Arabidopsis*. Curr Opin Plant Biol 14:66–73
11. Kelliher T, Egger RL, Zhang H et al (2014) Unresolved issues in pre-meiotic anther development. Front Plant Sci 5:347
12. Chaubal R, Anderson JR, Trimnell MR et al (2003) The transformation of anthers in the *msca1* mutant of maize. Planta 216:778–788
13. Xing S, Zachgo S (2008) *ROXY1* and *ROXY2*, two *Arabidopsis* glutaredoxin genes, are required for anther development. Plant J 53:790–801
14. Hong L, Tang D, Zhu K et al (2012) Somatic and reproductive cell development in rice anther is regulated by a putative glutaredoxin. Plant Cell 24:577–588
15. Kelliher T, Walbot V (2012) Hypoxia triggers meiotic fate acquisition in maize. Science 337:345–348
16. Hill JP, Lord EM (1989) Floral development in *Arabidopsis thaliana*: a comparison of the wild type and the homeotic *pistillata* mutant. Can J Bot 67:2922–2936
17. Owen HA, Makaroff CA (1995) Ultrastructure of microsporogenesis and microgametogenesis in *Arabidopsis thaliana* (L.) Heynh. ecotype Wassilewskija (Brassicaceae). Protoplasma 185:7–21
18. Zajac K (1997) Ultrastructural study of maturing pollen in *Arabidopsis thaliana* (L.) Heynh. (*Brassicaceae*). Acta Soc Bot Pol Pol Tow Bot 66:125–131
19. Yamamoto Y, Nishimura M, Hara-Nishimura I et al (2003) Behavior of vacuoles during microspore and pollen development in *Arabidopsis thaliana*. Plant Cell Physiol 44:1192–1201
20. Julian C, Herrero M, Rodrigo J (2014) Anther meiosis time is related to winter cold temperatures in apricot (*Prunus armeniaca* L.) Environ Exp Bot 100:20–25
21. Heslop-Harrison J (1966) Cytoplasmic connexions between angiosperm meiocytes. Ann Bot 30:221–222
22. Blackmore S, Wortley AH, Skvarla JJ et al (2007) Pollen wall development in flowering plants. New Phytol 174:483–498
23. Burgos MH, Fawcett DW (1955) Studies on the fine structure of the mammalian testis I. Differentiation of the spermatids in the cat (*Felis domestica*). J Biophys Biochem Cytol 1:287–300
24. Fawcett DW, Ito S, Slautterback D (1959) The occurrence of intercellular bridges in groups of cells exhibiting synchronous differentiation. J Biophys Biochem Cytol 5:453–460
25. King RC (1970) Ovarian development in *Drosophila melanogaster*. Academic, New York, NY
26. Feijó JA, Pais MSS (1988) Ultrastructural modifications of plastids and starch metabolism during the microsporogenesis of *Ophrys lutea* (Orchidaceae). Ann Bot 61:215–219
27. Feijó JA, Pais MSS (1989) Cytomixis in meiosis during the microsporogenesis of *Ophrys lutea*: an ultrastructural study. Caryologia 42:37–48
28. Arnoldy W (1900) Beiträge zur Morphologie der Gymnospermen. IV. Was sind die "Keimbläschen" oder "Hofmeisters-Körperchen". Eizelle Abietin Flora 87:194–204
29. Dawson J, Wilson ZA, Aarts MGM et al (1993) Microspore and pollen development in six male-sterile mutants of *Arabidopsis thaliana*. Can J Bot 71:629–638
30. Meyer NR, Yaroshevskaya AS (1976) Phylogenetic significance of the development of pollen grain walls in Liliaceae, Juncaceae and Cyperaceae. In: Ferguson IK, Muller J (eds) The evolutionary significance of the exine. Academic, London
31. Sampson FB (1977) Pollen tetrads of *Hedycarya arborea* (Monimiaceae). Grana 16:61–73
32. McGlone MS (1978) Pollen structure of the New Zealand members of the Styphelieae (Epacridaceae). N Z J Bot 16:91–101
33. Preuss D, Lemieux B, Yen G et al (1993) A conditional sterile mutation eliminates surface components from *Arabidopsis* pollen and disrupts cell signaling during fertilization. Genes Dev 7:974–985
34. Skvarla JJ, Larson DA (1966) Fine structural studies of Zea mays pollen I: cell membranes and exine ontogeny. Am J Bot 53:1112–1125

35. Paxson-Sowders DM, Owen HA, Makaroff CA (1997) A comparative ultrastructural analysis of exine pattern development in wild-type *Arabidopsis* and a mutant defective in pattern formation. Protoplasma 198:53–65
36. Dong X, Hong Z, Sivaramakrishnan M et al (2005) Callose synthase (CalS5) is required for exine formation during microgametogenesis and for pollen viability in *Arabidopsis*. Plant J 42:315–328
37. Heslop-Harrison J (1968) Pollen Wall Development. Science 161:230–237
38. Ahlers F, Thom I, Lambert J et al (1999) ^{1}H NMR analysis of sporopollenin from *Typha angustifolia*. Phytochemistry 50:1095–1098
39. Domínguez E, Mercado JA, Quesada MA et al (1999) Pollen sporopollenin: degradation and structural elucidation. Sex Plant Reprod 12:171–178
40. Kim SS, Douglas CJ (2013) Sporopollenin monomer biosynthesis in *Arabidopsis*. J Plant Biol 56:1–6
41. Wallace S, Fleming A, Wellman CH et al (2011) Evolutionary development of the plant and spore wall. AoB Plants 2011:plr027
42. Ross KJ, Fransz P, Jones GH (1996) A light microscopic atlas of meiosis in *Arabidopsis thaliana*. Chromosome Res 4:507–516
43. Zhou S, Wang Y, Li W et al (2011) Pollen *semi-sterility1* encodes a kinesin-1-like protein important for male meiosis, anther dehiscence, and fertility in rice. Plant Cell 23:111–129
44. Armstrong SJ, Caryl AP, Jones GH et al (2002) Asy1, a protein required for meiotic chromosome synapsis, localizes to axis-associated chromatin in *Arabidopsis* and *Brassica*. J Cell Sci 115:3645–3655
45. de Jong JH, Fransz P, Zabel P (1999) High resolution FISH in plants - techniques and applications. Trends Plant Sci 4:258–263
46. Armstrong SJ, Franklin FC, Jones GH (2001) Nucleolus-associated telomere clustering and pairing precede meiotic chromosome synapsis in *Arabidopsis thaliana*. J Cell Sci 114:4207–4217
47. Golubovskaya IN (1979) Genetic control of meiosis. Int Rev Cytol 58:247–290
48. Mercier R, Grelon M (2008) Meiosis in plants: ten years of gene discovery. Cytogenet Genome Res 120:281–290
49. Mercier R, Mézard C, Jenczewski E et al (2015) The molecular biology of meiosis in plants. Annu Rev Plant Biol 66:297–327
50. Koltunow A, Grossniklaus U (2003) Apomixis: a developmental perspective. Annu Rev Plant Biol 54:547–574
51. Wang CJ, Tseng CC (2014) Recent advances in understanding of meiosis initiation and the apomictic pathway in plants. Front Plant Sci 5:497
52. Bretagnolle F, Thompson JD (1995) Gametes with the somatic chromosome number: mechanisms of their formation and role in the evolution of autopolyploid plants. New Phytol 129:1–22
53. Brownfield L, Köhler C (2011) Unreduced gamete formation in plants: mechanisms and prospects. J Exp Bot 62:1659–1668
54. Spillane C, Curtis MD, Grossniklaus U (2004) Apomixis technology development-virgin births in farmers' fields. Nat Biotechnol 22:687–691
55. Sattler MC, Carvalho CR, Clarindo WR (2016) The polyploidy and its key role in plant breeding. Planta 243:281–296
56. Dresselhaus T, Franklin-Tong N (2013) Male-female crosstalk during pollen germination, tube growth and guidance, and double fertilization. Mol Plant 6:1018–1036
57. Kessler SA, Grossniklaus U (2011) She's the boss: signaling in pollen tube reception. Curr Opin Plant Biol 14:622–627
58. Li H, Yang WC (2016) RLKs orchestrate the signaling in plant male-female interaction. Sci China Life Sci 59:867–877
59. Hedhly A, Vogler H, Schmid MW et al (2016) Starch turnover and metabolism during flower and early embryo development. Plant Physiol 172:2388–2402
60. Pacini E, Hesse M (2004) Cytophysiology of pollen presentation and dispersal. Flora 199:273–285
61. Lora J, Herrero M, Hormaza JI (2009) The coexistence of bicellular and tricellular pollen in *Annona cherimola* (Annonaceae): Implications for pollen evolution. Am J Bot 96:802–808
62. Bolick MR (1981) Mechanics as an aid to interpreting pollen structure and function. Rev Palaeobot Palynol 35:61–79
63. Wodehouse RP (1935) Pollen grains: their structure, identification, and significance in science and medicine. McGraw-Hill, New York, NY
64. Heslop-Harrison J (1979) An interpretation of the hydrodynamics of pollen. Am J Bot 66:737
65. Katifori E, Alben S, Cerda E et al (2010) Foldable structures and the natural design of pollen grains. Proc Natl Acad Sci U S A 107:7635–7639
66. Volkova OA, Severova EE, Polevova SV (2013) Structural basis of harmomegathy: evidence

from Boraginaceae pollen. Plant Syst Evol 299:1769–1779

67. Müller J (1979) Form and function in angiosperm pollen. Ann Mo Bot Gard 66:593–632
68. Edlund AF, Zheng Q, Lowe N et al (2016) Pollen from *Arabidopsis thaliana* and other Brassicaceae are functionally omniaperturate. Am J Bot 103:1006–1019
69. Higashiyama T, Takeuchi H (2015) The mechanism and key molecules involved in pollen tube guidance. Annu Rev Plant Biol 66:393–413
70. Vogler H, Martinez-Bernardini A, Grossniklaus U (2016) Maybe she's NOT the boss: male-female crosstalk during sexual plant reproduction. Genome Biol 17:96
71. Chebli Y, Kaneda M, Zerzour R et al (2012) The cell wall of the *Arabidopsis* pollen tube--spatial distribution, recycling, and network formation of polysaccharides. Plant Physiol 160:1940–1955
72. Fayant P, Girlanda O, Chebli Y et al (2010) Finite element model of polar growth in pollen tubes. Plant Cell 22:2579–2593
73. Vogler H, Draeger C, Weber A et al (2013) The pollen tube: a soft shell with a hard core. Plant J 73:617–627
74. Hu C, Munglani G, Vogler H et al (2017) Characterization of size-dependent mechanical properties of tip-growing cells using a lab-on-chip device. Lab Chip 17:82–90
75. Yetisen AK, Jiang L, Cooper JR et al (2011) A microsystem-based assay for studying pollen tube guidance in plant reproduction. J Micromech Microeng 21:054018
76. Agudelo CG, Sanati Nezhad A, Ghanbari M et al (2013) TipChip: a modular, MEMS-based platform for experimentation and phenotyping of tip-growing cells. Plant J 73:1057–1068
77. Sanati Nezhad A, Naghavi M, Packirisamy M et al (2013) Quantification of cellular penetrative forces using lab-on-a-chip technology and finite element modeling. Proc Natl Acad Sci U S A 110:8093–8098
78. Horade M, Yanagisawa N, Mizuta Y et al (2014) Growth assay of individual pollen tubes arrayed by microchannel device. Microelectron Eng 118:25–28
79. Shamsudhin N, Laeubli N, Atakan HB et al (2016) Massively parallelized pollen tube guidance and mechanical measurements on a lab-on-a-chip platform. PLoS One 11:e0168138
80. Vogler H, Felekis D, Nelson BJ et al (2015) Measuring the mechanical properties of plant cell walls. Plants (Basel) 4:167–182

Chapter 2

Apomixis: Engineering the Ability to Harness Hybrid Vigor in Crop Plants

Joann A. Conner and Peggy Ozias-Akins

Abstract

Apomixis, commonly defined as asexual reproduction through seed, is a reproductive trait that occurs in only a few minor crops, but would be highly valuable in major crops. Apomixis results in seed-derived progenies that are genetically identical to their maternal parent. The advantage of apomixis would lie in seed propagation of elite food, feed, and biofuel crops that are heterozygous such as hybrid corn and switchgrass or self-pollinating crops for which no commercial-scale hybrid production system is available. While hybrid plants often outperform parental lines in growth and higher yields, production of hybrid seed is accomplished through carefully controlled, labor intensive crosses. Both small farmers in developing countries who produce their own seed and commercial companies that market hybrid seed could benefit from the establishment of engineered apomixis in plants. In this chapter, we review what has been learned from studying natural apomicts and mutations in sexual plants leading to apomixis-like development, plus discuss how the components of apomixis could be successfully engineered in plants.

Key words Apomixis, Hybrid vigor, Apomeiosis, Parthenogenesis, MiMe, PsASGR-BBML

1 Overview of Apomixis

Apomixis, simply defined as asexual reproduction through seed, was described as early as 1841 [1] and can be achieved through multiple developmental pathways. The three major developmental forms of apomixis are adventitious embryony, diplospory, and apospory (Fig. 1). Distributed by genera, adventitious embryony is most common (found in 148 genera) followed by apospory (110 genera) and diplospory (68 genera) [2].

Adventitious embryony is classified as a form of sporophytic apomixis since embryos develop directly through the mitotic division of somatic cells of the ovule. Sexual reproduction in a sporophytic apomict typically remains functional since the endosperm, which is essential for seed development, is derived from fertilization of the central cell of a meiotically derived female gametophyte (embryo sac) in the same ovule. The adventitious embryo competes

Anja Schmidt (ed.), *Plant Germline Development: Methods and Protocols*, Methods in Molecular Biology, vol. 1669, DOI 10.1007/978-1-4939-7286-9_2, © Springer Science+Business Media LLC 2017

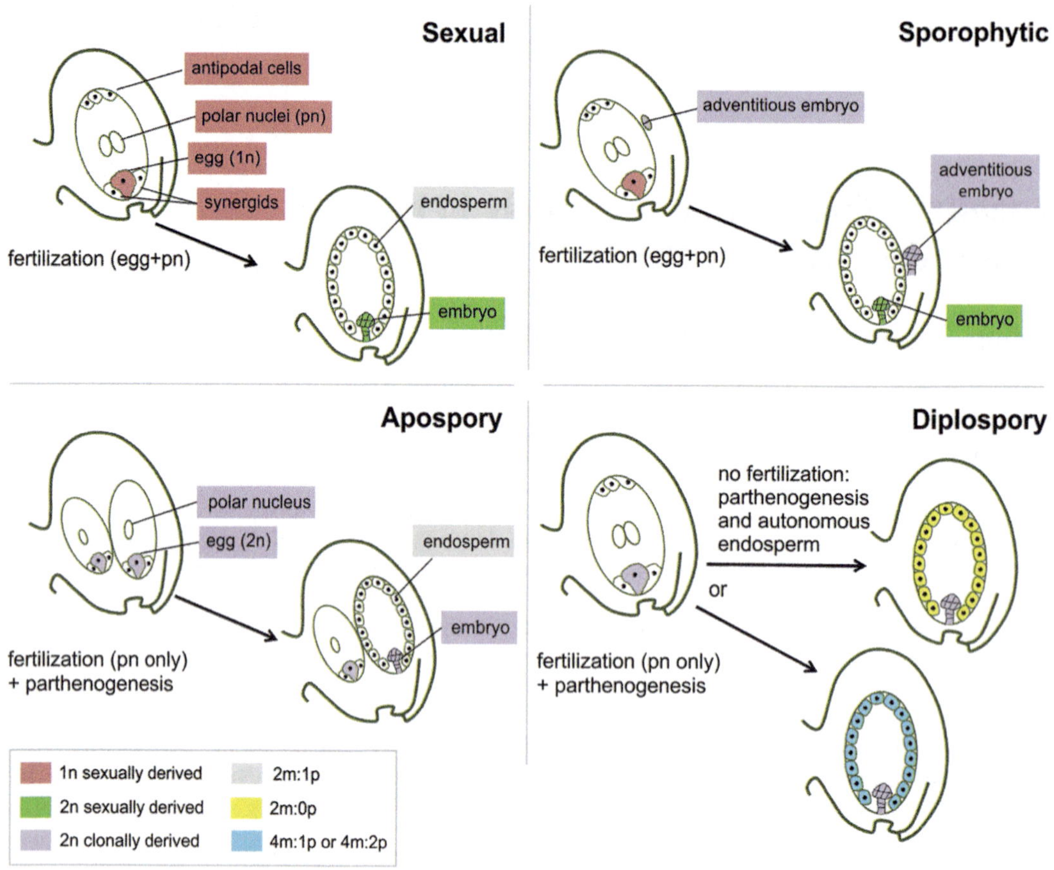

Fig. 1 Comparison between sexual and apomictic developmental pathways. Sexual and apomictic developmental pathways occur in the ovule. Depicted are mature embryo sacs prior to and after fertilization for the developmental pathways. Sexually and clonally derived egg and embryo development is shown. In the apospory pathway, only the pseudogamy pathway is presented; however, autonomous endosperm development can proceed in some species

with the sexually derived embryo for resources within the developing seed. Polyembryony within one seed, one genetically identical to the mother and the other a product of sexual reproduction, is a frequent outcome of adventitious embryony.

Both diplospory and apospory are classified as forms of gametophytic apomixis as the resulting embryo is derived not from sporophytic cells of the ovary but from the egg cell of an unreduced embryo sac. In diplospory, the unreduced embryo sac develops from the megaspore mother cell that undergoes restitutional meiosis or mitotic division to produce an unreduced embryo sac visually indistinguishable from a sexual embryo sac. In apospory, the unreduced embryo sac is formed from a nucellar cell that assumes the identity of an aposporous initial prior to mitotic divisions to produce an unreduced embryo sac. Apospory often is more easily identified visually due to formation of multiple aposporous embryo sacs or absence of antipodals in many species. Although multiple aposporous embryo sacs typically form in each ovule, usually only one embryo per seed is produced. The specification of an unreduced embryo sac initial is the first step in gametophytic apomixis and is termed apomeiosis. To create an embryo that is genetically identical to the mother plant, the egg cell of the unreduced embryo sac must take the identity of a zygote and initiate mitotic divisions to form an embryo in the absence of fertilization. This second step in gametophytic apomixis is termed parthenogenesis. To produce viable seed in gametophytic apomicts, endosperm formation must also proceed. In some apomicts, endosperm develops autonomously from the central cell. Most apomicts, however, require pollination and fertilization of the central cell (pseudogamy) for endosperm development.

The apomictic mode of reproduction occurs most often in polyploid genotypes, although the occurrences of natural and induced diploid populations that are apomictic indicate polyploidy is not a requirement for functional apomixis [3, 4]. Apomixis can be obligate, where all progenies of a plant are genetically identical because they are produced only through apomixis, or facultative, where progenies of a plant are a mixture of genotypes because the parent is capable of both apomictic and sexual reproduction either in the same ovule or in different ovules on the same plant.

2 Genetics of Apomixis

The genetic locus/loci for gametophytic apomixis have been mapped with molecular markers in multiple species. The polyploid nature of apomicts, the obligatory use of the apomict for the male parent, and the frequent lack of sexual genotypes at the same ploidy level as the apomictic male parent has made the genetic analysis of apomixis complex. A recombination- suppressed, single dominant locus has been found for *Tripsacum* (diplospory [5, 6]), *Brachiaria* (apospory [7–11]), *Pennisetum/Cenchrus* (apospory [12–16]), *Paspalum* (apospory [17–20]), *Panicum* (apospory [21]), and *Poa* (apospory [22, 23]). Recombination at the defined apomixis locus separating apomeiosis and parthenogenesis and associated molecular markers has been identified in *Poa* [24] and *Cenchrus* [25]. Flow cytometric seed screen (FCSS) identified the separation of apomeiosis and parthenogenesis in *Panicum* [26]. A deletion study in *Hieracium praealtum* (incorrectly identified as *H. caespitosum* in [27]) identified two unlinked dominant loci, LOSS OF APOMEIOSIS (LOA) and LOSS OF PARTHENOGENESIS (LOP), required for apomeiosis and parthenogenesis, respectively [27, 28]. In *Taraxacum* (diplospory [29]), and *Erigeron* (diplospory [30, 31]), loci controlling apomeiosis and parthenogenesis are genetically separate. Both molecular markers and FCSS studies identified two dominant loci for *Hypericum* (apospory [32]). Autonomous endosperm development is controlled by yet a third independent genetic locus (*AutE*) in *Hieracium/Pilosella* [33] and *Taraxacum* [34]. Given the data generated to date, a minimum of two genetic factors are required for functional apospory/diplospory with pseudogamy and three genetic factors for species with autonomous endosperm formation.

Genetic mapping for sporophytic apomixis has mostly been conducted in *Citrus* with a reported 308 kb region containing either the major gene associated with polyembryony or a major QTL affecting expression of polyembryony [35]. Additional loci, affecting the penetrance for the percentage of polyembryonic seed produced by a plant, complicates analysis of inheritance [36].

3 Identifying Genes Involved in Apomixis Using Natural Apomicts

Multiple differential expression analyses comparing apomictic and sexual ovaries from gametophytic and sporophytic apomicts have been conducted (Table 1). Studies have included differential display, cDNA-AFLP (amplified fragment length polymorphism), SSH (suppression subtractive hybridization), microarray analysis and next generation sequencing. Gene expression differences can range into the hundreds. Correlation of differentially expressed genes across studies was not readily found. These results may not be surprising given the different techniques used, the choice of multiple apomictic species with different developmental pathways and the multiple developmental stages and cell-type heterogeneity within the extracted RNA samples. These studies do demonstrate the complexity of transcriptional changes happening during development of apomictic and sexual ovaries. To reduce RNA tissue sample complexity, laser capture microdissection and next generation sequencing and/or microarray analysis is being used in apomictic species such as *Hieracium* [55] and *Boechera* [56].

APOLLO (*apo*mixis-*l*inked *lo*cus—from *Boechera*), *HpARI* (*ARIADNE7*—from *Hypericum*), Ps*ORC3a* (ORIGIN RECOGNITION COMPLEX—from *Paspalum*) and *ASGR-BBML* (Apospory-Specific Genomic Region-BabyBooM-Like—from *Pennisetum*) are candidate apomixis genes identified through tight genetic linkage to an apomixis locus [32, 57–59]. *APOLLO* contains a 20-nucleotide polymorphism present in the 5′ untranslated region of the gene and shows high expression in apomictic ovaries versus sexual ovaries [57]. *HpARI* is postulated to encode a truncated version of the sexual protein [32] and potentially interferes with assembly of a protein complex. *APOLLO* and *HpARI* are candidates for the apomeiosis genetic factor controlling diplospory and apospory, respectively. Ps*ORC3a* is classified as an expressed pseudogene. *PsORC3a* is

Table 1
Transcriptional studies of apomictic and sexual ovaries

Species	Apomictic type	Method of detection	Number of genes	Genes of interest noted by author	References
Brachiaria brizantha	Apospory	Macroarray	9	*BbrizHelic*, *BbrizRan*, *BbrizSec13* and *BbrizSti1*	[37]
		Differential display, reverse Northerns	11 (6 apomictic, 2 sexual, 3 different stages)		[38]
Cenchrus	Apospory	Differential display	8 (4 apomictic, 4 sexual)		[39]
		SSH, northern analysis	3 clones	*Pca21* and *Pca24*	[40]
Paspalum notum	Apospory	Differential display	12		[41]
		Differential display	65 (34 apomictic, 31 sexual)	N20 (GPI-anchored protein)	[42]
Paspalum simplex		cDNA-AFLP	78	auxin response factors	[43]
Panicum maximum	Apospory	Differential RNA screening	1	*ASG-1*	[44]
		Microarray analysis	394 clones		[45]
Boechera	Diplospory	SuperSAGE tags	1106 apomixis-specific tags, 237 sex-specific tags		[46]
		SuperSAGE, 454 transcriptome libraries	543 developmental time shifted at apomeiosis between sexual and apomictic	69 *B. holboellii* specific alleles	[47]
		454 transcriptome libraries		miR156/157 target squamosa promoter binding protein-like 11 (SPL11)	[48]
Pennisetum	Apospory	SSH, differential macroarray	26 upregulated in apomict		[49]
Poa pratensis	Apospory	cDNA-AFLP	179 fragments	*PpARM*, *PpAPK*, *PpRAB1*, *PpSERK*, *APOSTART*	[50, 51]

(continued)

Table 1
(continued)

Species	Apomictic type	Method of detection	Number of genes	Genes of interest noted by author	References
Hypericum perforatum L	Apospory	Roche 454 transcriptome sequencing	673 sexual and 552 apomictic unigenes		[52]
Citrus	Sporophytic	SSH, microarray	>200	orf45 (ankyrin repeat proteins), orf10 (SERK), orf37 (Alfin-like protein PHD-finger), orf19 (heat shock), orfs 21, 22, and 25 (unknown proteins)	[53]
		SSH-based microarray	399 consensus sequences	*psg-1*, *psg-2*, *msg-1*, *msg-2*, *msg-3*	[54]

postulated to downregulate sexual *PsORC* genes and considered to be an additional genetic factor required for pseudogamy in *Paspalum* apomicts that have adapted to tolerate a 4(maternal):1(paternal) genome ratio in endosperm to form viable seed through apomixis [58].

The *ASGR-BBML* gene was considered a good candidate gene for parthenogenesis based on its similarity to the *BABY BOOM* (*BBM*) genes of *Brassica napus* [59], its conservation and expression (unpublished data) in all tested apomictic *Pennisetum/Cenchrus* species [60], and the identification of an ASGR-recombinant plant that retained apomeiosis but lost parthenogenesis along with the *ASGR-BBML* genes [25]. *BABY BOOM* and *BABY BOOM-Like* genes have long been hypothesized to control aspects of embryogenesis. *BABY BOOM (BBM)* was originally discovered as a transcript (cDNA) induced in microspore cultures of *Brassica napus* (*BnBBM*) undergoing somatic embryogenesis. Overexpression of *BnBBM* in *Arabidopsis* leads to the formation of somatic embryos on leaf tissue [61]. *BBM* and *BBML* genes are members of a larger gene family of transcription factors which contain two conserved AP2 DNA binding domains [62] and a bbm-1 domain identified as important for somatic embryogenesis function [63]. Evidence for the role of *ASGR-BBML* in parthenogenesis was established when transgenic sexual pearl millet plants transformed with a genomic *PsASGR-BBML* transgene under control of its native promoter and terminator displayed parthenogenesis and haploid embryo development [64]. Interestingly, no paper has been published describing the successful use of *BBM* or *BBML* genes from non-apomictic species to promote parthenogenesis, suggesting that there is a unique feature or domain of *ASGR-BBML* that evolved and was selected for this function.

DNA and protein sequence analysis of additional *ASGR-BBML* genes along with *ASGR-BBML* genes not linked to the apomixis locus may help to elucidate the regions required for the unique function of this protein. It is unlikely that sequencing of additional *ASGR-BBML* genes from *Pennisetum/Cenchrus* species will yield much additional information as *P. squamulatum* and *C. ciliaris* *ASGR-BBML* genes are 98.6% conserved [59]. Additional information may be generated from sequencing *ASGR-BBML* genes from *Brachiaria spp.* (syn. *Urochloa*) and *Panicum* (syn. *Megathyrsus*), two more distantly related genera in the *Paniceae* tribe of the *Poaceae* with apomictic members. The *ASGR-BBML*-specific marker p779/p780 was found to be in complete linkage to the apomixis locus in a segregating population of *B. decumbens* and found to be predictive for reproductive phenotype in a wide range of *Brachiaria* species and in *P. maximum* [11]. While more data are needed, if the hypothesis of a common origin for aposporous apomixis in the *Paniceae* tribe is valid, conservation of candidate genes across species should help define the genetic factors underlying apospory.

The *PsASGR-BBML* transgene can be inherited through both male and female gametes. As the original pearl millet transformation was in a tetraploid, heterozygous background, backcrossing of a dihaploid line, originating from parthenogenesis of a reduced egg of the tetraploid, with a diploid homozygous inbred genotype is ongoing. The frequency of parthenogenesis identified by observing 45–60 cleared ovaries per plant from unpollinated *PsASGR-BBML* transgenic BC_1 (three lines composed of six plants) and BC_2 (two lines composed of seven plants) at 2 days after natural anthesis ranged from ~30 to 50% (Fig. 2). The average number of mature seed containing a haploid embryo based on FCSS data from a small number of individual lines is 0–20%. These data suggest that only a fraction of ovaries that initiate a parthenogenic embryo can complete development to a mature seed. Precocious embryo formation in ovaries of transgenic lines prior to pollen shed is more commonly observed than in natural apomicts, suggesting that parthenogenesis is being triggered in these lines days before it would be in natural apomicts. For strongly apomictic species *P. squamulatum* and *C. ciliaris*, embryo formation at or prior to pollen shed is almost never seen. One hypothesis for a decline in the number of haploid embryos that reach maturity is prevention of central cell fertilization due to excessive growth of the parthenogenetic embryo prior

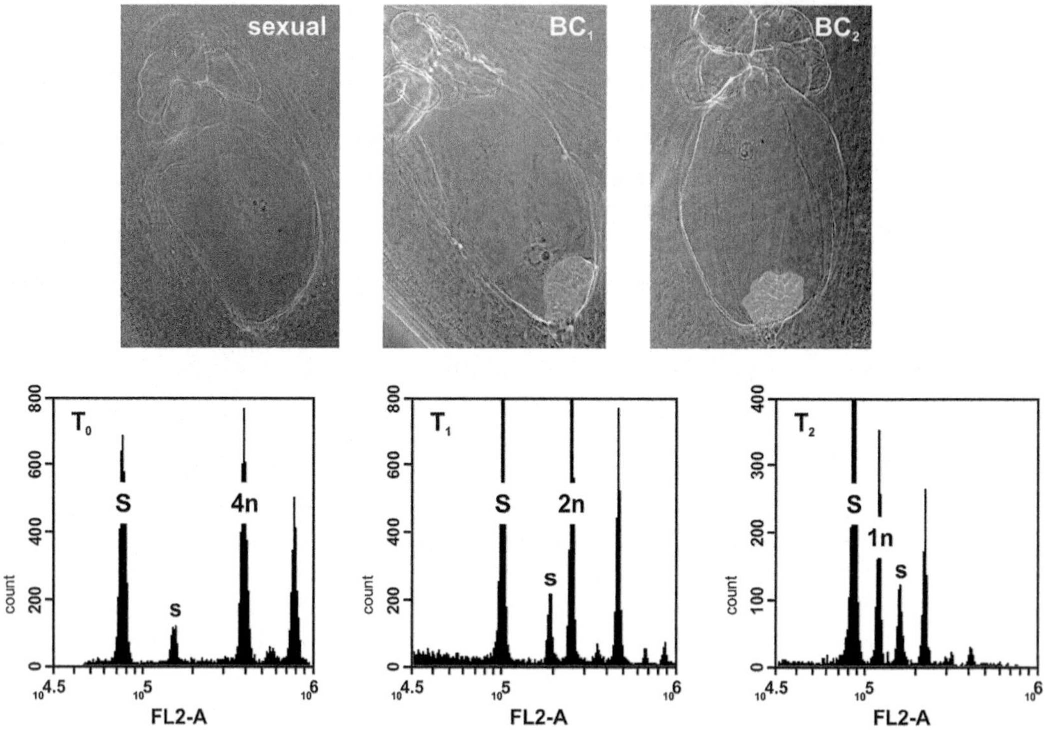

Fig. 2 Recurrent parthenogenesis in sexual millet carrying a *PsASGR-BBML* transgene. The *upper part* of the figure contains images of unfertilized sexual pearl millet ovaries two days post anthesis which have been cleared for imaging. The sexual ovary displays a mature sexual embryo sac without parthenogenesis. The BC_1 and BC_2 ovaries contain the *PsASGR-BBML* transgene and display a mature sexual embryo sac with parthenogenesis. The *lower part* of the figure displays flow cytometry from a T_0, T_1 and T_2 line carrying the *PsASGR-BBML* transgene. Sorghum leaf tissue ($S = 2n/2\times/2c$, $s = 2n/2\times/4c$) was used as a genome size standard in each sample. At T_0 a sexual millet plant with the *PsASGR-BBML* transgene is tetraploid (4*n*). In each successive generation, the ploidy level of some seed derived from the transgenic line was reduced ($T_1 = 2n$, $T_2 = 1n$)

to pollination. Another possibility is that the central cell is being fertilized, but because the embryo is developing out of synchrony with the endosperm, normal communication between endosperm and embryo is disrupted, arresting formation of seed containing haploid embryos.

Our goal in backcrossing is to produce a more homozygous, parthenogenetic line for use in transcriptional sequence analysis of the egg cell using laser capture microdissection and next generation sequencing. Identifying the transcripts regulated by *PsASGR-BBML* could lead to discovery of the gene network responsible for egg cell to zygote fate transition in the absence of fertilization. Transformation of other monocots with the *PsASGR-BBML* transgene is ongoing to investigate cross-genera function of the transgene and penetrance of the conferred phenotype.

4 Breeding of Apomictic Crop Plants and the Integration of Apomixis Loci

Crop plants that are natural apomicts include the turfgrass Kentucky bluegrass (*Poa pratensis*) and tropical forage grasses such as signalgrass (*Brachiaria decumbens*), Guinea grass (*Panicum maximum*), Bahia Grass (*Paspalum notatum*), buffelgrass (*Cenchrus ciliaris* syn. *Pennisetum ciliare*), and eastern gamagrass (*Tripsacum dactyloides*). Breeding programs at the International Center for Tropical Agriculture (CIAT) in Colombia and the Brazilian Agricultural Research Corporation (Embrapa) are using artificially doubled sexual *B. ruziziensis*, *B. brizantha*, and *B. decumbens* [65–68] to incorporate traits of interest along with the true-breeding seed advantage provided by apomixis [69].

Traditional breeding to integrate apomixis from a related wild species into crop plants such as pearl millet and maize has also been attempted with limited success. Pearl millet is a diploid, sexually reproducing drought tolerant crop grown as forage in the US and as an important cereal grain crop in some developing countries. A backcrossing program resulted in tetraploid apomictic pearl millet lines (BC_8) carrying a single chromosome from *P. squamulatum* conferring apomixis [70]. Apomictic pearl millet is not yet a reality, due to both lower seed set and an increase of facultativeness among the BC_8 lines. Reduction of fertility and/or the increase of sexual reproduction may be in part due to linkage drag from the presence of an entire *P. squamulatum* chromosome, epigenetic changes within the ASGR caused by backcrossing that could consequently alter gene expression of the apomixis "genetic factors" and/or modifying genes unlinked to the ASGR that increase penetrance of apomixis in the wild-type *P. squamulatum*. Similarly, introgression of the apomixis locus from *Tripsacum* to maize has been hindered by transmission and penetrance of the trait into the next generation with low seed fertility and poor development of viable adult plants [71].

5 Genetic Mutation(s) That Mimic the Apomictic Pathway

Mutational screens have identified sexual plants which mimic components of the apomictic pathway. For apomeiosis, the *dyad* recessive mutation in *Arabidopsis* alters the first meiotic division to a mitotic one, allowing for the formation of unreduced, nonrecombinant functional female gametophytes at a very low frequency [72]. The *dyad* mutation is a unique allele of the *SWITCH1 (SWI1)* gene of *Arabidopsis*, yielding an apomixis-like phenotype only in the female gametophyte. The *ameiotic1 (am1)* mutation in maize, an ortholog of *SWI1*, also displays a "mitosis-like" division instead of meiosis [73].

For parthenogenesis, the *multicopy suppressor of ira1* (*msi1*) mutation in *Arabidopsis* was identified to cause the initiation of haploid embryo and autonomous endosperm development within segregating *msi1* ovaries. The haploid embryo usually fails to proceed past the 2–4 cell stage [74]. Knockdown in expression of the single *Marchantia polymorpha* RWP-RK domain gene (*MpRKD*) leads to division of the egg cell without fertilization [75]. The authors propose that *MpRKD* establishes or helps maintain a quiescent state of the egg cell and lack of expression leads to parthenogenetic-like cell divisions.

For autonomous endosperm development, multiple mutants in *Arabidopsis* have been identified and are collectively known as the fertilization-independent seed (fis) class of mutants. *Fis* genes include *medea* [76, 77], *fis2* [78, 79], *fertilization-independent endosperm* [80, 81], and *msi1* [82, 83]. FIS class genes are components of the Polycomb Repressive Complex 2 and maintain repression of genes involved in endosperm development. Interestingly, FIE proteins in sexual *Hieracium* and *Arabidopsis* do not appear to be functionally conserved since downregulation in either sexual or apomictic forms of *Hieracium* does not promote autonomous endosperm development, and in vitro interaction with other FIS class proteins was not observed [84]. PRC2 genes have been identified in rice, maize, and barley. Studies of the monocot genes have noted differences between dicot and monocot endosperm regulation which will require more study [85].

Arabidopsis genes defective for the RNA-directed DNA methylation (RdDM) pathway such as *argonaute9 (ago9)*, *suppressor of gene silencing 3 (sgs3)*, *rna-dependent rna polymerase 2 (rdr2)*, *rna-dependent rna polymerase 6 (rdr6)*, and *dicer-like3 (dcl3)* create multiple aposporous-like initials in ovaries [86, 87]. Some of these cells continue to mature to unreduced functional female gametophytes that can develop both autonomous embryo and endosperm, although at low levels [88]. The production of aposporous-like initials in ovaries from DNA methyltransferase mutants of maize has also been identified [89].

6 Complexity of Engineering Apomixis into Crop Plants

Linkage drag and potential expression/epigenetic changes of the genetic factors from apomixis loci introgressed into sexual species genetic backgrounds will probably continue to impede the development of agronomically acceptable hybrids using traditional breeding methods except in the case of forage grasses. The most straightforward means to create an apomictic crop plant is to introduce a transgenic "apomixis cassette" containing genes conferring apomeiosis, parthenogenesis, and autonomous endosperm development, all which function in a dominant manner. This cassette

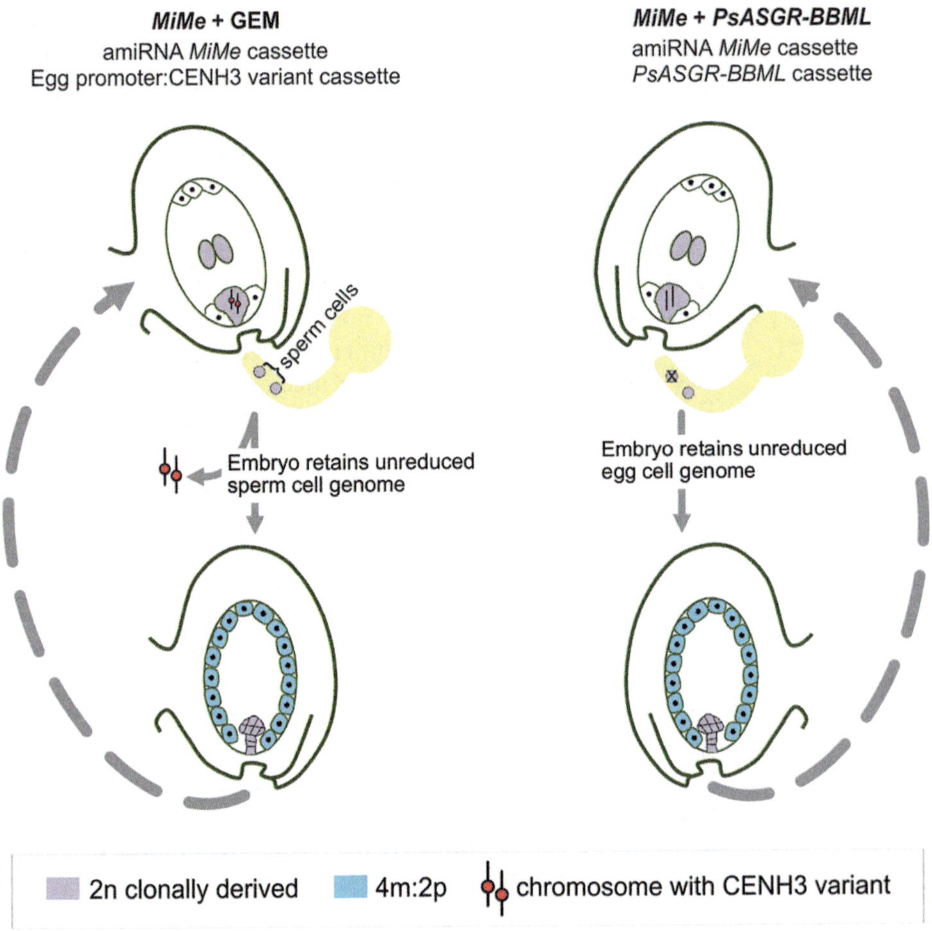

Fig. 3 Overview of potential methods of creating a self-reproducing apomictic pathway in plants using transgene cassettes. The *left figure* illustrates the use of both an amiRNA *MiMe* cassette and an egg-specific promoter fused with a weak CENH3 variant cassette to create a *MiMe* + GEM apomictic transgenic line. The *right figure* illustrates the use of both an amiRNA *MiMe* cassette and *PsASGR-BBML* cassette to create a MiMe + *PsASGR-BBML* apomictic transgenic line. The *gray arrows* denote the recurrent clonal propagation of the maternal genotype using this strategy

could be transformed into multiple genotypes and resulting apomictic lines could be used as pollen donors to create apomictic F_1 individuals. This cassette does not yet exist; however, other strategies to synthesize apomixis in crop plants through engineering of sexual pathways are being pursued [90, 91]. Below is a description of proposed engineering of apomixis in crop plants and how this strategy could be altered to incorporate the parthenogenesis gene *PsASGR-BBML* (Fig. 3). The discovery of new genes, from either apomictically or sexually reproducing organisms, will stimulate continued reexamination of potential ways to engineer apomixis in crops.

A diplospory-like, apomeiotic pathway was recovered in *Arabidopsis* using three recessive meiosis mutants. The *MiMe* and *MiMe-2* genotypes form unreduced male and female gametes that are genetically identical to their parent with high penetrance [92, 93]. Recessive alleles of *OSD1* (At3g57860), *ATREC8* (AT5G05490) and *ATSPO11–1* (AT3G13170) or *TAM* (*AT1G77390)*, *ATREC8*, and *ATSPO11* are responsible for the *MiMe* and *MiMe-2* phenotypes, respectively. To transfer the *MiMe* or *MiMe-2* phenotype to crop plants where duplicate genes may be present, the *MiMe* or *MiMe-2* orthologs from crop plants need to be identified and tested for function, as has recently been done for rice [94]. Genome sequence for many crop plants is available and therefore identification of orthologs and testing for tissue-specific expression should not be a limiting factor. An effective technology for creating targeted null alleles would be the CRISPR (*c*lustered *r*egularly *i*nterspaced *s*hort *p*alindromic *r*epeats)/Cas system for gene editing (plant-specific CRISPR/Cas review papers [95–97]). Given the recessive nature of the 3-gene *MiMe* phenotype, mutations in or knockdown of all three genes will be necessary to switch meiosis to mitosis. The creation of artificial microRNAs (amiRNAs) to post-transcriptionally silence the three genes is one possible approach that has been proposed [91]. Either monocot or dicot gene expression can be effectively knocked down using amiRNAs [98–100]. The amiRNAs could be expressed individually or in tandem to suppress expression of the three genes. The amiRNA approach is also dominant, paralleling the dominant nature of apomeiosis in natural apomicts. If a functional MiMe crop plant were created, parthenogenesis would still be required for apomictic reproduction and to prevent an increase in ploidy level that would result from fertilization of unreduced gametes.

A method to provide the requirement for parthenogenesis was demonstrated in *Arabidopsis MiMe* and *dyad* mutants through crosses with the GEM (*g*enome *e*limination induced by a *m*ix of CENH3 variants) line [101]. In this approach, fertilization of the egg cell with an unreduced male gamete occurs; however, either the male or female parental genome is programmed for elimination due to centromere conflicts. When an unaltered or "wild-type" genome is hybridized with a genome derived from a CENH3 variant line, the variant CENH3 chromosomes are lost, regardless of the sex of the parent. This method of producing clonal seed relies on recurrent crossing between two parental lines to perpetuate the clonal phenotype [101]. Crossing could be eliminated if the CENH3 variant were preferentially loaded onto the centromeres of the unreduced egg cell chromosomes due to egg-cell-targeted expression [91]. The alteration of only egg cell chromosomes would allow for the propagation of clonal seed by self-pollination (Fig. 3). Besides *Arabidopsis*, GEM lines with low penetrance have been engineered in maize [102].

Another method to induce parthenogenesis would be with *PsASGR-BBML* (Fig. 3). The *PsASGR-BBML* transgene also acts in a dominant manner and therefore a construct containing both amiRNA for the MiMe phenotype and *PsASGR-BBML* driven by an egg-cell-specific promoter should confer $2n$ clonal embryo formation. Endosperm initiation and development after self-pollination should be normal since the expected genome ratio of 4m:2p would result from unreduced *MiMe* pollen. This endosperm ratio, while the same 2m:1p ratio, but increased copy numbers, found during normal sexual reproduction in crop plants, does allow for plump seed formation in maize [103, 104] and apomictic *Boechera* [4]. Due to the dominant nature of a *MiMe* amiRNA and *PsASGR-BBML* phenotypes, synthesis of apomictic lines with these two genes would be dependent on transformation and regeneration of the elite F_1 hybrid of choice. With 100% penetrance, all progeny from the resulting T_0 plant would be genetically identical to the elite F_1 plant.

7 Conclusion

The search for additional apomixis genes and mutants of sexual genes which mimic apomictic phenotypes will continue, with the ultimate goal of creating a highly penetrant apomictic crop plant. The availability of technologies such as laser capture microdissection, targeted gene editing, and more efficient transformation systems for crop plants will advance our understanding of genes and gene networks governing clonal reproduction and ultimately lead to the installation of apomixis in crop plants.

References

1. Asker S, Jerling L (1992) Apomixis in plants. CRC Press, Boca Raton, FL
2. Hojsgaard D, Klatt S, Baier R, Carman JG, Hörandl E (2014) Taxonomy and biogeography of apomixis in angiosperms and associated biodiversity characteristics. Crit Rev Plant Sci 33(5):414–427
3. Noyes RD, Wagner JD (2014) Dihaploidy yields diploid apomicts and parthenogens in Erigeron (Asteraceae). Am J Bot 101 (5):865–874
4. Voigt-Zielinski M-L, Piwczyński M, Sharbel TF (2012) Differential effects of polyploidy and diploidy on fitness of apomictic Boechera. Sex Plant Reprod 25(2):97–109
5. Grimanelli D, Leblanc O, Espinosa E, Perotti E, Gonzalez De Leon D, Savidan Y (1998) Mapping diplosporous apomixis in tetraploid Tripsacum: one gene or several genes? Heredity 80(1):33–39
6. Leblanc O, Grimanelli D, Gonzalez-de-Leon D, Savidan Y (1995) Detection of the apomictic mode of reproduction in maize-Tripsacum hybrids using maize RFLP markers. Theor Appl Genet 90(7–8):1198–1203
7. Pessino S, Ortiz J, Leblanc O, Do Valle C, Evans C, Hayward M (1997) Identification of a maize linkage group related to apomixis in Brachiaria. Theor Appl Genet 94 (3–4):439–444
8. Pessino SC, Evans C, Ortiz JPA, Armstead I, Valle CBD, Hayward MD (1998) A genetic map of the apospory-region in Brachiaria hybrids: identification of two markers closely associated with the trait. Hereditas 128 (2):153–158

9. Zorzatto C, Chiari L, De Araújo Bitencourt G, Do Valle C, De Campos Leguizamón G, Schuster I, Pagliarini M (2010) Identification of a molecular marker linked to apomixis in Brachiaria humidicola (Poaceae). Plant Breed 129(6):734–736
10. Thaikua S, Ebina M, Yamanaka N, Shimoda K, Suenaga K, Kawamoto Y (2016) Tightly clustered markers linked to an apospory-related gene region and quantitative trait loci mapping for agronomic traits in Brachiaria hybrids. Grassl Sci 62(2):69–80
11. Worthington M, Heffelfinger C, Bernal D, Quintero C, Zapata YP, Perez JG, De Vega J, Miles J, Dellaporta S, Tohme J (2016) A parthenogenesis gene candidate and evidence for segmental allopolyploidy in apomictic Brachiaria decumbens. Genetics 116:190314
12. Gustine D, Sherwood R, Huff D (1997) Apospory-linked molecular markers in buffelgrass. Crop Sci 37(3):947–951
13. Ozias-Akins P, Roche D, Hanna WW (1998) Tight clustering and hemizygosity of apomixis-linked molecular markers in Pennisetum squamulatum implies genetic control of apospory by a divergent locus that may have no allelic form in sexual genotypes. Proc Natl Acad Sci 95(9):5127–5132
14. Roche D, Cong P, Chen Z, Hanna WW, Gustine DL, Sherwood RT, Ozias-Akins P (1999) An apospory-specific genomic region is conserved between Buffelgrass (Cenchrus ciliaris L.) and Pennisetum squamulatum Fresen. Plant J 19(2):203–208
15. Jessup R, Burson B, Burow G, Wang Y-W, Chang C, Li Z, Paterson A, Hussey M (2002) Disomic inheritance, suppressed recombination, and allelic interactions govern apospory in buffelgrass as revealed by genome mapping. Crop Sci 42(5):1688–1694
16. Goel S, Chen Z, Conner JA, Akiyama Y, Hanna WW, Ozias-Akins P (2003) Delineation by fluorescence in situ hybridization of a single hemizygous chromosomal region associated with aposporous embryo sac formation in Pennisetum squamulatum and Cenchrus ciliaris. Genetics 163(3):1069–1082
17. Pupilli F, Labombarda P, Caceres ME, Quarín CL, Arcioni S (2001) The chromosome segment related to apomixis in Paspalum simplex is homoeologous to the telomeric region of the long arm of rice chromosome 12. Mol Breed 8(1):53–61
18. Labombarda P, Busti A, Caceres ME, Pupilli F, Arcioni S (2002) An AFLP marker tightly linked to apomixis reveals hemizygosity in a portion of the apomixis-controlling locus in Paspalum simplex. Genome 45(3):513–519
19. Martínez EJ, Hopp HE, Stein J, Ortiz JP, Quarin CL (2003) Genetic characterization of apospory in tetraploid Paspalum notatum based on the identification of linked molecular markers. Mol Breed 12(4):319–327
20. Stein J, Pessino SC, Martínez EJ, Rodriguez MP, Siena LA, Quarin CL, Ortiz JPA (2007) A genetic map of tetraploid Paspalum notatum Flügge (bahiagrass) based on single-dose molecular markers. Mol Breed 20 (2):153–166
21. Ebina M, Nakagawa H, Yamamoto T, Araya H, Si T, Takahara M, Nakajima K (2005) Co-segregation of AFLP and RAPD markers to apospory in Guineagrass (Panicum maximum Jacq.) Grassl Sci 51(1):71–78
22. Barcaccia G, Mazzucato A, Albertini E, Zethof J, Gerats A, Pezzotti M, Falcinelli M (1998) Inheritance of parthenogenesis in Poa pratensis L.: auxin test and AFLP linkage analyses support monogenic control. Theor Appl Genet 97(1–2):74–82
23. Porceddu A, Albertini E, Barcaccia G, Falistocco E, Falcinelli M (2002) Linkage mapping in apomictic and sexual Kentucky bluegrass (Poa pratensis L.) genotypes using a two way pseudo-testcross strategy based on AFLP and SAMPL markers. Theor Appl Genet 104(2–3):273–280
24. Albertini E, Barcaccia G, Porceddu A, Sorbolini S, Falcinelli M (2001) Mode of reproduction is detected by Parth1 and Sex1 SCAR markers in a wide range of facultative apomictic Kentucky bluegrass varieties. Mol Breed 7 (4):293–300
25. Conner JA, Gunawan G, Ozias-Akins P (2013) Recombination within the apospory specific genomic region leads to the uncoupling of apomixis components in Cenchrus ciliaris. Planta 238(1):51–63
26. Kaushal P, Malaviya D, Roy A, Pathak S, Agrawal A, Khare A, Siddiqui S (2008) Reproductive pathways of seed development in apomictic guinea grass (Panicum maximum Jacq.) reveal uncoupling of apomixis components. Euphytica 164(1):81–92
27. Catanach AS, Erasmuson SK, Podivinsky E, Jordan BR, Bicknell R (2006) Deletion mapping of genetic regions associated with apomixis in Hieracium. Proc Natl Acad Sci 103(49):18650–18655
28. Koltunow AM, Johnson SD, Rodrigues J, Okada T, Hu Y, Tsuchiya T, Wilson S, Fletcher P, Ito K, Suzuki G (2011) Sexual reproduction is the default mode in apomictic Hieracium subgenus Pilosella, in which two dominant loci function to enable apomixis. Plant J 66(5):890–902

29. van Dijk PJ, Bakx-Schotman JT (2004) Formation of unreduced megaspores (diplospory) in apomictic dandelions (Taraxacum officinale, sl) is controlled by a sex-specific dominant locus. Genetics 166(1):483–492
30. Noyes RD, Rieseberg LH (2000) Two independent loci control agamospermy (apomixis) in the triploid flowering plant Erigeron annuus. Genetics 155(1):379–390
31. Noyes R, Baker R, Mai B (2007) Mendelian segregation for two-factor apomixis in Erigeron annuus (Asteraceae). Heredity 98(2):92–98
32. Schallau A, Arzenton F, Johnston AJ, Hähnel U, Koszegi D, Blattner FR, Altschmied L, Haberer G, Barcaccia G, Bäumlein H (2010) Identification and genetic analysis of the APOSPORY locus in Hypericum perforatum L. Plant J 62(5):773–784
33. Ogawa D, Johnson SD, Henderson ST, Koltunow AM (2013) Genetic separation of autonomous endosperm formation (AutE) from the two other components of apomixis in Hieracium. Plant Reprod 26(2):113–123
34. van Dijk PJ, van Baarlen P, De Jong JH (2003) The occurrence of phenotypically complementary apomixis-recombinants in crosses between sexual and apomictic dandelions (Taraxacum officinale). Sex Plant Reprod 16(2):71–76
35. Nakano M, Shimada T, Endo T, Fujii H, Nesumi H, Kita M, Ebina M, Shimizu T, Omura M (2012) Characterization of genomic sequence showing strong association with polyembryony among diverse citrus species and cultivars, and its synteny with Vitis and Populus. Plant Sci 183:131–142
36. Kepiro J, Roose M (2010) AFLP markers closely linked to a major gene essential for nucellar embryony (apomixis) in Citrus maxima× Poncirus trifoliata. Tree Genet Genomes 6(1):1–11
37. Silveira ÉD, Guimarães LA, de Alencar Dusi DM, Da Silva FR, Martins NF, do Carmo Costa MM, Alves-Ferreira M, de Campos Carneiro VT (2012) Expressed sequence-tag analysis of ovaries of Brachiaria brizantha reveals genes associated with the early steps of embryo sac differentiation of apomictic plants. Plant Cell Rep 31(2):403–416
38. Rodrigues JC, Cabral GB, Dusi DM, de Mello LV, Rigden DJ, Carneiro VT (2003) Identification of differentially expressed cDNA sequences in ovaries of sexual and apomictic plants of Brachiaria brizantha. Plant Mol Biol 53(6):745–757
39. Vielle-Calzada J-P, Nuccio ML, Budiman MA, Thomas TL, Burson BL, Hussey MA, Wing RA (1996) Comparative gene expression in sexual and apomictic ovaries of Pennisetum ciliare (L.) Link. Plant Mol Biol 32 (6):1085–1092
40. Singh M, Burson BL, Finlayson SA (2007) Isolation of candidate genes for apomictic development in buffelgrass (Pennisetum ciliare). Plant Mol Biol 64(6):673–682
41. Pessino SC, Espinoza F, Martinez EJ, Ortiz JPA, Valle EM, Quarin CL (2001) Isolation of cDNA clones differentially expressed in flowers of apomictic and sexual Paspalum notatum. Hereditas 134(1):35–42
42. Laspina NV, Vega T, Seijo JG, González AM, Martelotto LG, Stein J, Podio M, Ortiz JPA, Echenique VC, Quarin CL (2008) Gene expression analysis at the onset of aposporous apomixis in Paspalum notatum. Plant Mol Biol 67(6):615–628
43. Polegri L, Calderini O, Arcioni S, Pupilli F (2010) Specific expression of apomixis-linked alleles revealed by comparative transcriptomic analysis of sexual and apomictic Paspalum simplex Morong flowers. J Exp Bot 61 (6):1869–1883
44. Chen L, Miyazaki C, Kojimai A, Saito A, Adachi T (1999) Isolation and characterization of a gene expressed during early embryo sac development in apomictic guinea grass (Panicum maximum). J Plant Physiol 154 (1):55–62
45. Yamada-Akiyama H, Akiyama Y, Ebina M, Xu Q, Tsuruta S-i, Yazaki J, Kishimoto N, Kikuchi S, Takahara M, Takamizo T (2009) Analysis of expressed sequence tags in apomictic guineagrass (Panicum maximum). J Plant Physiol 166(7):750–761
46. Sharbel TF, Voigt M-L, Corral JM, Galla G, Kumlehn J, Klukas C, Schreiber F, Vogel H, Rotter B (2010) Apomictic and sexual ovules of Boechera display heterochronic global gene expression patterns. Plant Cell 22 (3):655–671
47. Sharbel TF, Voigt ML, Corral JM, Thiel T, Varshney A, Kumlehn J, Vogel H, Rotter B (2009) Molecular signatures of apomictic and sexual ovules in the Boechera holboellii complex. Plant J 58(5):870–882
48. Amiteye S, Corral JM, Vogel H, Sharbel TF (2011) Analysis of conserved microRNAs in floral tissues of sexual and apomictic Boechera species. BMC Genomics 12(1):1
49. Sahu PP, Gupta S, Malaviya D, Roy AK, Kaushal P, Prasad M (2012) Transcriptome analysis of differentially expressed genes during embryo sac development in apomeiotic non-parthenogenetic interspecific hybrid of

Pennisetum glaucum. Mol Biotechnol 51 (3):262–271

50. Albertini E, Marconi G, Barcaccia G, Raggi L, Falcinelli M (2004) Isolation of candidate genes for apomixis in Poa pratensis L. Plant Mol Biol 56(6):879–894
51. Albertini E, Marconi G, Reale L, Barcaccia G, Porceddu A, Ferranti F, Falcinelli M (2005) SERK and APOSTART. Candidate genes for apomixis in Poa pratensis. Plant Physiol 138 (4):2185–2199
52. Galla G, Vogel H, Sharbel TF, Barcaccia G (2015) De novo sequencing of the Hypericum perforatum L. flower transcriptome to identify potential genes that are related to plant reproduction sensu lato. BMC Genomics 16(1):1
53. Kumar V, Malik SK, Pal D, Srinivasan R, Bhat SR (2014) Comparative transcriptome analysis of ovules reveals stress related genes associated with nucellar polyembryony in citrus. Tree Genet Genomes 10(3):449–464
54. Nakano M, Kigoshi K, Shimizu T, Endo T, Shimada T, Fujii H, Omura M (2013) Characterization of genes associated with polyembryony and in vitro somatic embryogenesis in citrus. Tree Genet Genomes 9(3):795–803
55. Okada T, Hu Y, Tucker MR, Taylor JM, Johnson SD, Spriggs A, Tsuchiya T, Oelkers K, Rodrigues JC, Koltunow AM (2013) Enlarging cells initiating apomixis in Hieracium praealtum transition to an embryo sac program prior to entering mitosis. Plant Physiol 163(1):216–231
56. Schmidt A, Schmid MW, Klostermeier UC, Qi W, Guthörl D, Sailer C, Waller M, Rosenstiel P, Grossniklaus U (2014) Apomictic and sexual germline development differ with respect to cell cycle, transcriptional, hormonal and epigenetic regulation. PLoS Genet 10(7): e1004476
57. Corral JM, Vogel H, Aliyu OM, Hensel G, Thiel T, Kumlehn J, Sharbel TF (2013) A conserved apomixis-specific polymorphism is correlated with exclusive exonuclease expression in premeiotic ovules of apomictic Boechera species. Plant Physiol 163 (4):1660–1672
58. Siena LA, Ortiz JPA, Calderini O, Paolocci F, Cáceres ME, Kaushal P, Grisan S, Pessino SC, Pupilli F (2016) An apomixis-linked ORC3-like pseudogene is associated with silencing of its functional homolog in apomictic Paspalum simplex. J Exp Bot 67(6):1965–1978
59. Conner JA, Goel S, Gunawan G, Cordonnier-Pratt M-M, Johnson VE, Liang C, Wang H, Pratt LH, Mullet JE, DeBarry J (2008) Sequence analysis of bacterial artificial chromosome clones from the apospory-specific genomic region of Pennisetum and Cenchrus. Plant Physiol 147(3):1396–1411
60. Akiyama Y, Goel S, Conner JA, Hanna WW, Yamada-Akiyama H, Ozias-Akins P (2011) Evolution of the apomixis transmitting chromosome in Pennisetum. BMC Evol Biol 11(1):1
61. Boutilier K, Offringa R, Sharma VK, Kieft H, Ouellet T, Zhang L, Hattori J, Liu C-M, van Lammeren AA, Miki BL (2002) Ectopic expression of BABY BOOM triggers a conversion from vegetative to embryonic growth. Plant Cell 14(8):1737–1749
62. Kim S, Soltis PS, Wall K, Soltis DE (2006) Phylogeny and domain evolution in the APETALA2-like gene family. Mol Biol Evol 23(1):107–120
63. El Ouakfaoui S, Schnell J, Abdeen A, Colville A, Labbé H, Han S, Baum B, Laberge S, Miki B (2010) Control of somatic embryogenesis and embryo development by AP2 transcription factors. Plant Mol Biol 74(4–5):313–326
64. Conner JA, Mookkan M, Huo H, Chae K, Ozias-Akins P (2015) A parthenogenesis gene of apomict origin elicits embryo formation from unfertilized eggs in a sexual plant. Proc Natl Acad Sci 112(36):11205–11210
65. Lutts S, Ndikumana J, Louant B (1994) Male and female sporogenesis and gametogenesis in apomictic Brachiaria brizantha, Brachiaria decumbens and F1 hybrids with sexual colchicine induced tetraploid Brachiaria ruziziensis. Euphytica 78(1–2):19–25
66. Swenne A, Louant B, Dujardin M (1981) Induction par la colchicine de formes autotétraploïdes chez Brachiaria ruziziensis Germain et Evrard (Graminée). Agron Trop 36 (2):134–141
67. Pinheiro A, Pozzobon M, Do Valle C, Penteado M, Carneiro V (2000) Duplication of the chromosome number of diploid Brachiaria brizantha plants using colchicine. Plant Cell Rep 19(3):274–278
68. Simioni C, Cd V (2009) Chromosome duplication in Brachiaria (A. Rich.) Stapf allows intraspecific crosses. Crop Breed Appl Biotechnol 9(4):328–333
69. Miles JW (2007) Apomixis for cultivar development in tropical forage grasses. Crop Science 47(Suppl 3):S-238-S-249
70. Singh M, Conner J, Zeng Y-J, Hanna W, Johnson V, Ozias-Akins P (2010) Characterization of apomictic BC_7 and BC_8 pearl millet: meiotic chromosome behavior and construction of an ASGR-carrier chromosome-specific library. Crop Sci 50(3):892–902

71. Leblanc O, Grimanelli D, Hernandez-Rodriguez M, Galindo PA, Soriano-Martinez AM, Perotti E (2009) Seed development and inheritance studies in apomictic maize-Tripsacum hybrids reveal barriers for the transfer of apomixis into sexual crops. Int J Dev Biol 53(4):585–596
72. Ravi M, Marimuthu MP, Siddiqi I (2008) Gamete formation without meiosis in Arabidopsis. Nature 451(7182):1121–1124
73. Pawlowski WP, Wang C-JR, Golubovskaya IN, Szymaniak JM, Shi L, Hamant O, Zhu T, Harper L, Sheridan WF, Cande WZ (2009) Maize AMEIOTIC1 is essential for multiple early meiotic processes and likely required for the initiation of meiosis. Proc Natl Acad Sci 106(9):3603–3608
74. Guitton A-E, Berger F (2005) Loss of function of MULTICOPY SUPPRESSOR OF IRA 1 produces nonviable parthenogenetic embryos in Arabidopsis. Curr Biol 15 (8):750–754
75. Rövekamp M, Bowman JL, Grossniklaus U (2016) Marchantia MpRKD regulates the gametophyte-sporophyte transition by keeping egg cells quiescent in the absence of fertilization. Curr Biol 26(13):1782–1789
76. Grossniklaus U, Vielle-Calzada J-P, Hoeppner MA, Gagliano WB (1998) Maternal control of embryogenesis by MEDEA, a polycomb group gene in Arabidopsis. Science 280(5362):446–450
77. Kinoshita T, Yadegari R, Harada JJ, Goldberg RB, Fischer RL (1999) Imprinting of the MEDEA polycomb gene in the Arabidopsis endosperm. Plant Cell 11(10):1945–1952
78. Chaudhury AM, Ming L, Miller C, Craig S, Dennis ES, Peacock WJ (1997) Fertilization-independent seed development in Arabidopsis thaliana. Proc Natl Acad Sci 94 (8):4223–4228
79. Luo M, Bilodeau P, Koltunow A, Dennis ES, Peacock WJ, Chaudhury AM (1999) Genes controlling fertilization-independent seed development in Arabidopsis thaliana. Proc Natl Acad Sci 96(1):296–301
80. Ohad N, Margossian L, Hsu Y-C, Williams C, Repetti P, Fischer RL (1996) A mutation that allows endosperm development without fertilization. Proc Natl Acad Sci 93 (11):5319–5324
81. Ohad N, Yadegari R, Margossian L, Hannon M, Michaeli D, Harada JJ, Goldberg RB, Fischer RL (1999) Mutations in FIE, a WD polycomb group gene, allow endosperm development without fertilization. Plant Cell 11(3):407–415
82. Köhler C, Hennig L, Bouveret R, Gheyselinck J, Grossniklaus U, Gruissem W (2003) Arabidopsis MSI1 is a component of the MEA/FIE Polycomb group complex and required for seed development. EMBO J 22 (18):4804–4814
83. Guitton A-E, Page DR, Chambrier P, Lionnet C, Faure J-E, Grossniklaus U, Berger F (2004) Identification of new members of Fertilisation Independent Seed Polycomb Group pathway involved in the control of seed development in Arabidopsis thaliana. Development 131(12):2971–2981
84. Rodrigues JC, Tucker MR, Johnson SD, Hrmova M, Koltunow AM (2008) Sexual and apomictic seed formation in Hieracium requires the plant polycomb-group gene FERTILIZATION INDEPENDENT ENDOSPERM. Plant Cell 20(9):2372–2386
85. Tonosaki K, Kinoshita T (2015) Possible roles for polycomb repressive complex 2 in cereal endosperm. Front Plant Sci 6:144
86. Olmedo-Monfil V, Durán-Figueroa N, Arteaga-Vázquez M, Demesa-Arévalo E, Autran D, Grimanelli D, Slotkin RK, Martienssen RA, Vielle-Calzada J-P (2010) Control of female gamete formation by a small RNA pathway in Arabidopsis. Nature 464 (7288):628–632
87. Hernández-Lagana E, Rodríguez-Leal D, Lúa J, Vielle-Calzada J-P (2016) A multigenic network of ARGONAUTE4 clade members controls early megaspore formation in Arabidopsis. Genetics 116:188151
88. Lawit SJ, Albertsen MC, Fox T, Gordon-Kamm W, VAN AMM, CALZADA JPV, ROSILLO CB, AREVALO ED, CHAVEZ CG, LAGANA EH (2016) Methods for reproducing plants asexually and compositions thereof. WO 2016/048909 A1
89. Garcia-Aguilar M, Michaud C, Leblanc O, Grimanelli D (2010) Inactivation of a DNA methylation pathway in maize reproductive organs results in apomixis-like phenotypes. Plant Cell 22(10):3249–3267
90. Chan SR, Maruthachalam R, Mercier R, Nogue F (2014) Synthetic clonal reproduction through seeds. US20140298507 A1
91. Lawit SJ (2012) Self-reproducing hybrid plants. US20120266324 A1
92. d'Erfurth I, Jolivet S, Froger N, Catrice O, Novatchkova M, Mercier R (2009) Turning meiosis into mitosis. PLoS Biol 7(6): e1000124
93. d'Erfurth I, Cromer L, Jolivet S, Girard C, Horlow C, Sun Y, To JP, Berchowitz LE, Copenhaver GP, Mercier R (2010) The

cyclin-A CYCA1; 2/TAM is required for the meiosis I to meiosis II transition and cooperates with OSD1 for the prophase to first meiotic division transition. PLoS Genet 6(6): e1000989
94. Mieulet D, Jolivet S, Rivard M, Cromer L, Vernet A, Mayonove P, Pereira L, Droc G, Courtois B, Guiderdoni E (2016) Turning rice meiosis into mitosis. Cell Res 26 (11):1242–1254
95. Schiml S, Puchta H (2016) Revolutionizing plant biology: multiple ways of genome engineering by CRISPR/Cas. Plant Methods 12 (1):1
96. Liang G, Zhang H, Lou D, Yu D (2016) Selection of highly efficient sgRNAs for CRISPR/Cas9-based plant genome editing. Sci Rep 6:Article number: 21451
97. Belhaj K, Chaparro-Garcia A, Kamoun S, Patron NJ, Nekrasov V (2015) Editing plant genomes with CRISPR/Cas9. Curr Opin Biotechnol 32:76–84
98. Warthmann N, Chen H, Ossowski S, Weigel D, Hervé P (2008) Highly specific gene silencing by artificial miRNAs in rice. PLoS One 3(3):e1829
99. Schwab R, Ossowski S, Riester M, Warthmann N, Weigel D (2006) Highly specific gene silencing by artificial microRNAs in Arabidopsis. Plant Cell 18(5):1121–1133
100. Ossowski S, Schwab R, Weigel D (2008) Gene silencing in plants using artificial microRNAs and other small RNAs. Plant J 53(4):674–690. doi:10.1111/j.1365-313X.2007.03328.x
101. Marimuthu MP, Jolivet S, Ravi M, Pereira L, Davda JN, Cromer L, Wang L, Nogué F, Chan SW, Siddiqi I (2011) Synthetic clonal reproduction through seeds. Science 331 (6019):876–876
102. Kelliher T, Starr D, Wang W, McCuiston J, Zhong H, Nuccio ML, Martin B (2016) Maternal haploids are preferentially induced by CENH3-tailswap transgenic complementation in maize. Front Plant Sci 7:414
103. Lin B-Y (1984) Ploidy barrier to endosperm development in maize. Genetics 107 (1):103–115
104. Singh M, Goel S, Meeley RB, Dantec C, Parrinello H, Michaud C, Leblanc O, Grimanelli D (2011) Production of viable gametes without meiosis in maize deficient for an ARGONAUTE protein. Plant Cell 23(2):443–458

Part II

Clearing, Imaging Techniques, and Immunolocalization to Study Cell Lineage, Cell Type Specificity, Cytoskeleton, Cell Cycle, Signaling, and Hormones

Chapter 3

3D Imaging of Whole-Mount Ovules at Cellular Resolution to Study Female Germline Development in Rice

Ethel Mendocilla-Sato, Wenjing She, and Célia Baroux

Abstract

Recent advances in fluorescence-based staining of cellular compartments coupled with confocal microscopy imaging have allowed the visualization of three-dimensional (3D) structures with cellular resolution in various intact plant tissues and species. Such approaches are of particular interest for the analysis of the reproductive lineage in plants including the meiotic precursor cells deeply embedded within the ovary of the gynoecium enclosed in the flower. Yet, their relative inaccessibility and the lack of optical clarity of plant tissues prevent robust staining and imaging across several cell layers. Several whole-mount tissue staining and clearing techniques are available. One of them specifically allows staining of cellular boundaries in thick tissue samples while providing extreme optical clarity, using an acidic treatment followed by a modified Pseudo-Schiff propidium iodide (mPS-PI) method. While commonly used for *Arabidopsis* tissues, its application to other species like the model crop rice required protocol adaptations for obtaining robust staining that we present here. The procedure comprises six steps: (a) Material sampling; (b) Material fixation; (c) Tissue preparation; (d) Staining; (e) Sample mounting; and (d) Microscopy imaging. Particularly, we use ethanol and acetic anhydride as fixative reagents. A modified enzymatic treatment proved essential for starch degradation influencing optical clarity hence allowing acquisition of images at high resolution. This improved protocol is efficient for analyzing the megaspore mother cells in rice (*Oryza sativa*) ovary but is broadly applicable to other crop tissues of complex composition, without the need for tissue sectioning.

Key words Female germline, Megaspore mother cell, Ovary, Development, Rice, Whole-mount staining, 3D imaging, mPS-PI

1 Introduction

In angiosperms, the female germline develops in the ovule enclosed in the gynoecium of the flower. Forward and reverse genetics studies have contributed to shed light onto the genetic and molecular control of ovule development in terms of morphology and cell fate acquisition [1–3]. In both *Arabidopsis* and rice (*Oryza sativa*), model plant systems for eudicots and monocots, respectively, a single sub-epidermal cell develops into the archesporial cell, which in turn differentiates into a megaspore mother cell (MMC). The

Anja Schmidt (ed.), *Plant Germline Development: Methods and Protocols*, Methods in Molecular Biology, vol. 1669, DOI 10.1007/978-1-4939-7286-9_3, © Springer Science+Business Media LLC 2017

MMC is the precursor of the female reproductive lineage. Meiosis is followed by three rounds of mitoses, cellularization, and cellular differentiation, giving rise to the female gametophyte.

Noninvasive microscopy imaging of either fluorescent proteins tagging proteins of interests, or chemical dyes labeling cellular compartments, has allowed the analysis of three-dimensional (3D) cellular organization in intact plant tissues. Particularly, the analysis of 3D tissue patterning exploits the robust and cell-wall specific staining provided by propidium iodide suitable for high-resolution confocal microscopy imaging and 3D reconstruction of whole tissues or organs with cellular resolution. Most of the fixation, clearing, and staining procedures have been adapted on *Arabidopsis* leaves, flowers, embryos, anthers, ovules, and roots [4–8]. Such an approach provides the basis for tissue morphodynamic studies that relies on measuring tissue growth parameters such as cell proliferation and anisotropic cell expansion [9, 10]. These measurements are derived from 3D image reconstructionWhole-mount staining and digital segmentation of individual cells using open-source and commercial software which have been successfully used for plant tissue analyses [11–13].

For visualizing ovule tissue patterning and female germline development in rice, we have optimized the original modified Pseudo-Schiff propidium iodide staining technique [4]. We have used ethanol and acetic anhydride as fixative reagents and included a step of enzymatic treatment for starch degradation as described in [7]. This latter step has proven crucial for staining deeply embedded structures, and improving optical clarity hence reducing light scattering. Collectively, this modified protocol provides good optical clarity and imaging depth of rice ovaries using a conventional laser scanning confocal microscope.

2 Materials

2.1 Plant Material and Equipment

1. Flowering rice plants (*see* **Note 1**).
2. Stereomicroscope.
3. Benchtop incubator/mixer for microcentrifuge tubes.

2.2 Material Fixation

1. 2 mL microcentrifuge tubes.
2. Scalpel.
3. Tungsten dissecting needles (diameter: 0.75 and 0.35 mm); cleanup needles with 70% (v/v) ethanol before dissection.
4. Tweezers.
5. Fixative: three parts of 100% ethanol, one part of acetic anhydride; freshly prepared with sterile deionized water (*see* Subheading 3.2, **step 1**).

2.3 Tissue Preparation

1. 1% (w/v) SDS, 0.2 N NaOH (*see* **Note 2**).
2. Enzyme solution: 0.5% (w/v) α-amylase (from *Bacillus licheniformis*, Sigma-Aldrich, sterile 100 mg-vial), 0.5% (v/v) pullulanase (Sigma-Aldrich, sterile 50 mL flask) (*see* **Note 3**).
3. 80% (v/v) ethanol.
4. 1% (w/v) periodic acid. Periodic acid is weighed under the fume hood and dissolved in distilled water. Keep the solution under the fume hood.

2.4 Staining

1. Modified Pseudo-Schiff reagent with propidium iodide (mPS-PI): Prepare a stock solution of 1 M sodium metabisulfite ($Na_2S_2O_3$). Next, prepare a working solution of 10 mL of the 1 M $Na_2S_2O_3$ solution in 100 mL of 0.15 N HCl. Store solutions at room temperature for up to 4 months. Before use, add propidium iodide to reach a final concentration of 10 μg/mL.

2.5 Mounting Samples on Slides

All the solutions can be stored at room temperature. They are stable for several months.

1. Dust-free paper.
2. 70% (v/v) ethanol.
3. Diamond- or carbon-tip pen.
4. Cover slips: 18 × 18 mm, 0.17 ± 0.01 mm thickness.
5. Glass slides: 76 × 26 mm.
6. Chloral hydrate solution: 40 g chloral hydrate—dissolve with slight heating while stirring in 20 mL of water. When dissolved, add 10 mL glycerol. Keep at room temperature protected from light (stable for several months).
7. Hoyer's solution: Add 30 g gum arabic to 50 mL distilled water. Heat to 60 °C and stir overnight. Successively, add 200 g chloral hydrate. Once the chloral hydrate has dissolved, add 20 g glycerol. Filter the solution through glass wool (*see* **Note 4**).
8. Transparent nail polish.

2.6 Microscopy Imaging

1. Confocal scanning laser microscope, ideally with fast, resonant scanner (*see* **Note 5**).
2. Objective: 63× glycerol immersion NA 1.3 or more, 280 μm working distance or more, confocal grade (APO PL corrections) (*see* **Note 6**).

3 Methods

3.1 Plant Material Sampling

1. Cut off at the base of the leaf sheaths which enclose the young panicles.
2. With a scalpel, cut longitudinally the leaf sheath and open it up to expose the panicle with the spikelets (Fig. 1).
3. Under a stereomicroscope, take one spikelet and gently strip the lemma, palea with tweezers to isolate the ovary.
4. Further dissection of stamens and lodicules can be done with dissecting needles to avoid damaging the ovary. Unlike in *Arabidopsis*, one ovary bears a single ovule.

3.2 Tissue Fixation

1. Isolated ovaries are immediately transferred to a 2 mL microcentrifuge tube with 1 mL of fixative solution (*see* **Note** 7).
2. Fix the tissue overnight at 4 °C.
3. Wash in 70% (v/v) ethanol or store in 70% ethanol at 4 °C for up to 1 month.

3.3 Tissue Preparation

The steps for tissue preparation were adapted from protocols published by Truernit et al. [4], Yoshida et al. [8], and Nguyen and Mc Curdy [7].

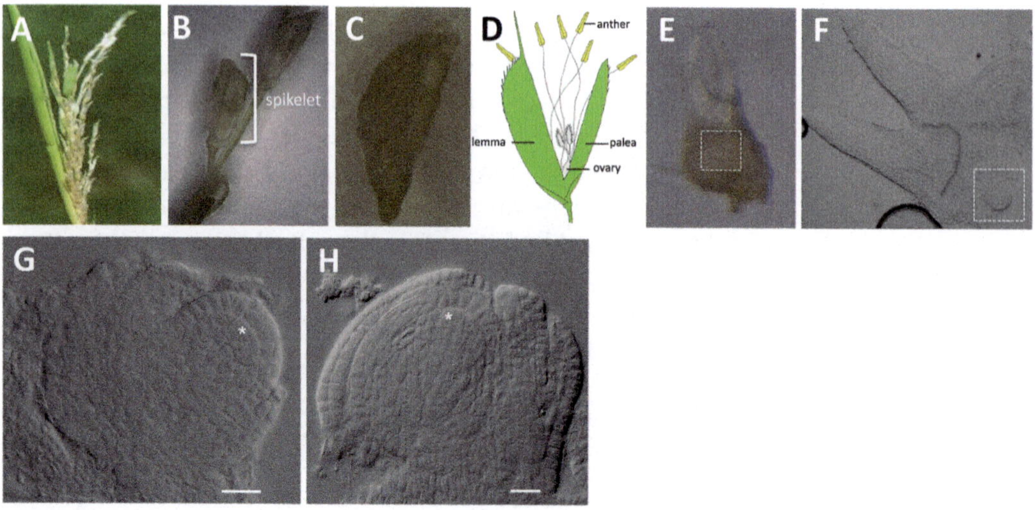

Fig. 1 Dissection of rice flowers to isolate ovule primordia in wild-type rice (*Oryza sativa*, var. Nipponbare). (**a–e**) An example for dissection of a rice inflorescence. (**a**) Morphology of the wild-type rice panicles. (**b**) Closeup of the panicle in (**a**), with spikelets growing along the stem in phyllotaxy. (**c**) A young spikelet collected from (**b**). (**d**) Morphology of a rice spikelet. (**e**) One pistil dissected from a spikelet. (**f**) A dissected pistil, with the ovule primordium visible (*dashed line*). (**g**, **h**) Developing ovule primordia at the megaspore mother cell (*asterisk*) stage (**g**) and functional megaspore (*asterisk*) stage (**h**, note the elongated integuments). Scale bar: 20 μm

1. Transfer the fixed tissues into 1 mL 1% SDS, 0.2 N NaOH solution in a 2 mL microcentrifuge tube. Incubate for 10 min at room temperature.
2. Remove the solution, rinse with sterile distilled water. Pipet out the solutions slowly with great care to avoid losing samples.
3. Remove the water and add 1 mL enzyme solution. Incubate at 37 °C for 6 h (*see* **Note 8**).
4. Rinse two times with 1 mL distilled water (pipet carefully the solution).
5. Under the fume hood: Remove the water and add 1 mL of 1% periodic acid. Incubate for 15 min.
6. Remove as much as possible the periodic acid solution and rinse the tissue with 1 mL sterile distilled water.

3.4 Staining

1. Incubate the tissues in mPS-PI for 1–2 h (*see* **Note 9**).
2. Rinse two times in water.
3. Remove as much water as possible. Add 300 μL of chloral hydrate solution and leave in darkness overnight at room temperature.

3.5 Sample Mounting

1. Clean a slide and a coverslip with 70% ethanol using dust-free paper. Allow to air-dry before use.
2. Cut a coverslip into 4–6 small pieces using a diamond tip pen.
3. Place two pieces of coverslips on the slide. Arrange them vertically and leave a space of approximately 10–12 mm in between.
4. Add one drop of Hoyer's solution on a slide between the two pieces of coverslip.
5. Using tungsten needles, remove the upper and lower parts of the ovary (*see* **Note 10**).
6. Gently place a coverslip on top of the preparation. Avoid bubbles and most importantly, do not squeeze the preparation (*see* **Note 11**).
7. If the mounting solution does not fill the entire area below the coverslip, add Hoyer's solution with a pipet from the side. Clean the excess with dust-free paper.
8. Seal the edges of the preparation with nail polish.
9. Allow the sample to settle in the Hoyer's solution for 2–3 days before imaging.

3.6 Microscopy Imaging

1. Add a drop of 80% glycerol on the coverslip.
2. Locate the ovule samples under a 10× or 20× (air or multi-immersion) objective using transmitted light or epifluorescence.

3. Switch to higher magnification objective (63× or 40×). Samples close to the coverslip are better suited for optimal resolution imaging.
4. The excitation and emission maxima of Propidium Iodide are 536 and 617 nm, respectively. Set up an acquisition profile using an appropriate excitation line (488, 514, 543, or 561 nm laser lines provide an efficient excitation source, with a preference for 561 nm) and collect emission between 570 and 620 nm. The fluorophore is very stable and allows using high laser transmission and several-fold averaging (to be balanced with acquisition time).
5. Use the Z-stack mode under optimal resolution to record serial sections of the whole ovule for 3D reconstruction (*see* **Note 12**).

3.7 Application: Image Reconstruction and 3D Segmentation

High-resolution 3D images provide material for studying the shape, size, and cellular arrangements following 3D reconstruction. Depending on the aim of the biological study, image segmentation can be performed where intensity signals from cell walls are used to define cellular boundaries. Several software are available: open-source (e.g., Fiji, MARS-ALT, MorphographX, and RACE) [11, 12, 14] or commercial (for instance Imaris, among others, as used in Fig. 2d, *see* **Note 13**). This approach ultimately gives a digital image composed of 3D objects corresponding to individual cells which can be labeled according to their identity (Fig. 2) and used to extract quantitative information on their size, shape, and position [8, 12, 13, 15, 16]. This quantitative information is valuable for the study of growth dynamics.

1. To visualize the generated volumetric data (e.g., Z-stacks), we recommend using either MorphographX or Imaris. Both interfaces are user-friendly.

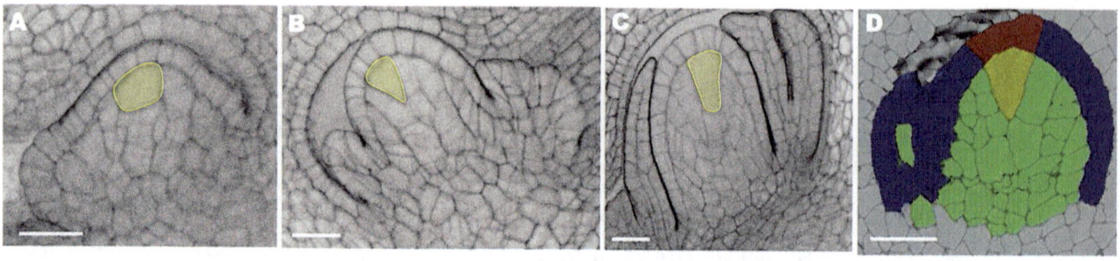

Fig. 2 Confocal images of wild-type rice ovules at three developmental stages and 3D segmentation. (**a–c**) Single optical sections through whole-mount rice ovaries showing developing ovule primordia. (**a**) A sub-epidermal cell acquires the identity of spore mother cell (shown in *yellow*). (**b**) Outer and inner integuments start growing and the megaspore mother cell elongates. (**c**) Integuments cover most of the nucellus, protecting the developing functional megaspore. (**d**) Segmentation of a 3D reconstructed image of an ovule primordium showing the L1 dome (in *red*), L1 layer (in *dark blue*), the MMC (*yellow*), and other L2/L3 cells (*light green*). 3d reconstruction and segmentation powered by Imaris (Bitplane, CH). Scale bars: 20 μm

2. Both software share the same workflow, which comprises the following steps: (a) image blurring—to reduce the noise; (b) segmentation; (c) segmentation correction; (d) extraction of geometric data.

4 Notes

1. The reproductive phase of *Oryza sativa* L. ssp. japonica var. Nipponbare starts with the booting stage—a bulge at the base of the leaf sheath suggests the panicle initiation and formation. At this stage, collect one leaf sheath to check for the presence of spikelets in the developing panicle (Fig. 1). In our conditions, Nipponbare plants are grown with a 10 h light:14 h darkness photoperiod at 28 °C. var. Nipponbare starts flowering on average 120 days after sowing. Optimizing temperature and day length will fasten flowering (https://ag.purdue.edu/hla/Hort/Greenhouse/Pages/Rice-Method.aspx) [17].
2. We recommend autoclaving this solution. Store at room temperature for up to 4 months.
3. The two enzymes dissolve well in distilled water. Use preferably freshly prepared enzyme solution.
4. To filter the Hoyer's solution, accommodate a thin layer of glass wool inside a conical glass/plastic funnel. Add subsequently the solution to remove the impurities from the gum arabic.
5. With a confocal scanning laser microscope, we obtained good depth resolution across 150 μm of sample thickness. Alternatively, multiphoton microscopy might provide better results.
6. Another option for imaging thick samples is to use a 40× objective with a high NA and long working distance. If the immersion medium of the objective is oil, replace then the Hoyer's solution with mineral oil.
7. As soon as the rice ovary is removed from the surrounding tissues, transfer it to the fixative solution so that cellular components are stabilized and preserved.
8. In our hands, the enzymatic treatment with α-amylase and pullulanase for 6 h is enough to effectively remove starch. çlpha-amylases initiate the hydrolysis of starch and pullulanases continue with further degradation of starch-related polysaccharides.
9. After adding propidium iodide to the modified Pseudo-Schiff reagent, cover microcentrifuge tubes with aluminum foil. Propidium iodide is light sensitive. We would recommend staining the tissues until they become light reddish.

10. The single ovule is positioned in the middle of the ovary and develops perpendicularly to the growth axis.
11. Hold one corner with a finger. Hold the opposite (diagonal) corner with another finger and slowly approach the coverslip from the preparation. Do not let the coverslip fall. This step is crucial to preserve the 3D structure of the tissues.
12. Set up the image format and z-step to produce voxel size 2–3× smaller than the theoretical resolution. For this, check the provider's description. For instance, if lateral and axial resolutions are of 150 and 210 nm, respectively, a 2× oversampling requires 75 nm-wide pixels (adjusted with the digital zoom and image format) and 100 nm z-step. A good solution is to set a cubic voxel of 80 nm size (80, 80, 80 nm in x, y, z).
13. In a separate protocol, we describe how to perform 3D image segmentation following mPS-PI staining, cell annotation, and export of quantitative data [13].

Acknowledgments

This work was supported and funded by the Commission for Technology and Innovation (CTI grant 16997), the Baugarten Stiftung Zürich, the University of Zürich, and the Swiss National Science Foundation (SNF grant 31003A_149974). We acknowledge Professor Ueli Grossniklaus (UG) for scientific support and insightful discussions and technical assistants of the department for organizational support and assistance with microscopy imaging (Christoph Eichenberger, Valeria Gagliardini, Arturo Bolaños, Peter Kopf).

References

1. Sieber P, Gheyselinck J, Gross-Hardt R, Laux T, Grossniklaus U, Schneitz K (2004) Pattern formation during early ovule development in Arabidopsis thaliana. Dev Biol 273(2):321–334. doi:10.1016/j.ydbio.2004.05.037
2. Colombo L, Battaglia R, Kater MM (2008) Arabidopsis ovule development and its evolutionary conservation. Trends Plant Sci 13(8):444–450. doi:10.1016/j.tplants.2008.04.011
3. Cucinotta M, Colombo L, Roig-Villanova I (2014) Ovule development, a new model for lateral organ formation. Front Plant Sci 5:117. doi:10.3389/fpls.2014.00117
4. Truernit E, Bauby H, Dubreucq B, Grandjean O, Runions J, Barthélémy J, Palauqui JC (2008) High-resolution whole-mount imaging of three-dimensional tissue organization and gene expression enables the study of Phloem development and structure in Arabidopsis. Plant Cell 20(6):1494–1503. doi:10.1105/tpc.107.056069
5. Kurihara D, Mizuta Y, Sato Y, Higashiyama T (2015) ClearSee: a rapid optical clearing reagent for whole-plant fluorescence imaging. Development 142(23):4168–4179. doi:10.1242/dev.127613
6. Musielak TJ, Schenkel L, Kolb M, Henschen A, Bayer M (2015) A simple and versatile cell wall staining protocol to study plant reproduction. Plant Reprod 28(3–4):161–169. doi:10.1007/s00497-015-0267-1
7. Nguyen ST, McCurdy DW (2015) High-resolution confocal imaging of wall ingrowth deposition in plant transfer cells: semi-quantitative analysis of phloem parenchyma transfer cell development in leaf minor veins

of Arabidopsis. BMC Plant Biol 15:109. doi:10.1186/s12870-015-0483-8

8. Yoshida S, Barbier de Reuille P, Lane B, Bassel GW, Prusinkiewicz P, Smith RS, Weijers D (2014) Genetic control of plant development by overriding a geometric division rule. Dev Cell 29(1):75–87. doi:10.1016/j.devcel.2014.02.002
9. Bassel GW, Smith RS (2016) Quantifying morphogenesis in plants in 4D. Curr Opin Plant Biol 29:87–94. doi:10.1016/j.pbi.2015.11.005
10. Coen E, Rolland-Lagan AG, Matthews M, Bangham JA, Prusinkiewicz P (2004) The genetics of geometry. Proc Natl Acad Sci U S A 101(14):4728–4735. doi:10.1073/pnas.0306308101
11. Fernandez R, Das P, Mirabet V, Moscardi E, Traas J, Verdeil JL, Malandain G, Godin C (2010) Imaging plant growth in 4D: robust tissue reconstruction and lineaging at cell resolution. Nat Methods 7(7):547–553. doi:10.1038/nmeth.1472
12. Barbier de Reuille P, Routier-Kierzkowska AL, Kierzkowski D, Bassel GW, Schüpbach T, Tauriello G, Bajpai N, Strauss S, Weber A, Kiss A, Burian A, Hofhuis H, Sapala A, Lipowczan M, Heimlicher MB, Robinson S, Bayer EM, Basler K, Koumoutsakos P, Roeder AH, Aegerter-Wilmsen T, Nakayama N, Tsiantis M, Hay A, Kwiatkowska D, Xenarios I, Kuhlemeier C, Smith RS (2015) MorphoGraphX: a platform for quantifying morphogenesis in 4D. elife 4:05864. doi:10.7554/eLife.05864
13. Mendocilla Sato E, Baroux C (2017) Analysis of 3D cellular organization of fixed plant tissues using a user-guided platform for image segmentation. Bio-Protocols 7(12):e2355
14. Stegmaier J, Amat F, Lemon WC, McDole K, Wan Y, Teodoro G, Mikut R, Keller PJ (2016) Real-time three-dimensional cell segmentation in large-scale microscopy data of developing embryos. Dev Cell 36(2):225–240. doi:10.1016/j.devcel.2015.12.028
15. Sankar M, Nieminen K, Ragni L, Xenarios I, Hardtke CS (2014) Automated quantitative histology reveals vascular morphodynamics during Arabidopsis hypocotyl secondary growth. elife 3:e01567. doi:10.7554/eLife.01567
16. Hervieux N, Dumond M, Sapala A, Routier-Kierzkowska AL, Kierzkowski D, Roeder AH, Smith RS, Boudaoud A, Hamant O (2016) A mechanical feedback restricts sepal growth and shape in Arabidopsis. Curr Biol. doi:10.1016/j.cub.2016.03.004
17. Summerfield RJ, Collinson ST, Ellis RH, Roberts EH, Penning de Vries FWT (1992) Photothermal responses of flowering in rice (*Oryza sativa*). Ann Bot 69(2):101–112. doi:10.1093/oxfordjournals.aob.a088314

Chapter 4

Live-Cell Imaging of F-Actin Dynamics During Fertilization in *Arabidopsis thaliana*

Daichi Susaki, Daisuke Maruyama, Ramesh Yelagandula, Frederic Berger, and Tomokazu Kawashima

Abstract

Fertilization comprises a complex series of cellular processes leading to the fusion of a male and female gamete. Many studies have been carried out to investigate each step of fertilization in plants; however, our comprehensive understanding of all the sequential events during fertilization is still limited. This is largely due to difficulty in investigating events in the female gametophyte, which is deeply embedded in the maternal tissue. Recent advances in confocal microcopy assisted by fluorescent marker lines have contributed to visualizing subcellular dynamics in real time during fertilization in vivo. In this chapter, we describe a method focusing on the investigation of F-actin dynamics in the central cell during male gamete nuclear migration. This method also allows the study of a wide range of early sexual reproduction events, from pollen tube guidance to the early stage of seed development.

Key words Arabidopsis, Fertilization, Gamete nuclear migration, F-actin, Confocal microscopy, Live-cell imaging

1 Introduction

Fertilization is the fundamental event required to initiate embryonic development in both animals and plants [1, 2]. In flowering plants, male gametes (sperm cells) are immotile and delivered to the female gamete via a pollen tube (pollen tube guidance). Each pollen tube releases two sperm cells between the two female gametes (egg and central cell). Each sperm cell recognizes the egg or central cell and this recognition phase is followed by gamete cell fusion (plasmogamy). After plasmogamy, the sperm nucleus migrates into the female gamete toward the nucleus. Gamete nuclear fusion (karyogamy) then occurs to complete fertilization and initiate embryogenesis [1, 2].

In the past decade, molecular mechanisms controlling male-female interaction, including pollen tube guidance, have been extensively investigated [3]. Monitoring events following gamete

Anja Schmidt (ed.), *Plant Germline Development: Methods and Protocols*, Methods in Molecular Biology, vol. 1669, DOI 10.1007/978-1-4939-7286-9_4, © Springer Science+Business Media LLC 2017

fusion that takes place inside the female gametophyte was enabled by the semi in vivo system of the *Arabidopsis thaliana* pollen tube attraction assay [4] together with advances in confocal microscopy and fluorescent marker lines [5]. Since the first time-lapse imaging of plant fertilization [6, 7], detailed real-time events of pollen tube-female gamete interaction [8–11], male gamete nuclear migration after plasmogamy [12], as well as the zygote elongation [13] were successfully visualized. Furthermore, the optimization of media for seed growth now allows for the monitoring of embryogenesis in real time [14]. Observation of sexual reproduction from intact plants is also possible with two-photon excitation microscopy [15].

Here, we describe the procedure for monitoring fertilization in a semi in vivo manner. In this chapter, we focus on visualizing F-actin dynamics in the central cell during fertilization. F-actin can be fluorescently labeled by expressing lifeact peptides, which bind specifically F-actin with minimum artifacts [16]. Using the sperm cell marker line (*proHTR10:H3.10-mRFP1*) [17] as well as the central cell F-actin marker line (*proFWA:lifeact-Venus*) [12, 18], time-lapse images of male gamete nuclear migration with F-actin dynamics in the central cell can be obtained. This procedure can also be applied to investigate pollen tube micropylar guidance as well as male-female gamete interaction, karyogamy, through the early phase of seed development.

2 Materials

2.1 Plant Material and Equipment

1. *Plant material. Arabidopsis thaliana* Col-0, wild-type, transgenic plants expressing lifeact-Venus in the central cell (*proFWA:lifeact-Venus*), and transgenic plants expressing H3.10-mRFP1 in the sperm cells (*proHTR10:H3.10-mRFP1*).
2. Plant growth chamber.
3. Dissection microscope.
4. Confocal microscope (40×/1.30 and 60×/1.40 oil immersion objective lens, e.g., Nikon Ti-E, Japan) equipped with a disc-scan confocal system (e.g., Yokogawa CSU-X1, Japan) and CMOS camera (e.g., Hamamatsu Photonics ORCA-Flash4.0, Japan).
5. 4-well glass-bottom dish.
6. Aluminum cooling/heating block.
7. 27-gauge needle.
8. Tweezers with fine tips (INOX No. 5).

2.2 Reagents

Use molecular biology grade chemicals for making all reagents. All the reagents are prepared using sterile double distilled Mono Q water.

1. 14% sucrose.
2. Pollen growth medium: 14% (w/v) sucrose, 0.001% (w/v) boric acid, 1.27 mM $Ca(NO_3)_2$, 0.4 mM $MgSO_4$, adjust pH to 7.0 with KOH. Add low gelling temperature agarose to a final concentration of 1.5% and melt the agarose in a microwave. The stock can be kept at 4 °C for up to 1 month.

3 Methods

3.1 Plant Growth

1. Sow seeds directly on soil and stratify 2–4 days in the dark at 4 °C.
2. Germinate seeds and grow plants for 2–3 weeks under short day conditions (16/22 °C and 8/16 h light/dark, respectively).
3. Afterward, transfer plants to long day conditions (22 °C, 16/8 h light/dark) for bolting and flowering (*see* **Note 1**).

3.2 Preparation of Pollen Growth Dish

1. Pour 1 mL of 14% sucrose around the well of a glass-bottom dish (*see* **Note 2**) and place the dish on an aluminum block at room temperature (Fig. 1a; *see* **Note 3**).
2. Pour 200 μL of agarose-melted pollen growth medium (~60 °C) into the well, and cover the entire surface with the medium.
3. Aspirate the medium by slowly pipetting to remove excess medium (*see* **Note 4**).
4. A few minutes later, aspirate all unsolidified medium by pipetting from the center of the dish. A very thin layer of the medium should be formed at the center of the dish (Fig. 1b).
5. Once completely solidified, remove the dish from the aluminum block, cover with a lid, and keep at room temperature until use.

3.3 Pollen Tube Growth

1. Emasculate closed flower buds of wild-type plants 1–2 days before live-cell imaging.

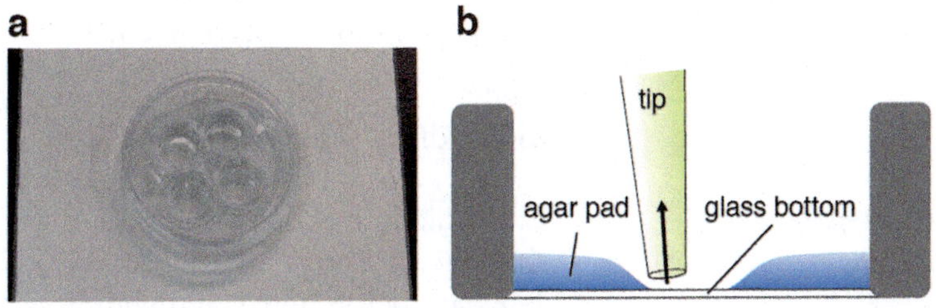

Fig. 1 Pollen growth dish setup. (**a**) Image of a dish placed on an aluminum block. (**b**) Schematic illustration of the pollen growth agarose medium in a well. Melted agarose is aspirated from the center to generate a very thin layer of the medium

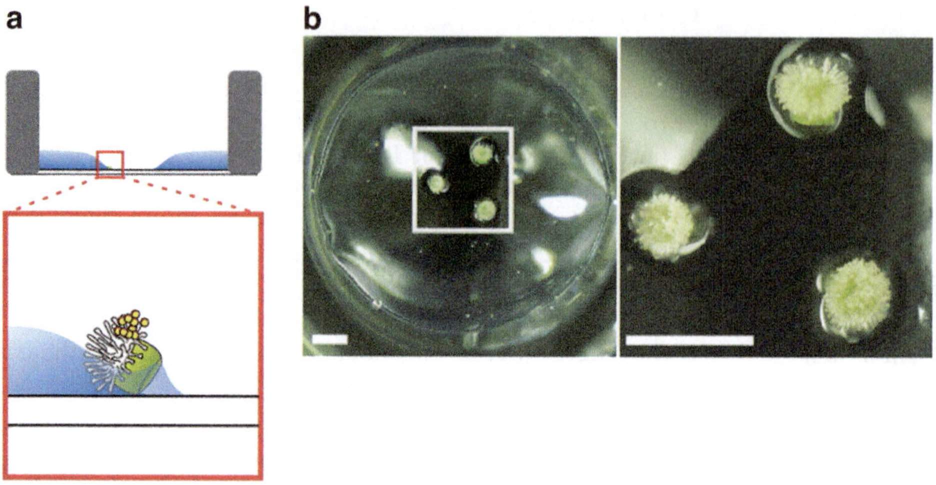

Fig. 2 (**a**) Schematic illustration of dissected stigma on the medium. (**b**) Image of the well with stigma placed on the medium. The enlarged image within the *white box* of the *left panel* is shown in the *right panel*. Bar = 1 mm

2. Under a dissection microscope, pollinate ~200 pollen grains from the *proHTR10:H3.10-mRFP1* transgenic plant onto the emasculated pistils in **step 1**.
3. Cut the top of a carpel (right beneath the stigma) with a 27-gauge needle, placing the back side of the needle toward the stigma (Fig. 2a). Usually, the stigma remains attached to the back side of the needle.
4. Transfer the stigma to the border between the thin and thick layers of the medium in the pollen growth dish prepared in Subheading 3.1 (Fig. 2a).
5. Adjust the position of the stigma so that the cut surface is facing the center of the dish (toward the thin layer) (Fig. 2b). Make sure the cut surface is covered by liquid coming from the medium; otherwise, pollen tubes do not grow out from the style (Fig. 2).
6. Place the lid back on the dish and seal with parafilm. Cover with aluminum foil to block the light.
7. Incubate at 22 °C for 2–2.5 h.
8. Under a dissection microscope, check pollen tube growth into the medium (Fig. 3a).

3.4 Preparation of Ovules

1. Emasculate un-opened flowers of the transgenic plants expressing lifeact-Venus in the central cell (*proFWA:lifeact-Venus*) 1–2 days before live-cell imaging.
2. Under a dissection microscope, collect ovules from pistils in **step 1** with a 27-gauge needle (*see* **Note 5**).

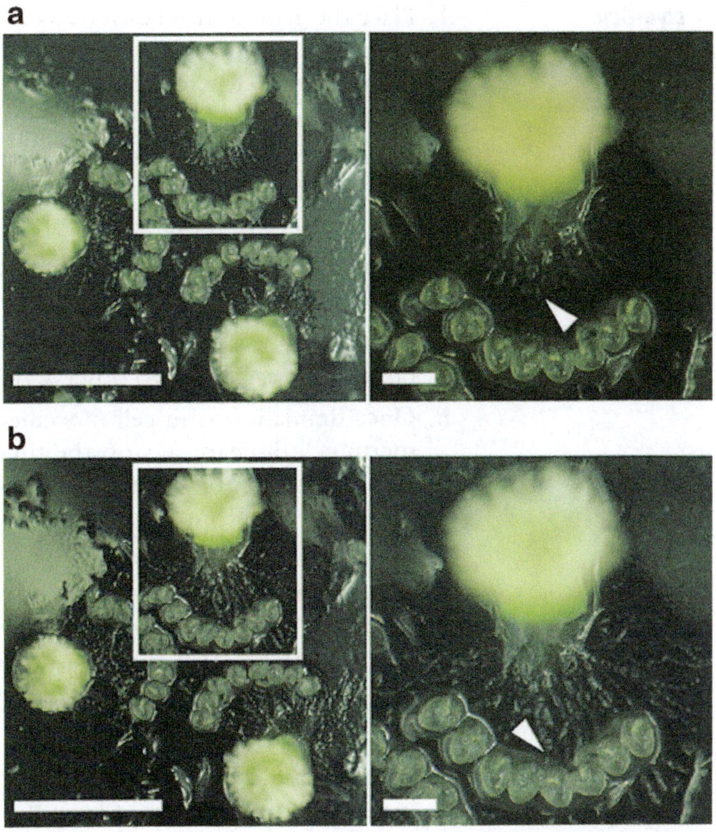

Fig. 3 Pollen tube growth on the medium. (**a**) Image of pollen tubes coming out from the stigma. The enlarged image within the *white box* of the *left panel* is shown in the *right panel*. (**b**) Image of pollen tubes growing toward dissected ovules on the medium. The enlarged image within the *white box* of the *left panel* is shown in the *right panel*. *Arrow heads* point to the pollen tube that reached the ovule. Bar = 1 mm

3. Place the collected ovules onto the center of the dish from Subheading 3.3, **step 8**.
4. Twist a kimwipe into a string and remove all liquid accumulated around the stigma using the kimwipe string.
5. Using a needle, position 8–10 ovules per stigma 50–100 μm away from the radially growing pollen tubes (Fig. 3a).
6. Place the lid back on the dish and seal with parafilm. Cover with aluminum foil to block the light.
7. Incubate at 22 °C for 2–2.5 h.
8. Under a dissection microscope, check pollen tube growth to the ovules (Fig. 3b).

3.5 Live-Cell Imaging

1. Place the glass-bottom dish on a confocal microscope stage and use a 40× objective lens and bright field illumination to focus on an ovule where pollen tubes are about to enter the micropyle.
2. Using the minimum laser power that barely produces a detectable signal, observe the mRFP1 signal (RFP channel) from the two sperm cells in the pollen tube (*see* **Note 6**).
3. Using a 60× objective lens, set the center of the Z-position to where the two sperm cells are visible.
4. Obtain 3 Z-section images (3 μm above and below the set Z-position) every 3 min to observe the sperm cell movement.
5. Once dramatic sperm cell movement is observed (the sign of sperm cell discharge), stop the time-lapse imaging in **step 4**, and reset the center of the Z-position to where two sperm cells are visible (*see* **Note 7**).
6. Adjust the 515 and 561 nm laser powers to the level that lifeact-Venus signal shows cable structures and the mRFP1 signal just becomes visible, respectively (*see* **Note 8**).
7. Using both YFP and RFP channels, obtain 7–10 Z-section images (the set Z-position as the center; 1–1.5 μm interval) every 30 s.
8. When the signals become weak, usually the lifeact-Venus signal, increase the laser power to keep the signal visible (*see* **Note 9**).
9. Analyze the acquired images using available software, e.g., Fiji [19] (*see* **Note 10**).

4 Notes

1. Large, healthy growing ovules usually give strong stable fluorescent signals. Plants can be grown first under short day condition so that plants grow bigger, generating large healthy ovules for better imaging.
2. In order to keep moisture within the dish during observation, this step is recommended. Water can be used instead of 14% sucrose.
3. An aluminum block can evenly capture the heat of the melting agarose from the bottom. This helps generate a smooth, thin agarose layer.
4. Aspirating excess agarose at this step speeds up the agarose solidifying process. When aspirating too much, the agarose no longer covers the entire bottom of the well due to surface tension.

5. For example, stabilizing dissected pistils on double-sided adhesive tape makes ovule dissection easier. Under a dissection microscope, slit each carpel wall along the margin with the replum using a needle and remove the carpel wall from the pistil. Cut the funiculi and transfer the dissected ovules to a dish.
6. Photo bleaching is an issue in time-lapse live-cell imaging. It is important to use the minimum laser power to check the sperm cell position (RFP signal) to prevent photo bleaching.
7. When sperm cells are discharged, there is a dramatic movement of sperm cells compared to the movement of sperm cells in the pollen tube toward the pollen tube tip before discharge [6].
8. In order to obtain time-lapse images, the laser power should be adjusted as low as possible due to the problem of photo bleaching. However, it is important to obtain cable structure images from Lifeact-Venus signal (YFP) for F-actin dynamics. The laser power and image quality depend on the confocal microscope system used. The appropriate conditions need to be identified for each system.
9. Based on the confocal microscope system mentioned in this chapter, the signal of Lifeact-Venus becomes weaker. The laser power can be increased when a reduction of the signal is noticed.
10. Z projection, adjustment of brightness and contrast, as well as movie generation can be performed with Fiji [19, 20].

Acknowledgments

D.M. is supported by Toyoaki Scholarship Foundation and JSPS KAKENHI grant, number 15K14541 (Grant-in-Aid for Challenging Exploratory Research) and 16H06173 (Grant-in-Aid for Young Scientists).

References

1. Kawashima T, Berger F (2011) Green love talks; cell-cell communication during double fertilization in flowering plants. AoB Plants 2011:plr015
2. Dresselhaus T, Sprunck S, Wessel GM (2016) Fertilization mechanisms in flowering plants. Curr Biol 26:R125–R139
3. Higashiyama T, Takeuchi H (2015) The mechanism and key molecules involved in pollen tube guidance. Annu Rev Plant Biol 66:393–413
4. Palanivelu R, Preuss D (2006) Distinct short-range ovule signals attract or repel Arabidopsis thaliana pollen tubes in vitro. BMC Plant Biol 6:7
5. Jeong S, Lukowitz W (2015) Embryos, camera, laser, action. Dev Cell 34:137–138
6. Hamamura Y, Saito C, Awai C et al (2011) Live-cell imaging reveals the dynamics of two sperm cells during double fertilization in Arabidopsis thaliana. Curr Biol 21:497–502
7. Rotman N, Rozier F, Boavida L et al (2003) Female control of male gamete delivery during fertilization in Arabidopsis thaliana. Curr Biol 13:432–436

8. Ngo QA, Vogler H, Lituiev DS et al (2014) A calcium dialog mediated by the FERONIA signal transduction pathway controls plant sperm delivery. Dev Cell 29:491–500
9. Iwano M, Ngo QA, Entani T et al (2012) Cytoplasmic Ca2+ changes dynamically during the interaction of the pollen tube with synergid cells. Development 139:4202–4209
10. Hamamura Y, Nishimaki M, Takeuchi H et al (2014) Live imaging of calcium spikes during double fertilization in Arabidopsis. Nat Commun 5:4722
11. Denninger P, Bleckmann A, Lausser A et al (2014) Male-female communication triggers calcium signatures during fertilization in Arabidopsis. Nat Commun 5:4645
12. Kawashima T, Maruyama D, Shagirov M et al (2014) Dynamic F-actin movement is essential for fertilization in Arabidopsis thaliana. Elife 3: e04501
13. Kimata Y, Higaki T, Kawashima T et al (2016) Cytoskeleton dynamics control the first asymmetric cell division in Arabidopsis zygote. Proc Natl Acad Sci U S A 113:14157–14162
14. Gooh K, Ueda M, Aruga K et al (2015) Live-cell imaging and optical manipulation of Arabidopsis early embryogenesis. Dev Cell 34:242–251
15. Mizuta Y, Kurihara D, Higashiyama T (2015) Two-photon imaging with longer wavelength excitation in intact Arabidopsis tissues. Protoplasma 252:1231–1240
16. Riedl J, Crevenna AH, Kessenbrock K et al (2008) Lifeact: a versatile marker to visualize F-actin. Nat Methods 5:605–607
17. Ingouff M, Hamamura Y, Gourgues M et al (2007) Distinct dynamics of HISTONE3 variants between the two fertilization products in plants. Curr Biol 17:1032–1037
18. Kawashima T, Berger F (2015) The central cell nuclear position at the micropylar end is maintained by the balance of F-actin dynamics, but dispensable for karyogamy in Arabidopsis. Plant Reprod 28:103–110
19. Schindelin J, Arganda-Carreras I, Frise E et al (2012) Fiji: an open-source platform for biological-image analysis. Nat Methods 9:676–682
20. Schneider CA, Rasband WS, Eliceiri KW (2012) NIH image to ImageJ: 25 years of image analysis. Nat Methods 9:671–675

Chapter 5

Development and Observation of Mature Megagametophyte Cell-Specific Fluorescent Markers

Mark A. Chamberlin and Shai J. Lawit

Abstract

Visualization of the intact embryo sac within the ovular/gynoecial tissues and clear identification of cell types can be logistically difficult and subject to interpretation. Cellular marker technologies have been available for the embryo sac, but have typically labeled only one cell type in a particular line. Here, we describe techniques for simultaneous labeling each cell type in the embryo sac and visualization methods for such in *Arabidopsis*, soybean, maize, and sorghum.

Key words Fluorescent reporter, Cell labeling, Megagametophyte, Embryo sac, Cell visualization

1 Introduction

The well-characterized *Polygonum*-type of embryo sac is the most prevalent among angiosperm species, including the model *Arabidopsis* as well as many agronomically important crops including maize, sorghum, rapeseed, and soybean. In the majority of the species, this results in the formation of an eight-nucleate, seven-celled mature female gametophyte. This comprises three accessory antipodal cells, two pollen tube attracting synergid cells, one egg cell, and one central cell with two polar nuclei that fuse prior to or during fertilization reviewed by [1–3]. The number of antipodal cells is typically three as observed in *Arabidopsis* and soybean, but can be variable between species. In sorghum we observed five and in maize we have observed 50 or more antipodal cells. The sheltered female gametophyte structures are relatively inaccessible, and therefore, difficult to observe in a minimally invasive manner. This plus the greater number of male reproductive cells relative to female reproductive cells have led to a preponderance of male reproductive studies relative to female. Due to the nature of ovule development into seeds, the gametophyte cells are embedded within several layers of ovular tissues, nucellus, and integuments. Traditional

Anja Schmidt (ed.), *Plant Germline Development: Methods and Protocols*, Methods in Molecular Biology, vol. 1669, DOI 10.1007/978-1-4939-7286-9_5, © Springer Science+Business Media LLC 2017

methods often are too time-consuming to allow efficient observations of large numbers of ovules. Even with relatively transparent integuments of the model system *Arabidopsis* the embryo sac is surrounded by five cell layers that make observation difficult without chemical fixation and clearing of whole ovules. Until recently, individual cell identity within the embryo sac facilitated through GUS marker lines has been the predominant methodology [4–6] often requiring crossing of mutants to multiple reporter lines to discern relevant biology. Genes with gametophyte cell-specific expression patterns have been identified by a number of studies [7–12], allowing for utilization of their promoters for differentially labeling of several unique cell/tissue types simultaneously. These and other cell-specific expression patterns are useful for fluorescent visualization of individual live cells to monitor development, interactions, and aberrations thereof.

We discuss methods for development of multi-color fluorescent reporter systems designed to specifically and simultaneously visualize individual cell types of the megagametophyte. Such systems facilitate visual tracking of each cell type; in this example, from embryo sac cellularization through fertilization and early embryo/endosperm development. Variants of this system can be useful in many tissues and organs to investigate development and the consequences of experimental manipulations.

2 Materials

2.1 Molecular Constructs and Transgenic Fluorescent Tissue

1. Cell-specific promoter sequences for reproductive tissues as in: [5, 7–13] (*see* **Note 1**).
2. Fluorescent protein coding sequences: e.g., Clontech (USA)/Takara (Japan); Evrogen (Russia) (*see* **Notes 2–4**). A very large selection of fluorescent proteins is available. It is critical to select proteins with nonoverlapping emission spectra.
3. Standard molecular cloning materials [14].
4. Transformation vectors appropriate to the species of interest.
5. Plant material grown to appropriate developmental stage (*see* Subheading 3.2).

2.2 Cytology

1. Gold Seal Microscope slides (Erie Scientific) for upright microscopes or Lab-Tek II Chambered Coverglass (Nunc) for inverted microscopes. Other brands may be used, but Gold Seal have been observed to be the cleanest in our experience.
2. Gold Seal Cover slips (Erie Scientific).
3. Phosphate-buffered saline (PBS; pH 7.2).
4. 2% paraformaldehyde:0.5% glutaraldehyde in PBS on ice.

5. Dissecting tools: ophthalmic scalpels (15° blade and 45° blade) and Dumont #5 and #7 forceps.
6. 2,2′-thiodiethanol (TDE) solutions as an optical clearing reagent.
7. FocusClear (CelExplorer Lab Co., Hsinchu, Taiwan).
8. Fluorescence capable compound microscope, e.g., Leica (Wetzlar, Germany) DMRXA epifluorescence microscope with a mercury light source using fluorescence, bright-field, and differential interference contrast (DIC) optics.
9. Appropriate fluorescent filter sets, e.g., from Omega Optical (Brattleboro, VT) or Chroma Technology (Bellows Falls, VT): Aqua #31036v2 (exc. 426–446, dichroic 455LP, em. 465–495); Alexa 488 #MF-105 (exc. 486–500, dichroic 505LP, em. 510–530); Alexa 532 #MF-106 (exc. 509–529, dichroic 535LP, em. 541–561); and Cy3 #C-106250 (exc. 541–551, dichroic 560LP, em. 565–605).
10. High sensitivity microscope compatible gray-scale camera, e.g., Photometrics (Tucson, AZ) CoolSNAP HQ CCD or Hamamatsu (Hamamatsu City, Japan) ORCA Flash 4.0 CMOS.
11. Image capture, analysis, and processing software, e.g., Molecular Devices MetaMorph (Downington, PA), ImageJ (https://imagej.nih.gov/ij), Carl Zeiss ZEN (Thornwood, NY), Leica Microsystems LAS X (Buffalo Grove, IL), etc.
12. Image editing software, e.g., Adobe Systems Photoshop CS (San Jose, CA), GIMP (https://www.gimp.org/), etc.

3 Methods

3.1 Transgenic Lines (See Note 5)

1. Prepare *Arabidopsis* lines as in Lawit et al. [13].
2. Transform rapeseed lines as in Chen et al. [15].
3. Transform soybean lines as in Li et al. [16].
4. Transform maize lines as in Cho et al. [17].
5. Transform sorghum lines as in Wu et al. [18].

3.2 Material Preparation—Staging, Harvesting, and Dissection

3.2.1 Floral Staging

1. *Arabidopsis*—Mature megagametophyte formation and pollination in *Arabidopsis* typically occurs concomitantly with floral opening.
2. Soybean—Fertilization often occurs prior to floral opening. In comparison with *Arabidopsis*, soybean gynoecia develop more rapidly than the ovules. Soybean gynoecia measure approximately 1.3 mm long relative to 270 μm long ovule at the mature megagametophyte stage.
3. Maize—Fertilization occurs within 24 h of pollination.

4. Sorghum—In the example of TX430, flowers at the mature embryo sac stage can be first found when the flag leaf initially appears. By the time the inflorescence (panicle) is obvious just above the flag leaf fertilization has already initiated.

3.2.2 Ovule Staging

In circumstances where a mature megagametophyte stage is desired, pollination may be prevented to ensure development does not progress beyond this stage. Various emasculation techniques are available. It may also be necessary to cover the stigma to prevent pollination by wind or insect mobilized pollen. Depending on the species, a maturation period may be allowed following emasculation to better synchronize development to the mature megagametophyte stage (in the case of *Arabidopsis* the maturation period is 24 h).

1. *Arabidopsis*—Gynoecia are easily peeled open with some dissection and the ovules removed and mounted in PBS or a clearing agent. Forty to fifty ovules may be observed per silique, divided evenly between the two valves.
2. Soybean—Gynoecia/pods are highly curved and the ovules are attached at the concave suture. Ovules are campylotropous, its micropylar end curved toward the stigma and the suture. Pods are easily peeled open and the ovules removed and mounted. One to four ovules may be observed per pod.
3. Maize—At 8 days after pollination maize embryos remain torpedo shaped and the endosperm is now fully cellularized and significantly fills the ovule, displacing 30–50% of the nucellus.
4. Sorghum—At a flower bud length of 3 mm the ovule is at the megaspore mother cell stage and pollen is at quartet to early uninucleate stage. At 5–6 mm the ovule is at the mature unfertilized embryo sac stage and the anther is near anthesis with mature starch-filled pollen.

3.2.3 Dissection

The developing ovules are contained within a gynoecium that must be dissected to make the ovule accessible. It is necessary to determine the correct developmental stage for observations; often this must be done after dissection. Dissected ovules from the gynoecium or siliques are mounted on well slides in PBS for microscopic observations.

1. *Arabidopsis*—Gynoecia may be removed from other floral organs under a dissecting microscope with the use of fine tweezers. Gynoecia may be opened by sliding a fine (33 gauge) needle along the replum. Ovules are easily removed from gynoecia by scraping along the septum or valves to free the ovules. *Arabidopsis* ovules are sufficiently transparent to allow for microscopic evaluation of fluorescent reporters

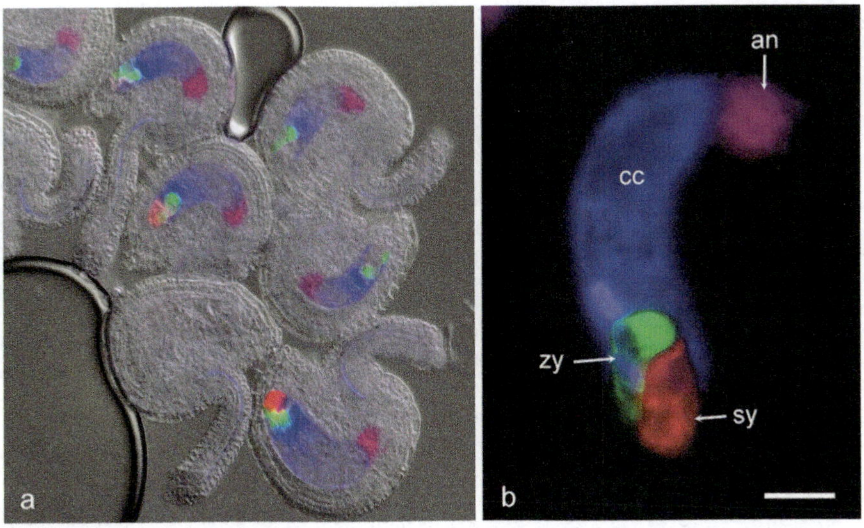

Fig. 1 Whole transgenic *Arabidopsis* ovules expressing a single construct of four embryo sac cell-specific promoters, each driving the expression of a different fluorescent protein (DD65 promoter-TagBFP (*blue*)—DD2 promoter-TagRFP (*red*)—DD1 promoter-mKate (*magenta*)—DD45 promoter-ZsGreen (*green*)). (**a**) Differential interference contrast—fluorescence overlay image of ovules expressing the four embryo sac-specific promoters. (**b**) Multi-color fluorescent image of a single post-fertilization ovule showing specificity of the promoter-fluorescent proteins to the embryo sac antipodals (an, mKate), central cell (cc, TagBFP), zygote (zy, ZsGreen) and persistent synergid (sy, TagRFP). Bar = 20 μm

without any further dissection (Fig. 1). Observation of fresh material is optimal.

2. Soybean—Gynoecia may be removed from other floral organs with the use of fine tweezers. Pods may be opened by peeling both valves with fine tweezers from the apical end. Ovules (Fig. 2) may be released by severing the funiculi.

3. Maize—Kernels at the appropriate stage are removed by their base from the ear, either by a scalpel or micro-spatula. Depending on the stage, glumes may need to be removed by their base before further dissection. The embryo sac is found at the base of the adaxial side of the young kernel (Fig. 3). To expose the embryo sac, the young kernel is held by forceps upright and horizontal longitudinal sections are made. First trim away 100 μm from the adaxial side of the kernel, then make a second section approximately 200 μm thick. This latter section will contain the intact embryo sac surrounded by the nucellus (Fig. 4a).

4. Sorghum—Like maize, sorghum contains a single ovule per gynoecium. The embryo sac is obliquely positioned at the basal end of the gynoecium. Ovules are best sectioned longitudinally removing approximately 200 μm of tissue to expose the embryo sac (Fig. 4b). Due to the angled position of the embryo sac within the ovule, trial and error is required to retrieve an intact embryo sac.

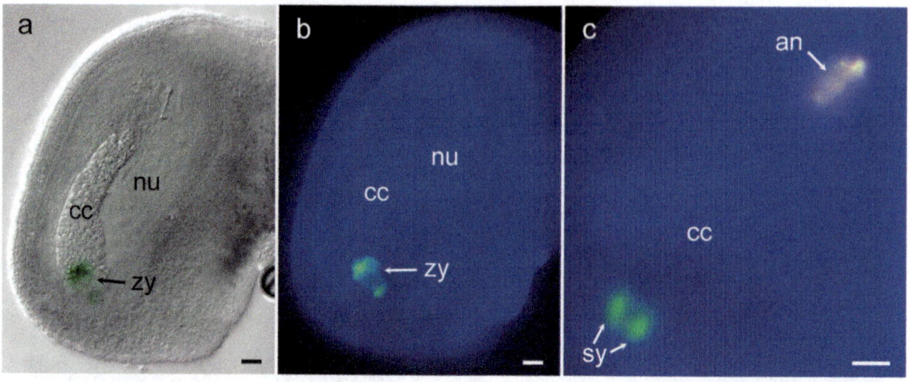

Fig. 2 Entire transgenic soybean ovules expressing embryo sac cell-specific promoters-fluorescent proteins. (**a** and **b**) Ovule expressing AT-DD45 promoter–ZsGreen in the zygote (zy) as shown in a DIC/fluorescence overlay (**a**) and fluorescence (**b**) images. (**a**) A common issue with early soybean embryo sacs is the abundance of large starch grains in the central cell (cc) that obscures clear fluorescence imaging of the zygote (zy) and other cells. (**c**) Unfertilized transgenic ovule transformed with a quad-labeled construct displaying expression of only two of the four cell types of the embryo sac, the micropylar two synergids (sy, AT-DD2-promoter-AcGFP1) and the chalazal antipodals (an, AT-DD1-promoter-ZsYellow1). *Blue* fluorescence is nonspecific autofluorescence from the ovular tissue in figs. **b** & **c**. nu = nucellus. Bars = 20 μm

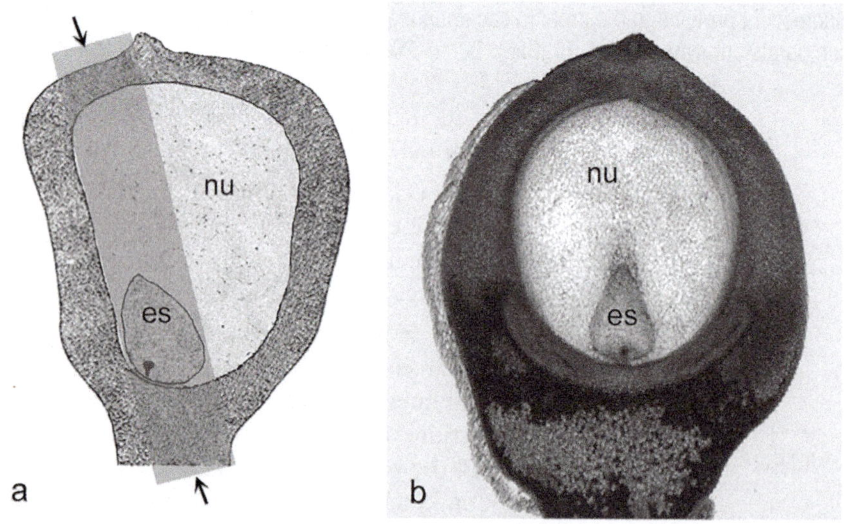

Fig. 3 Longitudinal sections of maize ovules at 8 days after pollination. (**a**) Vertical longitudinal diagram of an ovule depicting position of a dissected tissue slab, *gray* area with arrows. (**b**) *Bright-field* image of a hand dissected tissue slab, a horizontal longitudinal section that captured an intact embryo sac (es) and surrounding nucellar tissue (nu) and integuments-pericarp

3.3 Clearing and Fixing

1. FocusClear, and to a lesser degree TDE, were found to be compatible with fluorescent proteins (*see* **Notes 6–9**). Ovules may be mounted directly in FocusClear or run through a graded solution of TDE (10%, 25%, 50%, and 80%) [19]. Some ovule material may be recalcitrant to clearing (*see* **Note 10**).

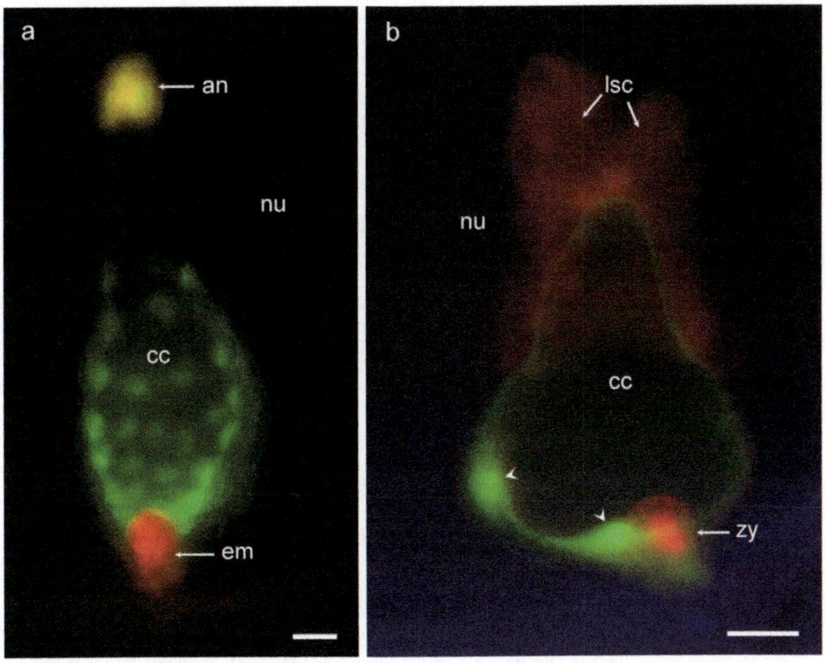

Fig. 4 Tissue slabs of transgenic ovules expressing a triple-labeled construct of embryo sac cell-specific promoter-fluorescent proteins. (**a**) Maize embryo sac, 2 days after pollination, showing the free-nuclear endosperm in the central cell (cc, *green*), a young embryo (em, *red*), and antipodal cells (an, *yellow*). (**b**) Sorghum embryo sac, shortly after fertilization, showing the zygote (zy, *red*) and a 2-nucleate endosperm (*arrowheads*) in the central cell (cc, *green*). Surprisingly, large stack cells (LSC) [20] in the nucellus (nu) become enlarged and express the egg/embryo-like marker ZmRKD2 promoter-DsRed2. Sorghum antipodals did not display any expression of the AtDD1 promoter-ZsYellow1 reporter dissimilar to that observed maize antipodals in (**a**). Bars = 50 μm

2. In cases when fixation is necessary to retain valuable material for several days, a mixture of 2% formaldehyde and 0.5% glutaraldehyde in PBS, pH 7.2 can be used. Glutaradehyde fixation may induce autofluorescence and mask the fluorescent proteins of interest, and therefore should be used at weak concentrations.
3. Fixation is accomplished at 4 °C for 1–4 h, depending on the thickness of the tissue.
4. Remove fixative with three washes of PBS and store tissue at 4 °C in PBS.
5. Because of the labile nature of dissected tissue and the florescent proteins, it is recommended to make observations on fresh, unfixed material when practical.

3.4 Cytology and Image Manipulations

The figures presented in this protocol were generated following **steps 1–3**.

1. Capture observations and images on fluorescence capable compound microscope using an attached camera designed for fluorescence (*see* **Notes 11** and **12**) controlled by microscope imaging software.
2. Images can be further manipulated using the microscope imaging software for merging and aligning grayscale images into multicolor overlay images.
3. Further image optimization, adjusting brightness, contrast, cropping, etc., can be performed using a downstream image processing-editing software, such as Photoshop or ImageJ.

4 Notes

1. Many embryo sac cell-specific promoter sequences are known, particularly for *Arabidopsis*. However, we have found that empirical knowledge is necessary to determine if a promoter functions as expected between species. With a limited sampling of promoters for the cell types of the embryo sac we have found that egg cell and antipodal cell promoters have tended to perform similarly from one species to another, even from dicots to monocots. However, synergid and central cell promoters have been much more species specific in their performance. For instance, the AtDD65 promoter expresses in the central cell and endosperm in *Arabidopsis*; however, it was not found to express in either soybean (Fig. 2c) or maize.
2. Close consideration should be made with multicolor fluorescence on a sample where the different fluorescent proteins would be expected to express in close proximity to one another, e.g., egg and synergids. When incorporating three or more fluorescent reporters expressing in overlapping or adjacent tissues, it is optimal to make selections from non-proximate ranges of the visible spectrum. Green fluorescent protein (GFP) works well with DsRed, but yellow fluorescent protein (YFP) and DsRed are not practical together due to spectral overlap of the emission and excitation wavelengths, respectively. The use of several colors and therefore multiple filter sets requires close scrutiny to the excitation and emission specs of each filter set. Avoid overlap of the spectral qualities of the sets. Band pass filter sets should be utilized in multi-color applications. Far-red fluorescent proteins are a more recent addition to the assortment of available proteins and can significantly expand the useful spectrum. Camera selection is important, only the newer scientific grade cameras are efficient in the far-red region. Spectra viewers (http://www.nightsea.com/sfa-sharing/fluorescence-spectra-viewers/) can be especially

helpful in making optimal fluorescent protein selections for your microscopy system.

3. Some overlap of excitation and emission spectra may be tolerated if expression is in non-adjacent cells or the signal is sufficiently strong enough to allow the use of narrow band pass filters to eliminate or attenuate non-desired signals.

4. Autofluorescence in plant tissue can be either problematic or useful for imaging. As an example, cell wall materials autofluoresce strongly when excited especially in the near-UV to blue wavelength range. This autofluorescence can either mask fluorescence from blue fluorescent protein (BFP) or add desired contrast when merged with other colors. It is advisable to observe the levels of autofluorescence in a non-transgenic sample of the tissue of interest prior to making fluorescent protein selections.

5. For transgenic material, it is best practice to perform qPCR of multiple regions of the transgenic cassette at early generations of material to identify simple, single insertion lines. Alternatively, Southern blotting with multiple probes or sequencing technologies could be used to the same effect.

   With suitably facile access to dissected ovules, microscopic screening of lines may be an efficient alternative way of screening transgenic lines for material with highly functional cell reporter activity. As specific and bright fluorescence of the reporters is the most critical transgenic quality of concern screening for these qualities in selected lines must be performed. Even with quality transgenic insertions, some reporters may not always function (for example AtDD45 promoter-DsRED Express in Fig. 2c). Functional analysis over several generations is advised.

6. Clearing solutions FocusClear, TDE, BABB (benzyl alcohol—benzyl benzoate), and Hoyer's solution were tested for compatibility with fluorescent proteins. Only FocusClear and TDE demonstrated compatibility, with TDE less compatible than FocusClear. Some materials may not require clearing; it is optimal for fluorescence to avoid clearing if possible.

7. Clearing causes softening (TDE) or hardening (BABB) of tissue depending on the methodology. This can cause difficulty in dissection. It is recommended to dissect and section material prior to clearing and mounting.

8. Chloral Hydrate may require licensing (such as US FDA) depending on jurisdiction.

9. Benzyl alcohol—benzyl benzoate (BABB) works very well for clearing tissue and has a refractive index (1.559) similar to glass slides and coverslips, as well as immersion oil. Although great

for clearing, we found it to be incompatible with fluorescence from fluorescent proteins.

10. In some ovules, subcellular bodies may be present which make clearing impossible. Soy embryo sacs are one example in that they are densely filled with starch granules by the mature megagametophyte stage (Fig. 2a). The large aggregate granules are opaque to fluorescence and therefore can impede observations.
11. The use of a wide-field microscope and a scientific-grade CCD (Photometrics CoolSNAP HQ, 1392 × 1040 imaging pixels, 6.45-μm pixels) or one of the newer generation CMOS (Hamamatsu ORCA Flash 4.0, 2048 × 2048, 6.5-μm pixels) cameras is the equipment of choice for low light multi-color fluorescence imaging. There is a limited amount of expressed fluorescent protein from the cells that we work with and these are therefore very labile to photobleaching. These low-light sensitive cameras reduce the chance of photobleaching and tissue damage.
12. Photobleaching issues may also be resolved with the newer generation confocal laser scanning microscopes (CLSMs) that use highly sensitive hybrid detectors, and thus, allow work at much lower laser intensities. This therefore greatly reduces the potential for photobleaching on sensitive fluorescent proteins with limited concentrations in cells or tissues of interest as well as reduces phototoxicity with living samples.

Acknowledgment

This work was developed at and supported by DuPont Pioneer. Sorghum analysis was supported by a sub-award from the CSIRO under the Capturing Heterosis for Smallholder Farmers grant from the Bill and Melinda Gates Foundation. We thank Katherine Thilges, Tim Fox, Brian Loveland, Eric S. Caswell, April Agee, Ping Che, and others for technical support of this work. We thank Kristin Haug-Collet, Susan D. Johnson, and Anna M. Koltunow for critical reading of this manuscript.

References

1. Sprunck S, Gross-Hardt R (2011) Nuclear behavior, cell polarity, and cell specification in the female gametophyte. Sex Plant Reprod 24 (2):123–136
2. Berger F, Twell D (2011) Germline specification and function in plants. Annu Rev Plant Biol 62(1):461–484
3. Sundaresan V, Alandete-Saez M (2010) Pattern formation in miniature: the female gametophyte of flowering plants. Development 137 (2):179–189
4. Vielle-Calzada JP, Baskar R, Grossniklaus U (2000) Delayed activation of the paternal genome during seed development. Nature 404(6773):91–94

5. Yang W, Jefferson RA, Huttner E, Moore JM, Gagliano WB, Grossniklaus U (2005) An egg apparatus-specific enhancer of Arabidopsis, identified by enhancer detection. Plant Physiol 139(3):1421–1432
6. Gross-Hardt R, Kagi C, Baumann N, Moore JM, Baskar R, Gagliano WB, Jurgens G, Grossniklaus U (2007) LACHESIS restricts gametic cell fate in the female gametophyte of Arabidopsis. PLoS Biol 5(3):e47
7. Sánchez-León N, Arteaga-Vázquez M, Alvarez-Mejía C, Mendiola-Soto J, Durán-Figueroa N, Rodríguez-Leal D, Rodríguez-Arévalo I, García-Campayo V, García-Aguilar M, Olmedo-Monfil V, Arteaga-Sánchez M, Martínez de la Vega O, Nobuta K, Vemaraju K, Meyers BC, Vielle-Calzada J-P (2012) Transcriptional analysis of the Arabidopsis ovule by massively parallel signature sequencing. J Exp Bot 63(10):3829–3842
8. Johnston A, Meier P, Gheyselinck J, Wuest S, Federer M, Schlagenhauf E, Becker J, Grossniklaus U (2007) Genetic subtraction profiling identifies genes essential for Arabidopsis reproduction and reveals interaction between the female gametophyte and the maternal sporophyte. Genome Biol 8(10):R204
9. Steffen JG, Kang IH, Macfarlane J, Drews GN (2007) Identification of genes expressed in the Arabidopsis female gametophyte. Plant J 51 (2):281–292
10. Wuest SE, Vijverberg K, Schmidt A, Weiss M, Gheyselinck J, Lohr M, Wellmer F, Rahnenführer J, von Mering C, Grossniklaus U (2010) Arabidopsis female gametophyte gene expression map reveals similarities between plant and animal gametes. Curr Biol 20 (6):506–512
11. Ohnishi T, Takanashi H, Mogi M, Takahashi H, Kikuchi S, Yano K, Okamoto T, Fujita M, Kurata N, Tsutsumi N (2011) Distinct gene expression profiles in egg and synergid cells of rice as revealed by cell type-specific microarrays. Plant Physiol 155(2):881–891
12. H-J Y, Hogan P, Sundaresan V (2005) Analysis of the female gametophyte transcriptome of arabidopsis by comparative expression profiling. Plant Physiol 139(4):1853–1869
13. Lawit SJ, Chamberlin MA, Agee A, Caswell ES, Albertsen MC (2013) Transgenic manipulation of plant embryo sacs tracked through cell-type-specific fluorescent markers: cell labeling, cell ablation, and adventitious embryos. Plant Reprod 26(2):125–137
14. Green M, Sambrook J (2012) Molecular cloning: a laboratory manual. Cold Spring Harbor Laboratory Press, New York
15. Chen W, Tulsieram L (2015) Microprojectile bombardment transformation of Brassica. US Patent No. 8,993,845
16. Li Z, Xing A, Moon BP, McCardell RP, Mills K, Falco SC (2009) Site-specific integration of transgenes in soybean via recombinase-mediated DNA cassette exchange. Plant Physiol 151(3):1087–1095
17. Cho M-J, Wu E, Kwan J, Yu M, Banh J, Linn W, Anand A, Li Z, TeRonde S, Register JC III (2014) Agrobacterium-mediated high-frequency transformation of an elite commercial maize (Zea mays L.) inbred line. Plant Cell Rep 33(10):1767–1777
18. Wu E, Lenderts B, Glassman K, Berezowska-Kaniewska M, Christensen H, Asmus T, Zhen S, Chu U, Cho M-J, Zhao Z-Y (2014) Optimized Agrobacterium-mediated sorghum transformation protocol and molecular data of transgenic sorghum plants. In Vitro Cell Dev Biol Plant 50(1):9–18
19. Staudt T, Lang MC, Medda R, Engelhardt J, Hell SW (2007) 2,2′-Thiodiethanol: a new water soluble mounting medium for high resolution optical microscopy. Microsc Res Tech 70 (1):1–9
20. Carman JG, Jamison M, Elliott E, Dwivedi KK, Naumova TN (2011) Apospory appears to accelerate onset of meiosis and sexual embryo sac formation in sorghum ovules. BMC Plant Biol 11(1):1

Chapter 6

Analysis of Fluorescent Reporter Activity in the Male Germline During Pollen Development by Confocal Microscopy

David Twell and Lynette Brownfield

Abstract

The male germline of flowering plants develops within the vegetative cell of the male gametophyte (pollen). The germline is established by asymmetric division of the microspore to form the generative cell. Mitotic division of the generative cell then produces the two sperm cells required for double fertilization. These differentiate to produce the proteins required for gamete attachment and fusion. An important aspect of understanding germline development is the characterization of germline gene expression. Here, we describe the use of a fluorescent reporter to study germline gene expression in developing pollen to assess the timing and specificity of expression.

Key words Fluorescent reporter, Male germline, Pollen, Sperm cells, Confocal laser scanning microscopy, Autofluorescence

1 Introduction

In flowering plants the germ cells are produced by the haploid male gametophyte (the pollen grain) that develops from unicellular microspores [1, 2]. During microspore development the nucleus migrates to produce a polarized cell that undergoes a highly asymmetric division. This results in a bicellular pollen grain comprising two distinct daughter cells with different fates. The large vegetative cell plays a supportive role and eventually forms the pollen tube. The smaller cell, the generative or germ cell, is the start of the male germline and its specification relies on the asymmetric nature of microspore division [3]. During development the generative cell is engulfed within the vegetative cell cytoplasm and begins to express the proteins required for gamete recognition and fusion. In species such as Arabidopsis (*Arabidopsis thaliana*) that release tricellular pollen, the generative cell divides before pollen release to produce the two sperm cells that are required for double fertilization.

Anja Schmidt (ed.), *Plant Germline Development: Methods and Protocols*, Methods in Molecular Biology, vol. 1669, DOI 10.1007/978-1-4939-7286-9_6,

In species that release bicellular pollen, this division occurs during pollen-tube growth. Identifying and characterizing the genes that are expressed in the generative and sperm cells (the male germline) is a key aspect of understanding flowering plant sexual reproduction and plant fertility.

An important consideration when characterizing male germline genes is to determine their expression patterns during pollen development, and whether expression is germline-specific within the male gametophyte (i.e., not in the microspore or vegetative cell). As the germline develops within the vegetative cell, techniques such as qRT-PCR to follow germline gene expression are difficult due to the large contribution of the vegetative cell to samples of isolated pollen or developing gametophytes. Instead, we have developed methods to use fluorescent reporters driven by promoters, or smaller regulatory regions, *cis*-regulatory modules [4], to follow expression of male germline genes from microspores to mature pollen. We have used these methods to characterize the expression of a number of genes such as *DUO POLLEN1* (*DUO1*) encoding a germline-specific transcription factor, selected DUO1 target (*DAT*) genes, non-germline-specific genes such as *DUO3* [5–8] and in the characterization of *ROD1* (*Regulatory region of DUO1*), a key *cis*-regulatory module of the *DUO1* promoter [9].

As fluorescent signals are often weak when genes are first developmentally expressed, and due to autofluorescence of the pollen cell wall, confocal laser scanning microscopy (CLSM) is used to enhance the sensitivity of detection. However, the enhanced sensitivity of CLSM means that the analysis becomes susceptible to induced autofluorescence artifacts caused by the pollen losing plasma membrane integrity as a result of stresses during sample preparation and viewing. The early stages of development (microspores and early bicellular pollen) are particularly sensitive to such autofluorescence artifacts.

Fluorescent reporter constructs can be built using a variety of cloning methods. We recommend the reporter consist of a promoter region driving the expression of Arabidopsis Histone H2B fused to the fluorescent reporter, with the H2B providing nuclear localization. This restricts the signal to nuclei aiding in the detection of weak fluorescence, and enables expression in germline cells and the vegetative cell to be easily distinguished. Using H2B also helps to distinguish induced autofluorescence from true signal when analyzing expression in early stages of pollen development.

Here, we describe methods for the analysis of male germline development by fluorescent reporters in Arabidopsis by CLSM, utilizing an H2B-fluorescent reporter driven by a promoter region from a male germline gene. In addition, we provide notes and advice on how to determine pollen developmental stages and how to recognize and avoid autofluorescence artifacts.

2 Materials

1. Flowering Arabidopsis plants expressing a fluorescent reporter for a male germline gene (preferably a promoter:H2B-fluorescent protein reporter).
2. Dissecting microscope.
3. Sterilized distilled water.
4. Fine needles and forceps.
5. Microscope slides and coverslips (number 1 thickness).
6. 0.3 M mannitol (autoclaved).
7. Nail varnish.
8. Confocal laser scanning microscope with transmitted light detector and image capturing software.

3 Methods

1. Select an inflorescence (bud cluster) from a plant to be analyzed for expression of a promoter:reporter (*see* **Note 1**).
2. Carefully cut the inflorescence from the plant and place it into a few drops of water on a microscope slide (*see* **Note 2**).
3. Under a dissecting microscope, remove each bud sequentially using fine needles and keep in a developmental order (from −1 to approximately −10). Keep buds in a small amount of water and do not allow to dry (*see* **Note 3**).
4. Select a bud for dissection and place in a drop of water on a new microscope slide (*see* **Note 4**).
5. Using the fine needles carefully open the bud by peeling back the sepals (and petals in later stages of development). Using the needle cut the filament of the stamen so the four medial anthers are excised (*see* **Note 5**).
6. Remove all non-anther material from the slide using the needles (or fine forceps) to leave four anthers in a small amount of water (*see* **Note 6**).
7. Pipette a small amount (2–5 μl) 0.3 M mannitol onto the anthers. While the precise volume is not important, keep the volume small as otherwise it is difficult to rip the anthers open in **step 8**. It is also important that the anthers do not dry out as this will lead to high levels of autofluorescence.
8. Using the needles rip open the anthers. Use one needle to hold the anther still and the other to rip the anther wall. Developing pollen should be released into the liquid. This is clearly visible

down a dissecting microscope (×10) for older stages. Continue ripping the anther until most pollen is released.

9. Remove as much anther debris as possible with the needles (*see* **Note 7**).
10. Add extra 10 μl of 0.3 M mannitol so there is sufficient liquid to prevent the coverslip pressing against and squashing pollen and prevent large air bubbles forming.
11. Gently place a cover slip on top evenly from above (*see* **Note 8**).
12. Seal the coverslip with nail varnish and view within 15 min.
13. View pollen using CLSM. This will require previous training/ use of the microscope. Ideally, the microscope will be equipped with a light source and detectors suitable for the fluorescent reporter and detectors for transmitted light.
14. Begin by finding pollen under low magnification and then increasing the magnification (×60 objective) using white light before converting to the CLSM setup.
15. Carry out initial scans using a short dwell time and a high gain setting to limit bleaching of the fluorescent reporter. Continually adjust the focus to scan through the pollen. In this quick scan view both the fluorescent signal and the transmitted light to give an indication of fluorescence intensity, developmental stage, pollen integrity and if there is significant autofluorescence (*see* **Note 9**).
16. Select a pollen grain and adjust the focus for detailed analysis and imaging. For this the dwell time is increased and the gain reduced. The precise settings will vary and need to be optimized for each developmental stage and reporter line. Use an average of four scans (Kalman 4) to reduce background. Image a number of pollen on each slide and determine developmental stage (*see* **Note 10** and Fig. 1) while avoiding pollen with autofluorescence (*see* **Note 9** and Fig. 1).
17. Repeat from **step 3** for each bud until reaching late microspore stage, or the stage when there is no longer a fluorescent signal, usually around −6/−9, although this varies with Arabidopsis accessions.

4 Notes

1. It is important that the inflorescence is healthy. Look for a full inflorescence with a full and consistent whorl of flower buds. Often the primary inflorescence on a stem is best, but only harvest after at least four flowers have opened.

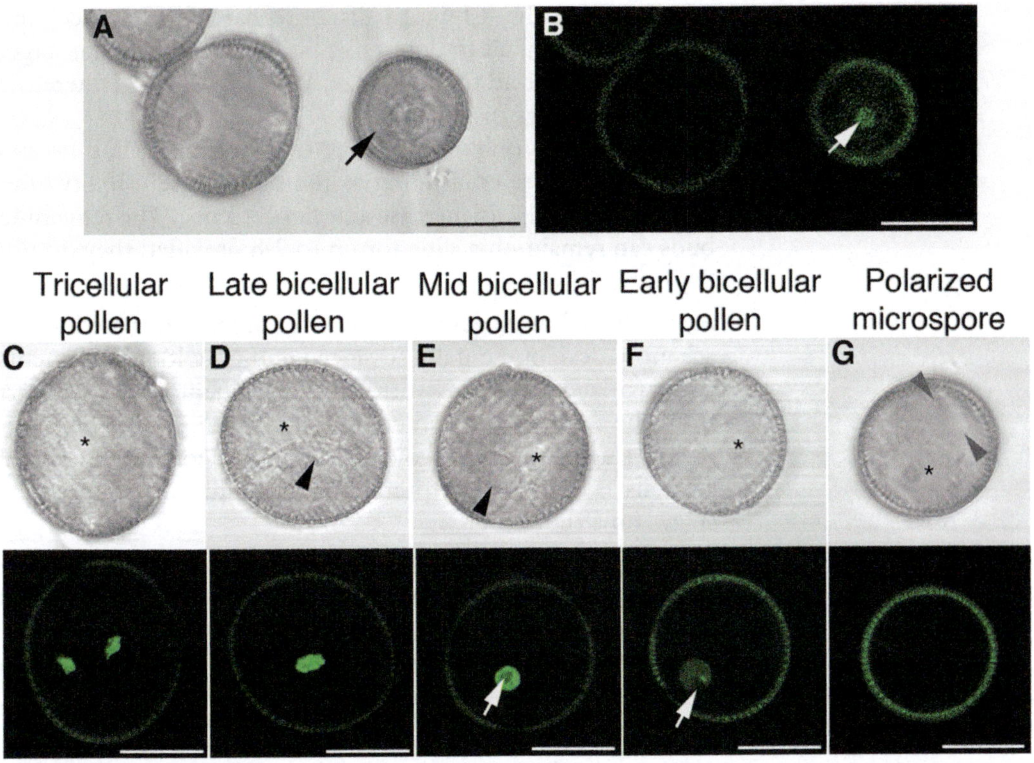

Fig. 1 Developmental stages of pollen as observed by CLSM. (**a**, **b**) Quick CLSM scan of microspores with no reporter fluorescence. (**a**) Transmitted light and (**b**) fluorescence with GFP settings. On the *left* is an intact microspore and on the *right*, a microspore that has lost plasma membrane integrity and has burst with the plasma membrane pulling away from the wall (*black arrow*), leading to an increase in autofluorescence from the pollen wall and nucleolus (*white arrow*). (**c**–**g**) CLSM images of pollen dissected in a series from the −1 bud to younger buds containing microspores. The top panel shows transmitted light and the lower panel a H2B-GFP fusion for a germline-specific reporter construct. For a detailed description of each stage *see* **Note 10**. In transmitted light images, *asterisks* (*) indicate the position of the vegetative cell or microspore nuclei, *black arrowheads* indicate the generative cell, and *gray arrowheads* show the extent of the vacuole. In fluorescence images, *white arrows* indicate reduced fluorescence in generative cell nucleoli compared to other regions of nucleoplasm. Scale bar = 10 μm

2. Place the microscope slide on an inverted lid of a petri dish, this makes it easy to move the slide and cover it by placing the bottom of the petri dish (also inverted) over the slide to prevent it drying out when not being used.
3. The −1 bud refers to the most developed, oldest unopened bud, −2 the second unopened bud etc. as described [10]. These are arranged in a whorl moving up the inflorescence stem and the size will decrease in the younger buds.
4. We usually begin with the −1 bud and work backward looking at every second bud (−1, −3, −5, etc.) until the earliest developmental stage of interest has been viewed. This is because the developmental stage when fluorescence is first detectable is

unknown and the −1 bud is likely to have fluorescence, and because it is difficult to determine pollen developmental stages based solely on bud number/size. Working from −1 therefore saves unnecessary dissections.

It is best to only prepare one or two slides at a time and view them before continuing as the pollen lose viability after dissection, which can increase autofluorescence. The remaining buds can remain on a slide for up to 2 h provided they do not dry out.

5. Arabidopsis flowers generally have six stamens, four medial that are more developmentally advanced and have longer filaments and two lateral stamens with shorter filaments [11]. Anthers from the four medial stamens should be at a similar developmental stage, but try to avoid including anthers from the lateral stamens as it makes interpretation more difficult with multiple stages on a slide.

6. It is important to keep the amount of water low. If a large amount of water is present, use needles to drag anthers away from most of the water and blot dry. Alternatively, use the needle to transfer the anthers to a new slide containing a small drop of 0.3 M mannitol.

7. Removing anther debris is relatively easy for older stages, but much harder for younger stages. If a small amount remains that is not a problem, as the purpose of removing the debris is to avoid overly thick material that will raise the cover slip.

8. Add the cover slip from above as placing it from an angle leads to many of the pollen moving with the meniscus and being lost from the viewing area. However, it is important to do this gently as otherwise the pollen can be squashed.

9. When analyzing fluorescence it is important to be confident signals are not due to autofluorescence. This is especially important in younger stages that are susceptible to autofluorescence when plasma membrane integrity is lost during the dissection and on the microscope slide. There are several things that can help limit this stress, such as observing slides immediately after dissection, ensuring buds and anthers never dry out and taking care when adding the coverslip. In addition, the use of 0.3 M mannitol provides osmotic support for pollen in the mounting medium.

   Even with care, some pollen grains lose integrity. Fortunately, there are visual indications in the transmitted light and fluorescent channels at the scanning stage that this has occurred (*see* Fig. 1a, b). Microspores that have lost plasma membrane integrity are generally smaller than viable microspores, and in some cases the membrane may be visibly pulled away from the wall (*see* Fig. 1a). Additionally, the nucleus is smaller, the nucleolus may appear darker and the vacuole is lost. Signs of a loss of integrity in the fluorescence channel include increased wall autofluorescence (*see* Fig. 1b). Additionally, the use of Histone H2B as the nuclear localization signal aids in distinguishing autofluorescence in microspores and developing pollen. Histone H2B-fluorescent fusion proteins in the male germline are usually partially excluded from the nucleolus (*see* Fig. 1e, f), but autofluorescence is commonly strongest in the nucleolus (*see* Fig. 1b). Thus, fluorescence signals in microspores and germ cells that are higher in the nucleolus should be ignored. In addition, using a line that is heterozygous for a single locus insertion of the reporter construct can be beneficial. Fluorescence from the reporter should be in approximately 50% of the pollen and a deviation from this ratio is also an indication that the fluorescence may be due to background.

10. A key aspect of the CLSM analysis is determining the developmental stage of the pollen. Here, we provide guidance on the key features of each stage.

    Triceullar pollen (buds −1 to −3, *see* Fig. 1c): Using transmitted light, this stage does not have many distinguishing features with the sperm cells being difficult to detect, although the vegetative nucleus may be visible as a smoother region. The presence of two sperm cells, indicating tricellular pollen, can be confirmed in the fluorescence channel. A number of pollen grains may need to be screened to find examples with the two sperm cells and the vegetative nuclei in the same focal plane. Increasing the pinhole diameter can help, but if set too wide, wall autofluorescence may increase. The nucleolus is not generally apparent in sperm cell nuclei.

    Late bicellular pollen (buds −3 to −5, *see* Fig. 1d): In late bicellular pollen the generative cell can be distinguished using transmitted light as it is surrounded by a slightly darker layer containing lipid droplets and is elongated. The generative cell is usually central and close to the vegetative nucleus that may be observed as a smoother region. At this stage, DNA in the generative cell nucleus is condensing in preparation for mitosis, and this can be observed when using Histone H2B fused to the reporter. If the gene promoter is only active late in pollen development a fluorescent signal may not be observed at this stage.

    Mid bicellular pollen (buds −4 to −6, *see* Fig. 1e): At this stage the generative cell can again be distinguished due to the surrounding dark lipid layer and is round and positioned away from the pollen wall. The vegetative nucleus can be observed as a smooth region with prominent nucleolus. When using Histone H2B fused to the reporter the fluorescence present is throughout the generative cell nucleoplasm, but is reduced in the nucleolus.

Early bicellular pollen (buds −5 to −7, *see* Fig. 1f): The pollen has just completed pollen mitosis I, and the generative cell is still located on the pollen wall. Both generative and larger vegetative cell nuclei can be seen, often with prominent nucleoli using transmitted light. When the generative cell is newly formed fluorescence from a germline specific gene is likely to be low and difficult to detect, especially during the CLSM scanning stage. Confocal settings, notably dwell and gain, may need to be increased to be confident that there is a true signal. If using a Histone H2B-reporter, fluorescence is normally lower in the generative cell nucleolus than the rest of the nucleoplasm.

Polarized microspores (buds −6 to −9, *see* Fig. 1g): The single nucleus is large with a prominent nucleolus and is located toward one side of the cell. A large vacuole on the opposite side of the cell may also be visible in microspores that have maintained plasma membrane integrity. Any fluorescence at this stage would indicate that the gene is not germline-specific.

Acknowledgments

This work was supported by a Fast Start grant from the Marsden Fund of the Royal Society of New Zealand (12-UOO-043) to L.B. and by research grants from the Biotechnology and Biological Sciences Research Council (BB/I011269/1; BB/N005090/1) to D.T.

References

1. Borg M, Brownfield L, Twell D (2009) Male gametophyte development: a molecular perspective. J Exp Bot 60:1465–1478
2. Twell D (2011) Male gametogenesis and germline specification in flowering plants. Sex Plant Reprod 24:149–160
3. Eady C, Lindsey K, Twell D (1995) The significance of microspore division and division asymmetry for the vegetative cell-specific transcription and generative cell differentiation. Plant Cell 7:65–74
4. Peters B, Aidley J, Cadzow M, Twell D, Brownfield L (2017) Identification of *cis*-regulatory modules that function in the male germline of flowering plants. In: Schmidt A (ed) Plant germline development, Methods in molecular biology. Springer, Dordrecht
5. Brownfield L, Hafidh S, Borg M, Sidorova A, Mori T, Twell D (2009) A plant germline-specific integrator of sperm specification and cell cycle progression. PLoS Genet 5: e10000430
6. Borg M, Brownfield L, Khatab H, Sidorova A, Lingaya M, Twell D (2011) The R2R3 MYB transcription factor DUO1 activates a male germline-specific regulon essential for sperm cell differentiation in Arabidopsis. Plant Cell 23:534–549
7. Brownfield L, Hafidh S, Durbarry A, Khatab H, Sidorova A, Doerner P, Twell D (2011) Arabidopsis DUO POLLEN 3 is a key regulator of male germline development and embryogenesis. Plant Cell 21:1940–1956

8. Borg M, Rutley N, Kagale S, Hamamura Y, Gherghinoiu M, Kumar S, Sari U, Esparza-Franco MA, Sakamoto W, Rozwadowski K, Higashiyama T, Twell D (2014) An EAR-dependent regulatory module promotes male germ cell division and sperm fertility in Arabidopsis. Plant Cell 26:2098–2113
9. Peters B, Casey J, Aidley J, Zohrab S, Borg M, Twell D, Brownfield L (2017) A conserved cis-regulatory module determines germline fate through activation of the transcription factor *DUO1* promoter. Plant Physiol 173:280–293
10. Lalanne E, Twell D (2002) Genetic control of male germ unit organization in Arabidopsis. Plant Physiol 129:865–875
11. Smyth DR, Bowman JL, Meyerowitz EM (1990) Early flower development in Arabidopsis. Plant Cell 2:755–767

Chapter 7

In Vivo Ploidy Determination of *Arabidopsis thaliana* Male and Female Gametophytes

Burcu Nur Keçeli, Nico De Storme, and Danny Geelen

Abstract

Organ- or tissue-specific ploidy level determination is often used for answering biological, molecular, genetic, or evolutionary questions in plant sciences. However, current techniques for ploidy determination either cannot provide information on single cell level, require destructive sample preparation, or are laborious and time-consuming. Here, we present a new approach developed in *Arabidopsis thaliana*, which is not only less labor intensive but also allows in vivo ploidy determination on single cell level. The technique is based on the incorporation of a transgenic construct, consisting of the centromere-specific protein CENH3 fused to the fluorescent reporter GFP that specifically labels centromeric regions and hence allows for an accurate visual determination of the cell's chromosome number. Moreover, by combining the construct with a gametophyte-specific promoter, the technique enables accurate chromosome quantification in all individual gametophytic cell types formed during micro- and megagametogenesis. As such, CENH3-based centromere visualization provides an easy and straightforward method to monitor meiotic cell division integrity, gametophytic chromosome dynamics, and reproductive ploidy stability.

Key words CENH3, Centromere, *Arabidopsis thaliana*, Ploidy analysis, Meiosis

1 Introduction

Genetics, molecular biology, and evolutionary research frequently use chromosome number determinations to answer questions on genome stability and cell division integrity. Depending on the research question, either organ-, tissue-, or even cell-specific ploidy assays are often required. As an important application, ploidy analyses in early gametophytic stages provide direct information on the stability of meiotic cell division and chromosome segregation, and thus may serve for both genetic and evolutionary studies [1]. More specifically, studies on unreduced and aneuploid gamete formation [2–4] and its evolutionary effects on speciation, or studies on intraspecific hybrid formation [5] and apomixis [6] that have both evolutionary and agronomic aspects might benefit from chromosome

Anja Schmidt (ed.), *Plant Germline Development: Methods and Protocols*, Methods in Molecular Biology, vol. 1669, DOI 10.1007/978-1-4939-7286-9_7, © Springer Science+Business Media LLC 2017

counts in male and female gametophytic stages. Additionally, techniques for gametophytic ploidy analysis may also be useful for studies focusing on endosperm-based triploid block [7] or parent-specific genome elimination [8] that are aiming to unravel the underlying molecular regulation of both phenomena in the search for putative applications for plant breeding.

DNA flow cytometry, chromosome spreads, and fluorescent in situ hybridization (FISH) are techniques that are frequently used for ploidy determination in plant research. These methods, however, are all destructive in vitro techniques that have several drawbacks. For instance, flow cytometry can be used for high-throughput analyses; however, it is not applicable for studies on single cell level [9]. Moreover, infrequent alterations in ploidy level are hard to detect because flow cytometry is an additive technique based on the analysis of large cell populations [10]. On the other hand, ploidy determination techniques that have cytological basis are labor intensive, time consuming, and often suffer from biological constraints. Karyotype analysis (i.e., chromosome spreads), for instance, is only informative when chromosomes are in a fully condensed state, which intrinsically requires the presence of cells that are actively dividing [11]. Lastly, FISH is a cytological technique in which one or more specific DNA sequences are tracked using fluorescently labeled probes that are complementary to the sequence of interest. Using the right combination of probes, FISH can thus also be applied for accurate chromosome counting on single cell level. On the other hand, FISH requires the availability of specific probes or probe combinations, making this method laborious and expensive [12]. To facilitate in vivo single cell ploidy analysis, we here present an alternative approach that uses recombinant CENH3, i.e., the centromere-specific histone 3 variant, fused to a GFP reporter to label the cell's centromeres, hence providing an easy and reliable visual assay for accurately determining the chromosome number in a broad range of cell types, including male and female gametophytic cells.

Centromeres are specific regions on the chromosomes that serve as docking centers for kinetochore proteins, hence playing a crucial role in metaphase spindle attachment, chromosome segregation, and cell division [13]. Centromeres are marked with a specific histone 3 (H3) variant, namely CENP-A or CENH3, which replaces some of the canonical H3 only at centromeric regions [14]. Depending on the species examined, CENH3-containing nucleosomes are either fully scattered throughout the whole chromosome (holocentric) or are concentrated in a single region on the chromosomes (monocentric). Most of the plant species studied carry monocentric chromosomes as opposed to a few that carry holocentric chromosomes. For example, *Arabidopsis thaliana*, rice (*Oryza sativa* L.), and maize (*Zea mays* L.) have monocentric chromosomes, whereas *Scirpus cernuus*, *Luzula sylvatica*, and *Isolepis prolifera* are species that carry holocentric chromosomes [15]. Considering the fact that each chromosome in monocentric species harbors one single centromeric domain, detection and/or visualization of centromeres provides a reliable alterative for accurately quantifying the chromosome number and thus ploidy of a single cell. Therefore, techniques based on the visualization of centromeres through detection of centromere-specific proteins, like CENH3, could be used as a novel in vivo ploidy determination strategy.

By recombinantly fusing green fluorescent protein (GFP) to the C-terminus of CENH3, centromeres can be visualized and chromosomes can be counted. Moreover, by using different promoters to drive the CENH3-GFP construct, tissue-specific expression can be achieved, while strongly reducing background fluorescence and largely facilitating the analysis of the ploidy level and centromere dynamics of the specific cell(s) of interest [16].

In order to assess the cell's ploidy level in gametophytic stages of *Arabidopsis thaliana*, we fused the CENH3-GFP reporter construct to the promoter of the *WUSCHEL-RELATED HOMEOBOX 2* (*WOX2*) gene and introduced this *in planta*. The *pWOX2*-driven CENH3-GFP protein labels centromeres in all cells from early mega- and microspore stages up till the fully matured gametophytic stage (e.g., pollen and embryo sacs), and thus allows easy visual assessment of the gametophytic ploidy level during all stages of micro- and mega-gametogenesis. Importantly, early micro- and megaspore stages represent the outcome of meiotic chromosome segregation, indicating that *pWOX2-CENH3-GFP*-based ploidy determination of young spores provides a reliable method to monitor the fidelity of meiotic cell division and chromosome segregation. Moreover, since *WOX2* is also expressed during *Arabidopsis thaliana* early embryo development, *pWOX2-CENH3-GFP* also serves useful for analyzing the ploidy of embryo's and embryonic cells [17]. In order to perform in vivo ploidy analysis in Arabidopsis somatic cells, we developed a second line that expresses CENH3-GFP driven by the cauliflower mosaic virus (CaMV) 35s promoter (*p35s*). This line shows expression of CENH3-GFP in almost all somatic cell types, strongly facilitating the detection and quantification of centromeres, and thus chromosomes, in any somatic organ, tissue, or cell of interest.

Besides the toolbox developed for *Arabidopsis thaliana* L., we also found that ectopic expression of *p35s-AtCENH3-GFP* in Tobacco Bright Yellow2 (BY2) suspension cells labels centromeric regions and thus allows for cell-specific ploidy analysis (*see* Fig. 1). These findings demonstrate that the use of recombinant CENH3 fused to a visual reporter to quantify the number of chromosomes on a single cell level can be applied in different organisms and in various tissue types.

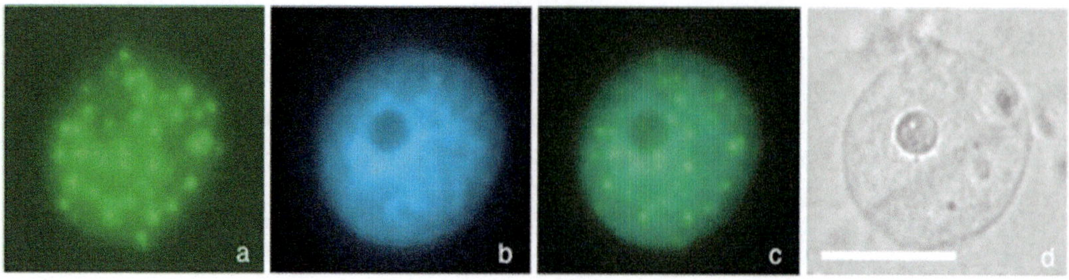

Fig. 1 CENH3-GFP driven by p35s labels centromeres in BY2 nuclei. (**a–d**) mitotic cells. (**a**) CENH3-GFP dots; (**b**) DAPI staining; (**c**) merged pictures of DAPI and GFP; (**d**) bright field. GFP images are processed z-stack files. (Scale bar: 10 μm)

In this chapter, we provide a protocol for germline-specific chromosome counts in *Arabidopsis thaliana* L. (Colombia ecotype) using the introgressed CENH3-GFP reporter construct driven by the *WOX2* promoter. Some additional recommendations and different strategies for which the construct could be used for are mentioned in Subheading 4.

2 Materials

1. *Arabidopsis thaliana* ecotype Col-0 flowering plants carrying the *pWOX2-CENH3-GFP* reporter construct (*see* **Note 1**).
2. Kl Medium: Dissolve 2.15 g Murashige-Skoog medium (without vitamins), 10 g D-(+)-sucrose, 0,1 g myo-inositol, and 0,5 g MES in 900 mL dH_2O. Adjust the pH of the solution to 5.7 with 1 M KOH. Add 8 g agar to the medium. Complete the volume of the solution to 1 L with dH_2O. Autoclave the medium. Under sterile conditions pour the medium into petri dishes.
3. DAPI (4′,6-diamidino-2-phenylindole) stock solutions: Dissolve 5 mg DAPI in 50 mL dH_2O with the help of 2–3 drops of methanol. Store this 0.01% (w/v) stock solution at −20 °C. (Final concentration of the stock solution: 3.6×10^{-4} M DAPI).
4. DAPI working solution: Mix 20 μL of DAPI stock solution with 980 μL Fluorescent Assay Buffer prior to use. Store this 2×10^{-4}% (w/v) working solution at 4–5 °C. (Final concentration of the working solution: 7.2×10^{-6} M DAPI).
5. Fluorescence Assay Buffer [18]: Dissolve 8.2 g Na_3PO_4 and 5 mL Triton X-100 (4-(1,1,3,3-tetramethylbutyl)phenyl-polyethylene glycol) in 950 mL dH_2O. The final concentration of the solution is 0.05 M Na_3PO_4 and 0.5% Triton X-100. Adjust the pH of the solution to 7.0 with 1 M HCl. Add dH_2O to complete the volume of the solution to 1 L.

6. Glass slides.
7. Cover slips (22 mm × 22 mm).
8. Needles.
9. Forceps.
10. Epi-fluorescent microscope or Confocal laser scanning microscope.
11. Stereo microscope.

3 Methods

3.1 Plant Cultivation

1. Sow seeds of *Arabidopsis thaliana* ecotype Col-0 carrying the *pWOX2-CENH3-GFP* construct in vitro on Kl medium and put the plates in a growth chamber at 20 °C with a photoperiod of 16 h day/8 h night after 2–3 days of vernalization period in the dark at 4 °C.
2. When seedlings are 6–9 days old, transfer them to soil.
3. Cultivate plants at 20 °C with a photoperiod of 16 h day/8 h night at 70% humidity.

3.2 Slide Preparation for Chromosome Counting in Male Gametogenesis

1. For in vivo chromosome quantification in nuclei at different microspore and pollen stages during male gametogenesis, select flower buds at stage 9 [19] and later for investigation (*see* **Note 2**).
2. Separate the flower bud of interest from the others with the help of needles and forceps.
3. Put the flower bud on a glass slide.
4. Add 10 μL DAPI working solution on the flower bud (*see* **Note 3**).
5. Put the cover slip on the bud and gently squash the flower bud in between the glass slide and the cover slip without applying extreme pressure.
6. Analyze the slides using epi-fluorescence or confocal laser scanning microscopy (*see* Fig. 2) (*see* **Notes 4** and **5**).

3.3 Slide Preparation for Chromosome Counting in Female Gametogenesis

1. For in vivo chromosome quantification in nuclei at different megaspore and embryo sac stages during female gametogenesis, flower buds at stage 11 [19] and later can be chosen.
2. For pistil dissection a stereo microscope can be used (*see* **Note 6**).
3. Separate the flower bud of interest from the whole inflorescence and put it on a glass slide. Add a drop of dH_2O on the top of the bud to protect the flower bud from drying out. Peel off petals, sepals, and anthers and remove them from the selected flower bud with the help of forceps and needles.

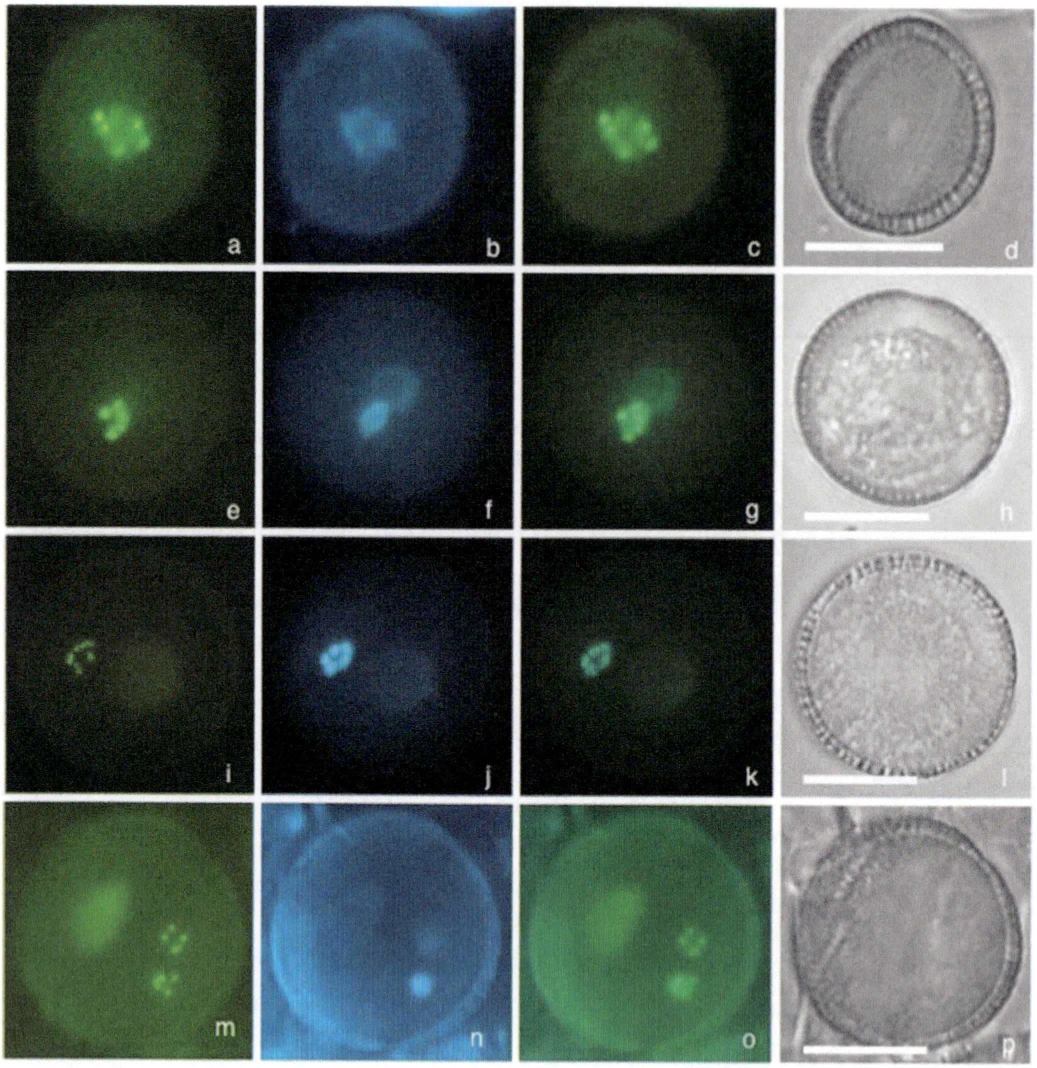

Fig. 2 CENH3-GFP driven by pWOX2 labels centromeres in developing microspores and pollen grains. (**a–d**) Uninuclear microspore stage. (**e–h**) Binuclear microspore stage. (**i–l**) Pollen mitosis II (PMII) or the mitotic division from binuclear to trinuclear microspore stage. (**m–p**) Trinuclear mature pollen grain (**a, e, i, m**) CENH3-GFP dots; (**b, f, j, n**) DAPI staining; (**c, g, k, o**) merged pictures of DAPI and GFP; (**d, h, l, p**) bright field. GFP images are processed z-stack files. (Scale bars: 10 μm)

4. Add 10 μL of DAPI working solution on a glass slide (*see* **Note 3**).
5. Gently put the naked pistil into the droplet of DAPI with the help of forceps.
6. Put the cover slip on the pistil and gently squash the pistil in between the glass slide and the cover slip. Analyze the slides using epi-fluorescence or confocal laser scanning (*see* Fig. 3) (*see* **Notes 4** and **5**).

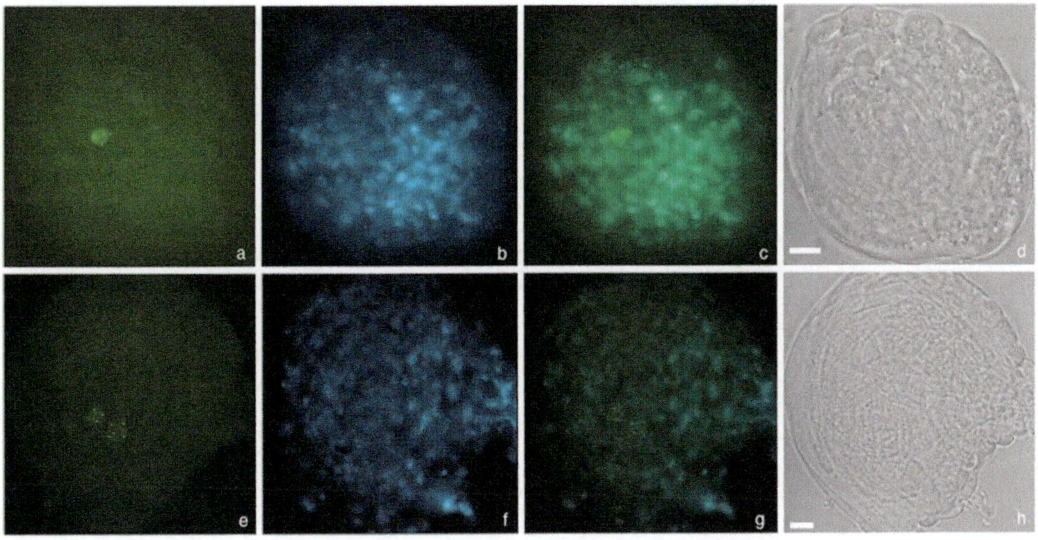

Fig. 3 CENH3-GFP driven by pWOX2 labels centromeres in megaspores and the developing embryo sac. (**a**–**d**) Uninuclear megaspore. (**e**–**h**) Binuclear embryo sac. (**a, e**) CENH3-GFP dots; (**b, f**) DAPI staining; (**c, g**) merged pictures of DAPI and GFP; (**d, h**) bright field. GFP images are processed z-stack files. (Scale bars: 10 μm)

3.4 Microscopy and Image Processing

1. Detect the gametophytic cell of interest using bright field microscopy. Start with a lower magnification (10× or 20×) and switch to a higher magnification lens (60× or 100×) to make it easier to focus on the cell of interest.
2. Take a picture of the cell of interest under bright field conditions.
3. Open the excitation filter for DAPI detection (381–393 nm band pass), focus on the nucleus, and take a picture of the nucleus (*see* **Note 7**).
4. Define the upper and lower Z-axis limits of the nucleus of the cell of interest using the DAPI channel.
5. Open the excitation filter for GFP detection (475–495 nm band pass). Record ~20 images at different focal planes in between the upper and lower Z-axis limits of the nucleus that was assigned under the DAPI channel (*see* **Note 7**). This process could be automatized using stage control features of your microscopy software (Z-stacking).
6. Select the pictures showing different centromeres.
7. Create Z-stacks of the selected pictures using picture-managing tools (*see* **Note 8**).

4 Notes

1. Seeds of *Arabidopsis thaliana* ecotype Col-0 carrying the *pWOX2-CENH3-GFP* construct and the plasmids used to create the line can be obtained upon request to the authors.

2. In case of mutant line investigation, the *pWOX2-CENH3-GFP* construct can be introduced in the mutant by intercrossing the mutant line with an *Arabidopsis thaliana* line carrying the construct encoding for the recombinant CENH3-GFP protein, or by directly transforming the mutant line with the construct. In case of intercrossing, the *pWOX2-CENH3-GFP*-harboring *Arabidopsis thaliana* line can be used as either pollen donor or acceptor. Importantly, since the line is not male sterile, emasculation should be performed when the line is used as a pollen acceptor.
3. For faster and cheaper investigations, DAPI staining could be omitted, and sample preparation can be solely performed in Fluorescence Assay Buffer. The Fluorescence Assay Buffer enhances the fluorescent signal and reduces photo bleaching [18]; however, dH_2O can alternatively be used for faster investigations.
4. In some stages of micro- or megasporogenesis, the different nuclei in the microspore, pollen, or developing embryo sac can be located at different microscopic focal planes. In those cases, Z-stack imaging is performed, in which different images are systematically taken at different focal planes, after which they are stacked together and overlapped in a single combined or projected image.
5. In Arabidopsis nuclei that have undergone one or more endoreduplication events, the number of CENH3-GFP foci does not represent the actual number of chromosomes and thus the endoploidy level of the cell [20].
6. For the female part, pistil dissection can be omitted and a whole flower bud squeeze can be used as explained in Subheading 3.2.
7. Depending on the application, fluorescent imaging can be performed using either an automatically optimized or a fixed exposure time. When the application requires quantitative analysis or comparison of fluorescence intensities, variables for the image recording should be kept fixed.
8. Stacking all GFP images without pre-selection could result in images showing blurry centromeric dots. In cases where all centromeres can be recorded in one focal plane, Z-stack imaging can be omitted.

References

1. Henry IM, Dilkes BP, Young K, Watson B, Wu H, Comai L et al (2005) Aneuploidy and genetic variation in the Arabidopsis thaliana triploid response. Genetics 170(4): 1979–1988
2. Mursalimov S, Deineko E et al (2016) Cytomixis in tobacco microsporogenesis: are there any genome parts predisposed to migration? Protoplasma 254(3):1379–1384. doi:10.1007/s00709-016-1028-1

3. De Storme N, Geelen D et al (2011) The arabidopsis mutant jason produces unreduced first division restitution male gametes through a parallel/fused spindle mechanism in meiosis II. Plant Physiol 155(3):1403–1415
4. De Storme N, Copenhaver GP, Geelen D et al (2012) Production of diploid male gametes in arabidopsis by cold-induced destabilization of postmeiotic radial microtubule arrays. Plant Physiol 160(4):1808–1826
5. Ramsey J, Douglas W et al (1998) Pathways, mechanisms, and rates of polyploid formation in flowering plants. Annu Rev Ecol Evol Syst 29:467–501
6. Kotani Y, Henderson ST, Suzuki G, Johnson SD, Okada T, Siddons H, Mukai Y, Koltunow AM et al (2014) The LOSS OF APOMEIOSIS (LOA) locus in Hieracium praealtum can function independently of the associated large-scale repetitive chromosomal structure. New Phytol 201(3):973–981. doi:10.1111/nph.12574
7. Garmendia A, Ferriol M, Juarez J, Zając A, Kałużny K, Merle H et al (2015) A rare case of a natural contact zone in Morocco between an autopolyploid and an allopolyploid of Centaurea aspera with sterile tetraploid hybrids. Plant Biol 17(3):746–757. doi:10.1111/plb.12284
8. Ravi M, Shibata F, Ramahi JS, Nagaki K, Chen C, Murata M, Chan SW et al (2011) Meiosis-specific loading of the centromere-specific histone CENH3 in Arabidopsis thaliana. PLoS Genet 7(6):e1002121. doi:10.1371/journal.pgen.1002121
9. Dolezel J, Bartos J et al (2005) Plant DNA flow cytometry and estimation of nuclear genome size. Ann Bot 95(1):99–110
10. Dolezel J, Greilhuber J, Suda J et al (2007) Estimation of nuclear DNA content in plants using flow cytometry. Nat Protoc 2 (9):2233–2244
11. Busch W, Herrmann RG, Houben A, Martin R et al (1996) Efficient preparation of plant metaphase spreads. Plant Mol Biol Rep 14 (2):149–155
12. Jiang JM, Gill BS et al (2006) Current status and the future of fluorescence in situ hybridization (FISH) in plant genome research. Genome 49(9):1057–1068
13. Cleveland DW, Mao Y, Sullivan KF et al (2003) Centromeres and kinetochores: from epigenetics to mitotic checkpoint signaling. Cell 112(4):407–421
14. Miell MDD, Fuller CJ, Guse A et al (2013) CENP-A confers a reduction in height on octameric nucleosomes. Nat Struct Mol Biol 20 (6):763–765. doi:10.1038/nsmb.2574
15. Zedek F, Veselý P, Horová L, Bureš P et al (2016) Flow cytometry may allow microscope-independent detection of holocentric chromosomes in plants. Sci Rep 6:27161. doi:10.1038/srep27161
16. De Storme N, Keçeli BN, Zamariola L, Angenon G, Geelen D et al (2016) CENH3-GFP: a visual marker for gametophytic and somatic ploidy determination in *Arabidopsis thaliana*. BMC Plant Biol 16(1). doi:10.1186/s12870-015-0700-5
17. Breuninger H, Rikirsch E, Hermann M, Ueda M, Laux T et al (2008) Differential expression of WOX genes mediates apical-basal axis formation in the Arabidopsis embryo. Dev Cell 14(6): 867–876. doi:10.1016/j.devcel.2008.03.008
18. Engel ML, Holmes-Davis R, McCormick S et al (2005) Green sperm. Identification of male gamete promoters in arabidopsis. Plant Physiol 138(4):2124–2133. doi:10.1104/pp.104.054213
19. Smyth DR, Bowman JL, Meyerowitz EM et al (1990) Early flower development in Arabidopsis. Plant Cell 2(8):755–767
20. Lermontova I, Koroleva O, Rutten T, Fuchs J, Schubert V, Moraes I et al (2011) Knockdown of CENH3 in Arabidopsis reduces mitotic divisions and causes sterility by disturbed meiotic chromosome segregation. Plant J 68(1):40–50

Chapter 8

Staining and Clearing of Arabidopsis Reproductive Tissue for Imaging of Fluorescent Proteins

Daniel Slane, Patrick Bürgel, and Martin Bayer

Abstract

Imaging of fluorescent proteins in whole-mount tissue is a powerful tool to understand growth and developmental processes, not only in plants. With the advent of genetically encoded fluorescent reporters, which specifically label reproductive cells in Arabidopsis, deep tissue imaging has become increasingly important for the study of plant reproduction. To penetrate the surrounding layers of maternal tissue, however, the tissue has to be cleared by homogenizing the refractive index of the sample, often leading to inactivation of fluorescent proteins. 2,2′-thiodiethanol (TDE) has recently been introduced as a clearing agent that allows the imaging of fluorescent proteins in a cleared plant tissue. Here, we describe a simple protocol that combines TDE-based tissue clearing with cell wall staining to outline cells that enable deep tissue imaging in reproductive structures of *Arabidopsis thaliana*.

Key words *Arabidopsis thaliana*, 2,2′-Thiodiethanol, Optical clearing, Deep tissue imaging, Renaissance SR2200, Embryogenesis, Pollen tube, Confocal microscopy, 2-Photon microscopy

1 Introduction

The study of plant reproduction has greatly benefited from the advent of genetically encoded fluorescent markers that enable the visualization of many plant reproductive processes [1, 2]. The visualization of these relatively weak fluorescent signals, however, can be technically challenging since sexual reproduction in flowering plants takes place deep inside maternal tissue. As for other plant tissues, imaging inside the ovule is notoriously difficult due to inhomogeneous refractive indexes of the cell wall, membranes, and cytoplasmic content that cause scattering of light and render the tissue nontransparent [3]. Consequently, there is a poor signal-to-noise ratio and the obtained images appear blurred. Tissue clearing can overcome this principle problem by adjusting the refractive index (RI) of the sample as uniformly as possible. For plant tissue, this is commonly a RI around 1.42—the RI of the cell wall. This can be readily achieved with Chloral hydrate-based

Anja Schmidt (ed.), *Plant Germline Development: Methods and Protocols*, Methods in Molecular Biology, vol. 1669, DOI 10.1007/978-1-4939-7286-9_8, © Springer Science+Business Media LLC 2017

clearing agents if tissue transparency is the only requirement; however, most of these clearing methods denature proteins and are therefore not compatible with the detection of fluorescent proteins [3, 4].

2,2′-Thiodiethanol (TDE) has been used successfully as a clearing agent due to its concentration-dependent RI of 1.33–1.52 in aqueous solutions [5]. Recently, TDE has been introduced as a clearing agent for microscopic analyses in the field of plant biology [3, 6]. While fluorescent proteins are stable and functional even in high TDE concentrations, the tissue is softened by the TDE treatment and therefore care has to be taken to preserve tissue integrity and morphology [3].

While more laborious protocols might be able to preserve morphological details better or might lead to stronger tissue clearing [7], this protocol was developed to be simple and reasonably fast. To understand the origin of fluorescent signals in the context of the reproductive structures, this protocol combines tissue-clearing with cell wall staining to give an outline of cells. For this reason, the fixative solution contains Renaissance SR2200 [8], a cell wall dye that has been successfully used to study plant reproduction and outlines of cells by staining beta-1,3- and beta-1,4-glucans in the cell wall [9]. In combination, TDE seems to help this dye to penetrate deep into the tissue, whereby a more uniform staining can be achieved. Combining tissue staining with the fixation step results in a simple two-step protocol. The cleared tissue is transparent enough to readily detect cells of the mature female gametophyte as well as the developing embryo (up to early globular stage) in whole-mount ovules (by single photon excitation) and in whole-mount siliques (by 2-photon microscopy). Furthermore, it is possible to use this clearing protocol to image pollen in whole-mount anthers or the growing pollen tube on the placenta surface in whole-mount siliques.

2 Materials

2.1 Tissue Fixation and Staining

1. Fresh Arabidopsis reproductive tissue (*see* **Note 1**).
2. Fixation and staining solution: 0.1% (v/v) SR2200 (*see* **Note 2**), 1% (v/v) DMSO, 0.05% (w/v) Triton-X 100, 3.95% (w/v) para-formaldehyde in PBS buffer [pH 7.4] (*see* **Note 3**).
3. PBS buffer: 0.9% (w/v) NaCl, 0.0795% (w/v) Na_2HPO_4, 0.0144% (w/v) NaH_2PO_4, pH 7.4.
4. 1.5 ml microcentrifuge tubes.
5. Dissecting microscope/binocular.
6. Vacuum centrifuge.
7. Double-sided adhesive tape.

8. Razor blade.
9. Syringe needle.
10. Forceps.

2.2 Tissue Clearing

1. Deionized water.
2. 2,2′-thiodiethanol (TDE; diluted in deionized water; for concentrations *see* Subheading 3.2 and **Note 4**).
3. 24-well cell culture plate.
4. Low-speed orbital platform shaker.

2.3 Mounting and Microscopy

1. Liquid blocker, "PAP pen."
2. Double-sided adhesive tape.
3. Microscope slides.
4. Microscope cover slips, 22 mm square.
5. Confocal laser scanning microscope (*see* **Notes 1**, **5** and **6**).

3 Methods

If not otherwise indicated, all the steps are carried out at room temperature.

3.1 Tissue Collection, Staining, and Fixation

3.1.1 Protocol for Young Ovules

1. Keep a 500 μl aliquot of fixation and staining solution in a 1.5 ml microcentrifuge tube ready.
2. Remove a silique containing ovules of the desired developmental stage and place it on a strip of double-sided adhesive tape residing on a microscope slide.
3. Under a dissecting microscope, open the silique alongside by carefully cutting a valve next to the replum with a syringe needle and pushing the valve to the side. Special care has to be taken, not to damage ovules inside the silique (Fig. 1).
4. Carefully transfer the now accessible ovules with the syringe needle tip to the fixation and staining solution.
5. To increase penetration depth of the fixative and dye, the microcentrifuge tube containing ovules is incubated at approx. 250 × *g* for 30 min at RT in a vacuum centrifuge. Afterward the tissue is further fixated and stained overnight at 4 °C and standard pressure (*see* **Note 7**).
6. Remove the fixation and staining solution without picking up the tissue and briefly wash the sample with 1 ml deionized water.

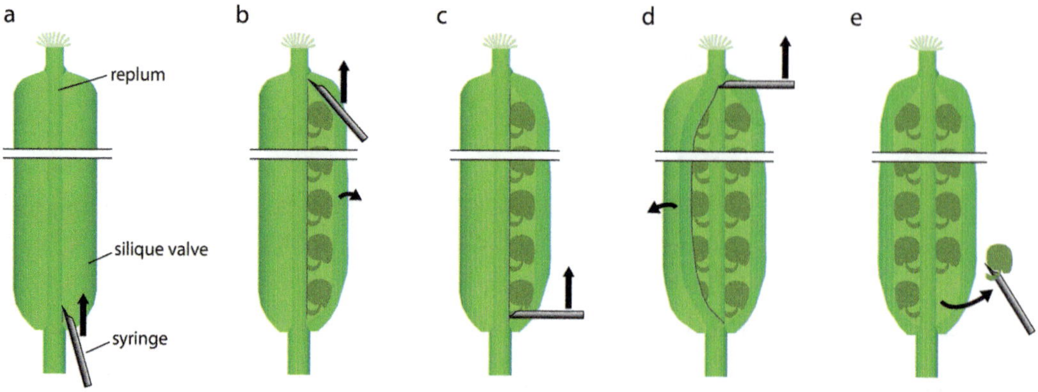

Fig. 1 Collecting ovules from manually dissected siliques. (**a**) Mounted on a double-sided adhesive tape, the silique is carefully cut with a syringe needle along the valve margin next to the replum without damaging the ovules. (**b**) The valve is carefully flipped to the side and fixed on the tape. (**c**) The septum is carefully cut right under the replum without damaging the ovules. (**d**) The second valve is carefully flipped to the side and fixed on the tape. (**e**) The ovules are now accessible and can be carefully picked up and transferred with the tip of the syringe needle

3.1.2 Protocol for Whole Siliques

1. Keep a 500 μl aliquot of fixation and staining solution in a 1.5 ml microcentrifuge tube ready.
2. Remove a silique containing ovules of the desired developmental stage and place it on a microscope slide.
3. Under a dissecting microscope carefully cut the silique at the bottom at a right angle to its main axis with a razor blade.
4. Transfer silique to the fixation and staining solution.
5. To increase penetration depth of the fixative and dye, the microcentrifuge tube containing siliques is incubated at approx. 250 × g for 30 min at RT in a vacuum centrifuge. Afterward the tissue is further fixated and stained overnight at 4 °C and standard pressure (*see* **Note 7**).
6. Remove the fixation and staining solution without damaging the tissue and briefly wash the sample with 1 ml deionized water.

3.2 Tissue Clearing

1. Transfer the tissue carefully to a 24-well cell culture plate and submerge in a 1 ml TDE solution (*see* **Note 4**). Special care has to be taken, not to damage the tissue. Ovules can be transferred with a 1000 μl pipet with the lower portion of the pipet tip cut off to increase the diameter of the opening. Make sure to transfer ovules in as little volume as possible. Whole siliques can be transferred by gently picking them up with fine forceps.
2. Place the cell culture plate on a low-speed platform shaker and agitate the samples in the TDE solution for 10 min.

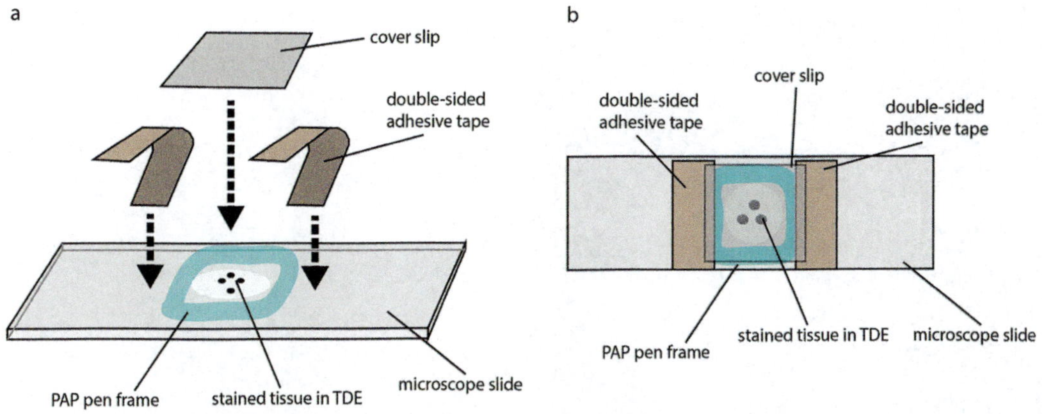

Fig. 2 Tissue mounting and microscope slide assembly. (**a**) A square of roughly 20 mm side length is marked on the microscope slide with a PAP pen. This square is then flanked by two layers of a double-sided adhesive tape. The cleared sample is transferred to the center of the marked square, mounted in a drop of the TDE solution, and covered with a coverslip. The coverslip is thereby resting on the two tape strips. (**b**) Schematic depiction of assembled slide (*top view*)

3. Replace the TDE solution with a fresh aliquot and incubate samples overnight at RT (*see* **Notes 7** and **8**).

3.3 Mounting and Microscopy

1. Draw a square of approximately 20 mm side length on the microscope slide with a "PAP" liquid blocker pen (Fig. 2).
2. Adjacent to the sides of the square, place two layers of a double-sided adhesive tape. This can be conveniently achieved by using tape strips of 40 mm length that are flipped over to form a double layer (Fig. 2).
3. Mount the cleared specimen in the 50 μl TDE solution pipetted to the center of the marked square.
4. Cover the sample with a 22 mm cover slip resting on the two strips of the adhesive tape (*see* **Note 9**).
5. The sample can directly be used for standard confocal or multi-photon imaging (Fig. 3; *see* **Notes 5** and **6**).

4 Notes

1. This protocol was optimized for plants grown under long-day conditions in walk-in chambers at 23 °C and 65% relative humidity on commercial potting mix (described in more detail in [11]).

   The clearing protocol was developed with a focus on young ovules containing developing female gametophytes and young embryos up to early globular stage. For later stages, 2-photon microscopy might be necessary to image fluorescent proteins

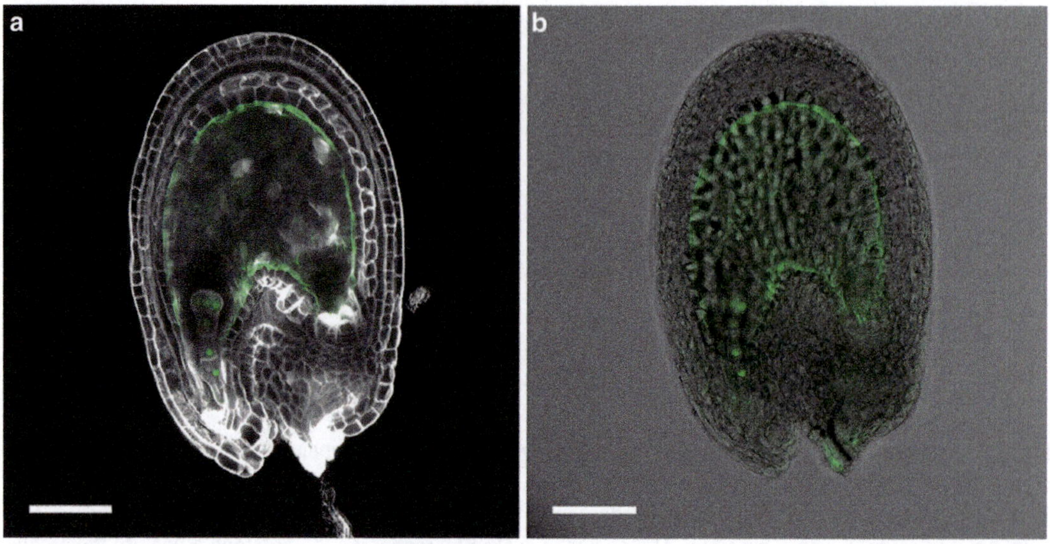

Fig. 3 Expression of nuclear-localized GFP in stained and cleared ovule. (**a**) Whole-mount ovule with nuclear-localized GFP (*green*) expressed in an embryo at the 4-cell stage (*pS4:NLS-GFP* [10]). Cell walls are stained by Renaissance SR2200 (*white*). (**b**) GFP channel (*green*) shown in (**a**) overlaid with DIC image (*grey scale*). Scale bar = 50 μm

inside the developing seed. Whole siliques can also be cleared and—depending on signal strength—growing pollen tubes can be imaged inside whole-mount siliques by standard confocal microscopy, but 2-photon microscopy might give superior results.

2. SR2200 can be ordered from: Renaissance Chemicals Ltd., Unit 1 Blackwood Hall Business Park, North Duffield, Selby, UK. Contact: Howard Weaver, enquiries@renchem.co.uk. The stock solution provided by supplier is considered 100% in this protocol. If background staining is not desired, the fixation solution can be prepared without SR2200 dye.

3. The fixation and staining solution is initially prepared as 4% (w/v) para-formaldehyde (PFA) solution by dissolving PFA powder in gently heated 1× PBS buffer (pH 7.4). After cooling of the PFA solution, the remaining ingredients are added to final volume. The fixation and staining solution can be stored in aliquots at −20 °C. Always use fresh aliquots for fixation and staining.

4. The concentration of the TDE solution depends on tissue type and clearing demand. In general, higher TDE concentrations make the tissue more transparent but at the same time soften it to an extent that might not be tolerable. The following TDE concentrations are compromises that enable good clearing of the tissue but also preserve sample morphology as much as

possible: 20% TDE, for young ovules containing female gametophytes or early-stage embryos (up to 8-cell stage); 50% TDE, for ovules at later stages of development (8-cell embryo and older) and for imaging of pollen grains in whole-mount anthers; 70% TDE, for pollen tubes in whole-mount siliques.

5. The cell wall dye Renaissance SR2200 used for tissue staining can be excited by a 405 nm laser line and shows broad emission in the range of 420–580 nm [9].
6. The cell wall dye SR2200 used in this protocol to outline cells is difficult to image by multi-photon excitation (780–810 nm) due to low fluorescence intensity under these conditions. If the outline of cells is essential, single-photon confocal microscopy might be the better choice.
7. Fixation and clearing time can be reduced to 1 h each if speed is the main priority but this will result in less pronounced tissue clearing and might lead to tissue deformation in higher TDE concentrations.
8. Overnight clearing of whole siliques in 70% TDE renders the tissue transparent enough to image pollen tubes growing on the placenta surface by standard confocal microscopy. However, depending on the fluorophore used and expression strength of pollen marker lines, 2-photon microscopy might be necessary to detect weak signals.
9. The two layers of the double-sided adhesive tape act as a platform to support the cover slip to prevent excessive weight on the sample. The plant tissue is softened by TDE clearing and would be deformed if the coverslip rested directly on the sample.

Acknowledgments

Our research is funded by the German Science Foundation [Deutsche Forschungsgemeinschaft, DFG—SFB1101/B01 to M. B.] and the Max Planck Society. This protocol is based on previously published work [3, 9]. We like to thank Christian Liebig and Thomas Musielak for support and help in establishing the initial TDE-based clearing protocol.

References

1. Sprunck S, Gross-Hardt R (2011) Nuclear behavior, cell polarity, and cell specification in the female gametophyte. Sex Plant Reprod 24(2):123–136. doi:10.1007/s00497-011-0161-4
2. Hamamura Y, Nagahara S, Higashiyama T (2012) Double fertilization on the move. Curr Opin Plant Biol 15(1):70–77. doi:10.1016/j.pbi.2011.11.001
3. Musielak TJ, Slane D, Liebig C, Bayer M (2016) A versatile optical clearing protocol for deep tissue imaging of fluorescent proteins in arabidopsis thaliana. PLoS One 11(8):

e0161107. doi:10.1371/journal.pone.0161107

4. McBryde MC (1936) A method of demonstrating rust hyphae and haustoria in unsectioned leaf tissue. Am J Bot 23(10):686–689. doi:10.2307/2436351
5. Staudt T, Lang MC, Medda R, Engelhardt J, Hell SW (2007) 2,2′-thiodiethanol: a new water soluble mounting medium for high resolution optical microscopy. Microsc Res Tech 70(1):1–9. doi:10.1002/jemt.20396
6. Hasegawa J, Sakamoto Y, Nakagami S, Aida M, Sawa S, Matsunaga S (2016) Three-dimensional imaging of plant organs using a simple and rapid transparency technique. Plant Cell Physiol 57(3):462–472. doi:10.1093/pcp/pcw027
7. Kurihara D, Mizuta Y, Sato Y, Higashiyama T (2015) ClearSee: a rapid optical clearing reagent for whole-plant fluorescence imaging. Development 142(23):4168–4179. doi:10.1242/dev.127613
8. Harris K, Crabb D, Young IM, Weaver H, Gilligan CA, Otten W, Ritz K (2002) In situ visualisation of fungi in soil thin sections: problems with crystallisation of the fluorochrome FB 28 (Calcofluor M2R) and improved staining by SCRI Renaissance 2200. Mycol Res 106:293–297. doi:10.1017/S0953756202005749
9. Musielak TJ, Schenkel L, Kolb M, Henschen A, Bayer M (2015) A simple and versatile cell wall staining protocol to study plant reproduction. Plant Reprod 28(3–4):161–169. doi:10.1007/s00497-015-0267-1
10. Slane D, Kong J, Berendzen KW, Kilian J, Henschen A, Kolb M, Schmid M, Harter K, Mayer U, De Smet I, Bayer M, Jurgens G (2014) Cell type-specific transcriptome analysis in the early Arabidopsis thaliana embryo. Development 141(24):4831–4840. doi:10.1242/dev.116459
11. Babu Y, Musielak T, Henschen A, Bayer M (2013) Suspensor length determines developmental progression of the embryo in arabidopsis. Plant Physiol 162(3):1448–1458. doi:10.1104/pp.113.217166

Chapter 9

Live-Cell Imaging of Auxin and Cytokinin Signaling in Maize Female Gametophytes

Antony M. Chettoor and Matthew M. S. Evans

Abstract

The plant life cycle is characterized by the alternation of generations between genetically active diploid sporophytes and haploid gametophytes. The gametophytes of flowering plants are sexually dimorphic. While the male gametophyte consists of only three cells (two sperm and a vegetative cell) and is released by the parent sporophyte, the female gametophyte (or embryo sac) is more complex and remains imbedded within diploid sporophyte tissues. In maize, the female gametophyte is embedded in a large ovule surrounded with multiple nucellar cell layers impeding live-cell imaging approaches to study embryo sac functions. Here, we describe a simple protocol to visualize embryo sacs with hormonal fluorescent reporters by increasing accessibility of the female gametophyte. The method described is applicable for visualization of any fluorescent embryo sac reporter. The embryo sacs visualization method developed for maize could be extended to facilitate visualization of embryos sac in other important cereals like wheat, rice, and oats.

Key words Maize-female gametophyte, Fluorescent proteins, Live-cell imaging, Auxin, Cytokinin

1 Introduction

Plants exhibit alternation of generations during their life cycle between a multicellular haploid generation called the gametophyte and a multicellular diploid generation called the sporophyte. Sexual reproduction is initiated in specialized sporophytic diploid cells (mother cells) that undergo meiosis and give rise to haploid spores. The spores undergo proliferation and differentiation to develop into multicellular haploid organisms, the gametophytes [1]. In lower plants these gametophytes are free-living, but in flowering plants they developed deeply imbedded within sporophytic tissues. The angiosperm female gametophyte, called the embryo sac, produces the two female gametes, the egg and central cells, as well as the synergids, and antipodals [2]. The male gametophyte contains a vegetative cell and two sperm cells. The two sperm cells of the male gametophyte fertilize the egg and central cell, producing the embryo and endosperm, respectively. Recent results demonstrate

Anja Schmidt (ed.), *Plant Germline Development: Methods and Protocols*, Methods in Molecular Biology, vol. 1669, DOI 10.1007/978-1-4939-7286-9_9, © Springer Science+Business Media LLC 2017

that hormone signaling pathways, notably auxin and cytokinin, play important roles in embryo sac development [3–5].

In most angiosperms, the embryo sac is located inside the ovule and is covered with multiple cell layers of the nucellus and integuments. Because the embryo sac develops imbedded within several cell layers of the ovule, it has been difficult to observe important events that occur during megasporogenesis, megagametogenesis, fertilization, and early seed development. A number of techniques have been used to visualize these cells in fixed tissue [4, 6–9]. However, these techniques are not compatible with live-cell imaging necessary to view dynamic processes in real time, such as female gametophyte development and fertilization. Toward this objective, attempts have been made in a number of plant model systems to improve resolution and sensitivity of imaging in living cells. The first model system successfully developed for imaging of live embryos sacs was *Torenia fournieri*. In *T. fournieri* the embryo sacs grow through the micropyle and protrude out of the ovule. The egg cell, synergids, and half of the central cell are located outside the ovule facilitating imaging of the embryo sac [10]. Arabidopsis embryo sacs with a few nucellar cell layers and integuments have also been amenable to live-cell imaging of dynamic processes, such as fertilization (reviewed in [11]). In monocot cereal flowers, imaging embryo sacs has been a challenge because of the large nucellus, floral bracts, and integuments that all cover the embryo sac. Techniques to overcome these obstacles have recently been applied in maize to visualize auxin signaling in the embryo sac [3, 4].

Here, we present a simple protocol for live imaging of the signaling pathways for the plant hormones auxin and cytokinin in maize embryo sacs using confocal microscopy. Removing the glumes, floral organs, and ovary wall and bisecting the ovule along the proximal-distal axis of the ear enables live-cell imaging of fluorescent proteins in maize embryo sacs. This protocol should facilitate live-cell imaging of any fluorescent proteins in maize embryo sacs, allowing the visualization of a variety of dynamic cellular processes such as hormone signaling, fertilization, and cytoskeletal function. Other monocot cereals like rice, wheat, and oats share a similar ovule and female gametophyte structure with maize, so the method for maize embryo sac visualization described here should be applicable for visualization of embryo sac development in other cereal crops.

2 Materials

1. Ears of the desired stage of transgenic *Zea mays* lines carrying reporters for auxin or cytokinin: An auxin-responsive promoter fusion (*proDR5::RFP*) and a translational fusion reporter for

the auxin efflux carrier PINFORMED1a (*proPIN1a::PIN1a-YFP*) and a two-component signaling Sensor (*proTCSv2::NLS-tdTomato*). These lines can be obtained from the Maize Cell Genomics resource (http://maize.jcvi.org/cellgenomics/index.php) [12]. Transgenic lines expressing fluorescent protein fusions for other aspects of maize cell biology are also available (including reporters for other hormones, such as ABA and GA, markers for the microtubule skeleton or actin filaments, nuclear markers, and markers for other organelles).

2. Forceps (fine tip), dissecting needles, and a scalpel.
3. Cover glass #1 (48 × 65 mm and 24 × 40 mm).
4. Confocal laser-scanning microscope (CSLM): Leica SP8 confocal laser-scanning microscope with LasX software with an inverted microscope stand.
5. 0.53 M Mannitol adjusted to pH 5.0 with 1 M Monobasic Potassium Phosphate and 0.1 M Potassium Hydroxide.

3 Methods

3.1 Harvesting Maize Ears

1. Harvest maize ears complete with husk leaves and the ear shank from greenhouse grown maize plants.
2. Keep the cut end of the ear shank in water before use and in between sessions of dissecting out ovules.

3.2 Preparing Maize Ears for Female Gametophyte Visualization

1. Using a scalpel make a 2.0 cm × 4.0 cm cut through the husk leaves (*see* **Note 1**).
2. Remove the husk leaves to expose a subset of the florets (about 40–50 at a time).
3. Gently cut silks above the ovary to expose 2–3 rows of florets. Each row should contain ~8–10 florets (each with a single ovule) (Fig. 1a, b).
4. Using fine forceps remove the outer glume, the inner glume-Glumen, the lemma, and the palea at the base of the floret (*see* **Note 2**) (Fig. 1c).
5. Next, the ovary wall needs to be removed to access the ovule. Using the needle and fine forceps, cut an arc halfway or more around the base of the silk on the side of the floret toward the base of the ear. Insert the side of the forceps into this slit and lift the silk like a cap away from the floret to expose the ovule (*see* **Note 3**) (Fig. 1d).
6. The embryo sac is located on the apical-basal midline of the ovule on the ear tip side and in the mature ovule is located tight against the cob side. With the scalpel make a longitudinal cut through the middle of the nucellus down to the base (along the

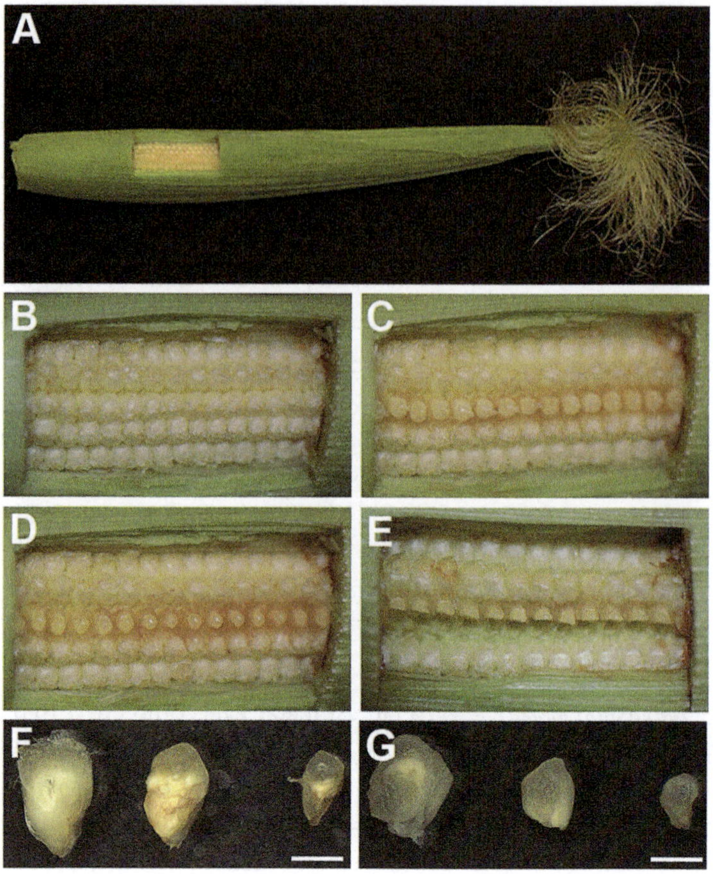

Fig. 1 Maize ovule dissections for embryo sac visualization. (**a**) Mature ear (silks emerged) with florets exposed and silks removed. (**b**) Closeup of the cut section in (**a**). (**c**) Row of florets with the glumes, palea, and lemma removed to expose the ovary. (**d**) Row of florets with the ovary wall removed to expose the ovule. (**e**) Row of exposed ovules bisected along the midline to expose the embryo sac. (**f**) Comparison of a whole floret with the glumes present (*left*), the glumes removed (*middle*), and the ovary wall removed (*right*). (**g**) Cut surface of a single ovule with the glumes present (*left*), the glumes removed (*middle*), and the ovary wall removed (*right*). Scale bar = 0.1 cm (**f**, **g**)

long axis of the ear) resulting in the ovule being split into two halves along the midline (Fig. 1e). This cut produces two ovule halves with the embryo sac within a few cell layers of the cut surface of one of the halves in the majority of the ovules. It is much easier to cut the ovules in half while they are still attached to the cob.

7. Remove the halves of the ovules from the cob (*see* **Note 4**) and float the ovule halves containing the embryo sacs with the cut surface down in a drop of water on a 48 mm × 68 mm cover glass when using an inverted microscope (*see* **Note 5**). For

longer term viability of the ovules, 0.55 M Mannitol pH 5.0 can be used instead of water. Place a second, smaller cover glass on the top of the ovule halves to press them against the cover glass. Be sure to exclude any bubbles and keep the ovule pieces away from the edges of the liquid to avoid image distortion.

3.3 Image Acquisition and Analysis

Follow the recommended standard procedures for your microscope.

1. In the case of CSLM on the Leica SP8X, record single slices and Z stacks with the white light laser using the following settings: YFP Emission (527–650 nm), RFP Emission (583–680), tdTomato (581–680), XY field size of 1024 × 1024 pixels, line average of three times, Z series interval of 1 μm, and color depth of 12 bits.
2. Image files can then be opened and analyzed using imageJ (NIH). For visualizing weaker signal, it is often helpful to sum several adjacent optical sections from the Z-stack within imageJ to increase the signal-to-noise ratio. The *proPIN1a::PIN1a-YFP* reporter should be localized to the cell surface in wild type, *proDR5::RFP* in the cytoplasm, and *proTCSv2::NLS-tdTomato* in the nucleus (Fig. 2). The PIN1 and TCS reporters, in particular, can be combined in the same plant to simultaneously visualize both auxin and cytokinin-related signaling. The strongest expression of *proDR5::RFP* is in the chalazal-most cell of the antipodal cell cluster, the micropylar nucellus and the integuments. Expression of *proPIN1a::PINa-YFP* is typically throughout the antipodal cell cluster as well as the micropylar nucellus and the integuments. *proTCSv2::NLS-tdTomato* is expressed in the nucellus surrounding the antipodal cells and the integuments.

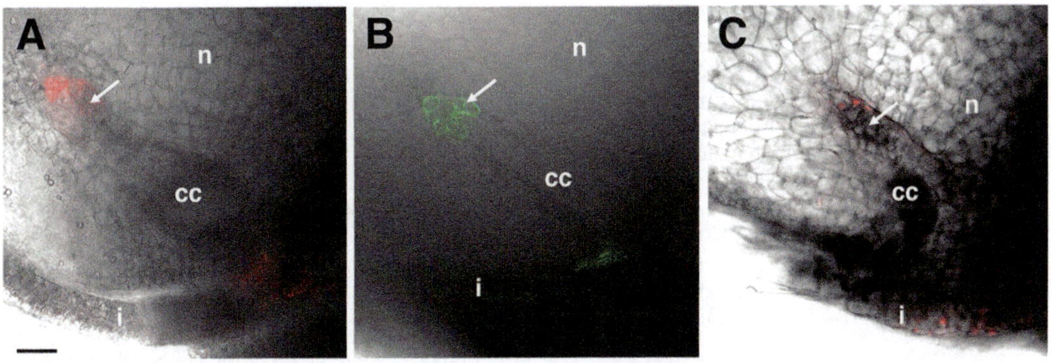

Fig. 2 Live-cell imaging of hormone signaling in maize embryo sacs. (**a**) *proDR5::RFP* reporter. (**b**) *proPIN1a::PIN1a-YFP* reporter. (**c**) *proTCSv2::NLS-tdTomato* reporter. The embryo sacs are oriented with the micropylar end at the lower right and the chalazal end at the upper left. Images are the sum of 5–20 optical sections. *cc* central cell, *n* = nucellus, *i* integument. *Arrows* indicate the antipodal cell cluster. Scale bar = 50 μm

4 Notes

1. Having larger areas cut on the husk leaves results in the ovules rapidly dehydrating and losing turgor. By exposing only a few florets at a time, more ovules can be processed per ear over the course of several hours, since the intact portions of the husk leaves prevent the remaining florets from desiccating.
2. Working on florets in the same row makes access and removal of glumes, lemma, and palea easier.
3. Scoring the ovary wall from the silk to the base of ovule with the needle prior to using the fine forceps helps in their removal.
4. When removing the ovule halves from the cob be sure to cut as close to the cob in the base of the ovule as possible so that the embryo sac is not left behind.
5. Using a cover glass instead of a glass slide helps in visualization of embryo sacs.

Acknowledgments

This work was supported by NSF Plant Genome Program Grant DBI-1340050 to M.M.S.E. All transgenic lines were generously supplied by the Maize Cell Genomics Project (http://maize.jcvi.org/cellgenomics/index.php).

References

1. Walbot V, Evans MMS (2003) Unique features of the plant life cycle and their consequences. Nat Rev Genet 4(5):369–379
2. Drews GN, Yadegari R (2002) Development and function of the angiosperm female gametophyte. Annu Rev Genet 36:99–124
3. Lituiev DS, Krohn NG, Müller B, Jackson D, Hellriegel B, Dresselhaus T, Grossniklaus U (2013) Theoretical and experimental evidence indicates that there is no detectable auxin gradient in the angiosperm female gametophyte. Development 140(22):4544–4553. doi:10.1242/dev.098301
4. Chettoor AM, Evans MMS (2015) Correlation between a loss of auxin signaling and a loss of proliferation in maize antipodal cells. Front Plant Sci 6:187. doi:10.3389/fpls.2015.00187
5. Yuan L, Liu Z, Song X, Johnson C, Yu X, Sundaresan V (2016) The CKI1 histidine kinase specifies the female gametic precursor of the endosperm. Dev Cell 37(1):34–46. doi:10.1016/j.devcel.2016.03.009
6. Faure JE, Rotman N, Fortune P, Dumas C (2002) Fertilization in Arabidopsis thaliana wild type: developmental stages and time course. Plant J 30(4):481–488
7. Christensen CA, King EJ, Jordan JR, Drews GN (1997) Megagametogenesis in Arabidopsis wild type and the Gf mutant. Sex Plant Reprod 10:49–64
8. Vollbrecht E, Hake S (1995) Deficiency analysis of female gametogenesis in maize. Dev Genet 16:44–63
9. Phillips AR, Evans MM (2011) Analysis of stunter1, a maize mutant with reduced gametophyte size and maternal effects on seed development. Genetics 187(4):1085–1097
10. Higashiyama T, Kuroiwa H, Kawano S, Kuroiwa T (1997) Kinetics of double fertilization in Torenia fournieri based on direct

observations of the naked embryo sac. Planta 203(1):101–110. doi:10.1007/s00050170

11. Dresselhaus T, Sprunck S, Wessel GM (2016) Fertilization mechanisms in flowering plants. Curr Biol 26(3):R125–R139. doi:10.1016/j.cub.2015.12.032

12. Mohanty A, Luo A, DeBlasio S, Ling X, Yang Y, Tuthill DE, Williams KE, Hill D, Zadrozny T, Chan A, Sylvester AW, Jackson D (2009) Advancing cell biology and functional genomics in maize using fluorescent protein-tagged lines. Plant Physiol 149(2):601–605. doi:10.1104/pp.108.130146

Chapter 10

Imaging Ca^{2+} Dynamics in Wild-Type and NADPH Oxidase-Deficient Mutant Pollen Tubes with Yellow Cameleon and Confocal Laser Scanning Microscopy

Christina Maria Franck, Jens Westermann, and Aurélien Boisson-Dernier

Abstract

While cytosolic calcium (Ca^{2+}) plays a central role in a myriad of signaling pathways as a secondary messenger, how dynamic changes of cytosolic calcium relate to cell growth control remains poorly understood. The engineering and continuous improvements of genetically encoded calcium sensors such as the Yellow Cameleon (YC) sensors combined with advances in microscopy have allowed imaging with great resolution of the spatiotemporal characteristics of cytosolic $[Ca^{2+}]_{cyt}$ in individual cells. An exciting new step consists therefore in cautiously studying calcium dynamics in mutant backgrounds that display disturbed cellular growth behavior to further enhance our understanding on growth-related processes. Here, we describe methods to perform imaging of $[Ca^{2+}]_{cyt}$ dynamics in growing *Arabidopsis thaliana* wild-type and NADPH-oxidase deficient *rbohH rbohJ* pollen tubes stably expressing YC3.6 using confocal laser scanning microscopy. We also present different ways to extract meaningful qualitative and quantitative information about calcium dynamics during growth.

Key words *Arabidopsis thaliana*, Pollen tube, Ca^{2+}, YC3.6, Ratio-imaging, Cell growth, NADPH oxidase, Cell wall integrity

1 Introduction

Cell growth in plants is a complex process and requires a close coordination between the internal growth machinery and the maintenance of cell wall integrity. One of the key factors in this context is Ca^{2+}, a key signaling constituent ubiquitously shared by eucaryots [1]. However, how calcium dynamics precisely relate to cell growth is still poorly understood. To study growth, tip-growing cells such as pollen tubes (PTs) are very apt, since they can be cultivated in vitro and solely elongate at one location, thus greatly facilitating live-imaging. Moreover, PTs are amenable to genetics,

Electronic Supplementary Material: The online version of this chapter (doi:10.1007/978-1-4939-7286-9_10) contains supplementary material, which is available to authorized users.

Anja Schmidt (ed.), *Plant Germline Development: Methods and Protocols*, Methods in Molecular Biology, vol. 1669, DOI 10.1007/978-1-4939-7286-9_10, © Springer Science+Business Media LLC 2017

transcriptomics, proteomics, pharmacological treatments, and mechanical stimulation. They rapidly cover extreme distances during sexual reproduction, as slow-growing tubes will less likely reach an unfertilized ovule and contribute to the next generation. PT growth rates can reach values of 16.7 μm/min in case of Lilium [2] or even 240 μm/min for fast-growing genera such as *Tradescantia* or *Hemerocallis* [3]. For steady PT growth, as observed in the plant model *Arabidopsis thaliana* [4], a stable tip-focused cytosolic calcium gradient is required. As reviewed by [5], disruption of this tip-focussed Ca^{2+} gradient by treatment with chemical agents such as Ca^{2+} channel blocker nifedipine [6] or actin polymerization preventing LatrunculinB [7] triggers pollen germination problems, PT growth irregularities, and ultimately complete cessation of growth. Unsteady Ca^{2+} oscillations at the tip of growing PTs have been shown to coincide with drastic growth oscillations, often followed by bursting. Examples here are knock-out double mutants of the partially redundant NADPH oxidases *Respiratory burst oxidase Homologue H* and *J* (*RbohH* and *RbohJ*) with a precocious bursting rate of up to 80% in vitro [8, 9].

For imaging cytosolic calcium, indicators are commonly used. These are either chemical indicators such as dyes or genetically encoded calcium sensors. There are many fluorescent dyes available, such as fura-2 [10], indo-1 [11], fluo-4 [12], and Calcium Green-1 [13, 14]. However, these dyes have drawbacks: they need to be introduced into the plant cell first and have been shown to leak into other cellular compartments. Genetically encoded indicators have the advantage of being stably expressed by the cell and staying located to the cytosolCytosolic. They can also be successfully targeted to other cellular compartments as shown, e.g., by [15]. Here, Förster resonance energy transfer (FRET)-based ratiometric Yellow Cameleon (YC) sensors are widely used [16].

YC3.6, the most popular FRET-based sensor to date, consists of a cyan fluorescent protein (CFP) residue linked to a yellow fluorescent protein (YFP) residue via a Calmodulin (CaM) domain and the M13 myosin light chain kinase peptide (Fig. 1) [17]. The difference between the original YC2.1 and the enhanced YC3.6 sensor is the circularly permutated YFP form in the YC3.6 known as Venus. The advantage of Venus lies in the enhanced brightness of the YFP signal and in the expanded dynamic range of the sensor [17]. At low [Ca^{2+}], CFP is excited at 440 nm and will emit light at 480 nm. The distance between the two fluorophores prevents a high FRET signal. Under high [Ca^{2+}] conditions, the CaM domain is activated by binding Ca^{2+} and thus associates with the CaM binding M13 domain. The subsequent conformational change of the YC3.6 brings the YFP and CFP domains in closer proximity to each other and the YFP can now be excited through FRET (Fig. 1). As the FRET signal increases, the CFP signal decreases. Recording both FRET and CFP signals overtime enables accurate

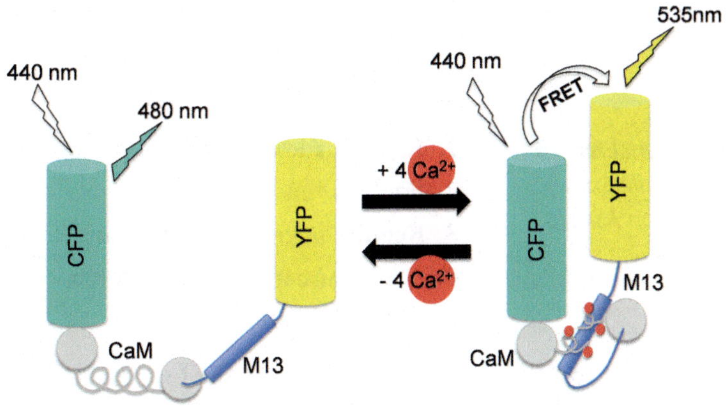

Fig. 1 Reversible conformational change of the YC3.6 ratiometric sensor in the presence/absence of Ca^{2+} ions

measurement of [Ca^{2+}] dynamics. Measuring and analyzing the ratio of FRET signal/CFP signal over time corrects for many artifacts independent of [Ca^{2+}], such as the difference in expression levels of the sensor or photobleaching during imaging [14].

For the methods we describe hereafter, the confocal laser scanning microscope Leica SP8 and its Leica Application Suite X (LASX) software (http://www.leica-microsystems.com) were used to image [Ca^{2+}] dynamics in growing *Arabidopsis thaliana* PTs stably expressing YC3.6 [4]. We present several ways to subsequently extract meaningful information from the acquired data and to analyze it. This includes ratio-imaging, kymograph creation, growth rate determination, and tip-localized [Ca^{2+}] dynamics quantification using mostly the free open source software ImageJ (https://imagej.nih.gov/ij/). We point out differences in cytosolic tip-focussed Ca^{2+} concentration and growth behavior of wild-type (WT) and NADPH oxidase-deficient *rbohH-3 rbohJ-3* mutant PTs to illustrate the potential of the presented methods.

2 Materials

2.1 Plant Material and Equipment

1. Wild-type and mutant plants stably expressing YC3.6 under a pollen-preferential promoter such as pACA9 [18], pLAT52 [19], or pACT1 [4].
2. Plant growth chamber set at 22 °C and 60% humidity.
3. 30 °C incubator.
4. Dissecting microscope.
5. Confocal laser scanning microscope.
6. Image processing program such as ImageJ (https://imagej.nih.gov/ij/), MetaFluor (https://www.moleculardevices.com/

systems/metamorph-research-imaging), Imaris (http://www.bitplane.com/imaris/imaris), or LASX (http://www.leica-microsystems.com).

2.2 Material for In Vitro Pollen Germination Assays

1. Razor blade.
2. Forceps.
3. Regular microscopy slides and cover slips.
4. A moisture incubation box made from a plastic box containing wet paper towel.

2.3 Reagents for In Vitro Pollen Germination Assays

Stock solutions should be prepared with MilliQ H_2O and filter sterilized.

1. Liquid Pollen Germination Medium (PGM): 5 mM KCl, 1 mM $MgSO_4$, 0.01% (w/v) H_3BO_3, 5 mM $CaCl_2$ and 10% (w/v) Sucrose, all dissolved in MilliQ H_2O and added up to a volume of 50 mL according to [20] (*see* **Note 1**). Adjust the pH to 7.5 with 0.1 N NaOH.
2. Solid PGM medium for slides: PGM as in **step 1** with the addition of 1.5% (w/v) low-melting point agarose.

3 Methods

3.1 Obtaining Mutant Plants Expressing YC3.6

1. Introduce the transgenes in your mutant background of choice by either crossing with an already established WT line expressing the transgene or by Agrobacterium-mediated transformation.
2. When picking T1 transformants, choose lines with no pollen abortion that display a strong and homogeneous fluorescent signal in about half of the pollen grains (*see* **Note 2**).

3.2 In Vitro Pollen Germination Assay

1. For your pollen germination assays, make sure your plants are at the correct stage (10–15 cm high stem) and have not been recently treated with pesticides as it usually impairs pollen development.
2. Collect between 20 and 30 nicely opened flowers for each genotype in the moisture incubation box. Incubate the flowers at 22 °C for at least 20 min to allow the pollen to hydrate.
3. Meanwhile, prepare 50 mL of fresh liquid PGM and use half to prepare the slides with solid medium. Bring the medium to a boil until it gets clear and homogeneous. Spread 450 μL on a glass slide and distribute it evenly using a razor blade to form an agarose pad. Allow the medium to solidify for 5 min and transfer the slide to a moisture incubation box.

Fig. 2 Pollen germination assay. (**a**) Brush the flowers on the solid medium in the center of the agarose pad. Arrange the flowers around the pollen in a small circle. (**b**) Three to six hours after incubation, PTs are growing in the middle of the circle. (**c**) Bright-field overview image of growing PTs

4. Grab a flower petiole with your forceps and under a dissecting scope gently brush the open flower on the solidified medium at the center to release the pollen. Then, place the flower around the pollen with the pistil facing inward. This will improve pollen germination. Brush the pollen of the remaining flowers in the middle of the flower circle to achieve higher pollen density (Fig. 2). Low pollen density leads to drastically decreased pollen germination rate.
5. Transfer the slides to the moisture incubation box and incubate it for 35–45 min at 30 °C to boost pollen metabolism.
6. Transfer the slides to 22 °C and let them incubate for another 3–4 h (*see* **Note 3**).
7. Gently remove the flowers from the slides and spread 80–100 μL of liquid PGM on the top of the germinating pollen. Gently apply a cover slip and avoid air bubbles. The samples are ready for microscopy.

3.3 Confocal Microscope Configuration

1. Excite the CFP part of the YC3.6 by using the argon-ion laser at 458 nm or the 405 nm emission of a diode laser. *See* Fig. 3 for microscope configuration and laser paths.
2. Minimize the excitation energy of the laser; adjust it to 20% power at maximum to minimize alteration or damage of the cellular properties of your sample.
3. Make sure that you use the proper primary dichroic mirror to separate the emitted fluorescence signal from the laser. It should match the laser line selected, thus either 458 nm for the argon-ion laser or 405 nm for the diode laser.
4. Enable two channels for CFP and FRET acquisition and set the Gain for both to 100 V and the Offset to 0% for quantitative imaging (*see* **Note 4**). If you use Hybrid detectors (HyDs) this is automatically set up.

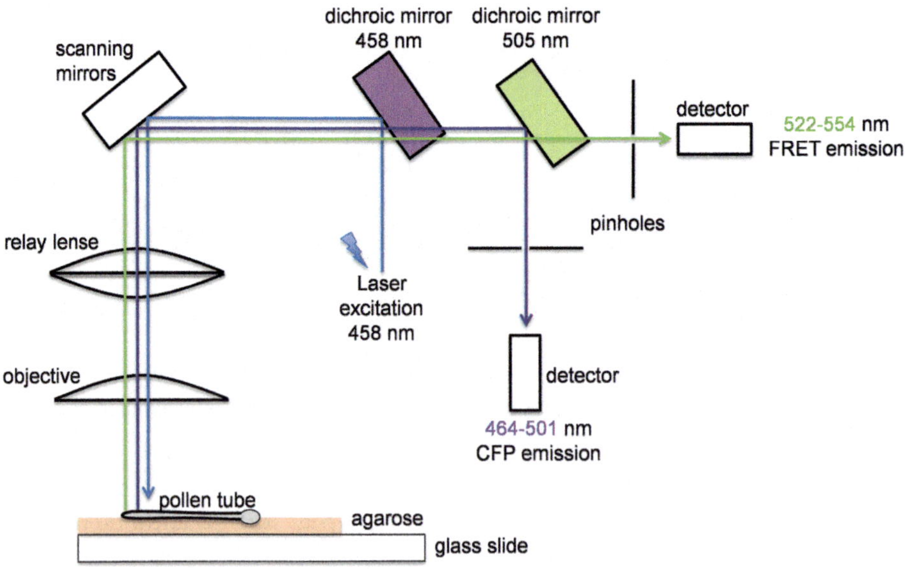

Fig. 3 Confocal microscopy setup for YC3.6 settings. The scanning mirrors determine which spot of the PT is focused on. The YC3.6 proteins expressed by the PT are excited and light emitted by the CFP and YFP fluorophores is reflected back at the scanning mirrors. At the first dichroic mirror, light at 458 nm wavelength is reflected, light with higher wavelengths passes through. The second dichroic mirror reflects light at wavelengths up to 505 nm which are guided through a pinhole and collected at a detector. This yields the CFP signal. Light at higher wavelengths passes through the mirror and is collected separately at another detector for the FRET signal

5. Adjust the fluorescence wavelengths to be collected in your separate FRET and CFP channels, also known as gating. For the CFP channel, set the gate values to 464–501 nm, for the FRET channel to 525–554 nm for example.
6. If possible, enable another channel to acquire a transmission image (bright-field) or, even better, a differential interference contrast (DIC) image (*see* **Note 5**). This will facilitate staying in focus during the time course.
7. Assign pseudo-colors to your channels: blue for CFP, yellow for FRET, and gray for bright-field or DIC for example.

3.4 Setting Up the Acquisition Parameters

1. For acquisition mode, choose a xyt stack mode to image PT [Ca^{2+}] dynamics over time.
2. For imaging, you want to find the best compromise between resolution and acquisition speed. A high spatial resolution will coincide with a low temporal resolution, likely causing a loss of information, as the time intervals between your separate images will be long. This is to be avoided when looking at dynamic [Ca^{2+}] changes within the cell that tend to occur rapidly. There are several parameters that greatly impact the relationship between spatial resolution and acquisition speed that will be discussed in the following lines.

3. "Format" limits the amount of pixels captured per image. A large format improves the quality of the picture but also slows the scanning process. Thus, set the Format to 512 × 512 or lower to minimize the frame interval. "Speed" defines the scan speed for each pixel. A slow scan speed improves the signal-to-noise ratio but at the same time increases photobleaching of your sample. Set "Speed" to 400 Hz or higher to minimize the frame interval.
4. "Line Average", "Line Accumulation", "Frame Average", and "Frame Accumulation" are ways to further increase the signal-to-noise ratio. Incrementing the settings to more than the default value 1 will improve the resolution of your images but again increase the acquisition time intervals. Set the "Line Average" to 2 or lower it to 1 to minimize the frame interval and leave "Line Accumulation", "Frame Average", and "Frame Accumulation" at 1.
5. "Pinhole" determines the diameter of the opening that the emitted light from the fluorophores will be passing through before reaching the detector. Larger pinhole diameter will increase the brightness of your signal; however, the detectors will pick up more diffuse light. Here, choose AU and *Airy 1.*
6. In our hands, with these parameters, the spatial resolution of the images is relatively high without increasing the minimal time interval in between the acquisition of the separate images to more than 3 s. Save your settings to recall them and use them for every sample, genotype, or treatment so as to permit a meaningful comparison of the results.

3.5 Data Acquisition

1. Put your sample on the stage and focus on the pollen grains with the 10x Dry objective.
2. Once in focus, switch to the 63× immersion objective. Select a PT of interest. Make sure it grows. Information gathered on nongrowing PTs is meaningless toward understanding relationships between calcium dynamics and growth in different backgrounds. Go to Live Scan.
3. Load your previously saved parameters.
4. Make sure the PT tip is at the center of the image and adjust the Zoom Factor to 4. Focus on the median plane of the PT tip as we assume symmetry for the cylindrical-shaped PT.
5. Confirm that the minimal Time Interval does not exceed 3 s. Set the Frame count, e.g., to 40 or 100 respectively to limit the number of images taken per time course.
6. Start the time-course acquisition. While it is running, continuously check if the PT grows out of focus and slightly readjust the Z position, if this is the case. If you do not have a third

channel for DIC or bright-field available, you can use "Quick-LUT" to ensure that you are still in focus. The black outline of the PT tip should be as thin as possible, i.e., the "red" signal of the PT should be as close as possible to the "green" background. Once it is done, screen your slide for another growing PT and repeat from **step 3** onward.

3.6 Ratio Imaging

1. Once the raw data are acquired and saved, several image processing programs (e.g., ImageJ with its RatioPlus plugin, MetaFluor, Imaris, or LASX) allow you to generate the ratio images or movies. In LASX, go to "Process", "Noise Reduction", and select "Median". Apply Radius 5.0 and Iteration 1 to each of your series and save the smoothed version.
2. Pick "Stack profile" and select "Calculator" and "Calcium imaging".
3. Calculate the fluorescence intensity (F) ratio (R) for each series:

   $R = F(\text{FRET channel})/F(\text{CFP channel})$.
4. Set the "Image Scaling" minimum to 0 and the maximum to 8. You can readjust these parameters until you find the ones that match all or most of your samples. Press calculate. The settings are now applied for all series.
5. To save the series with fluorescence intensity ratio, press export as .AVI and select "Export View", "Overlay Channels", and "Use compression". Select the amount of frames/second, e.g., 6. An example of a growing WT, as well as growing and bursting *rbohH-3 rbohJ-3* PTs with simultaneous FRET, CFP, DIC, and Ratio images can be seen in the Movies S1, S2, and S3 online, respectively.

3.7 Kymograph Creation in ImageJ

Kymographs are a great way to analyze cell growth by plotting distance grown over time. Unlike in a movie, you can follow the growth process in only one image and still visualize growth irregularities. Also, $[Ca^{2+}]$ oscillations can be observed in a ratio kymograph.

1. To create a ratio kymograph, open the ratio movie .AVI file (*see* **Note 6**) of your PT of choice in ImageJ.
2. Draw a segmented line in the middle of the path of the growing PT. Start in the middle of the tube on the first frame and end a bit further than the final position of the PT tip on the last frame.
3. Then go to "Image", "Stack", and "Reslice" the stack starting at the Top with Output spacing of 1 pixel. Tick "Avoid interpolation". A kymograph is created. Figures 4 and 5 depict kymographs of growing WT (left) and *rbohH-3 rbohJ-3* (right) PTs imaged during 2 and 5 min, respectively.

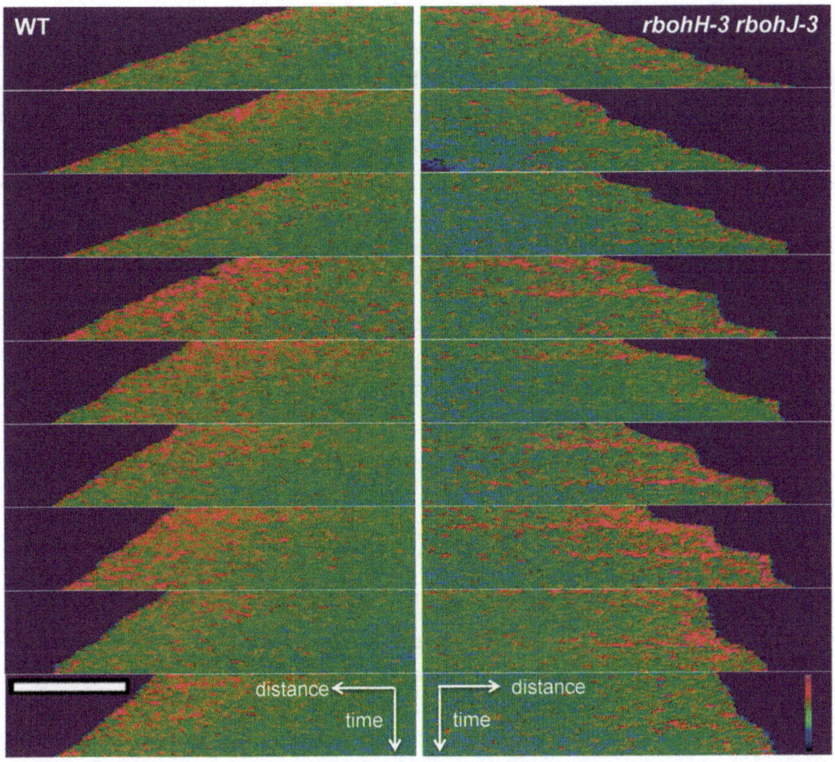

Fig. 4 Kymographs of nine growing WT (*left*) and *rbohH-3 rbohJ-3* PTs (*right*) imaged over 2 min with $\Delta t = 3$ s. Kymographs measure the changes in the relative position of the PT tip over time. The *X*-axis depicts distance covered, the *Y*-axis the time. The *color* code represents the relative Ca^{2+} concentration. Low Ca^{2+} levels are shown in dark blue whereas high Ca^{2+} levels are shown in *red*. The scale is 5 μm. Note how the WT PTs show steady growth, whereas the *rbohH-3 rbohJ-3* PTs depict irregular growth oscillations. Periods of stalled growth are followed by rapid cellular expansion

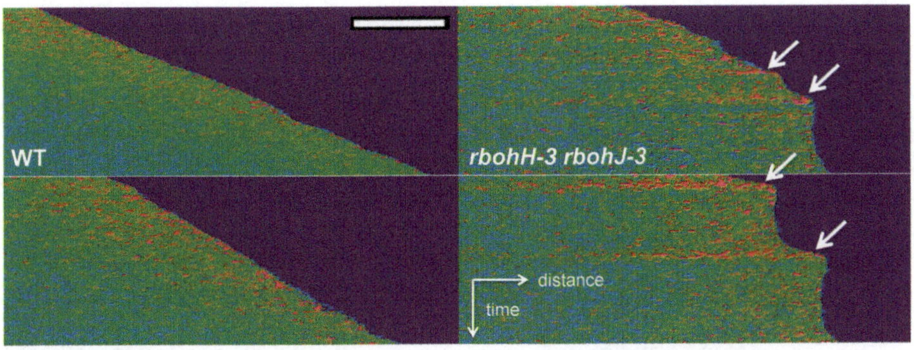

Fig. 5 Kymographs of growing WT (*left*) and *rbohH-3 rbohJ-3* PTs (*right*) imaged over 5 min with $\Delta t = 3$ s. Note how for *rbohH-3 rbohJ-3* PTs there is a [Ca^{2+}] maximum occurring at the tip in the late phase of the rapid expansion (*white arrows*) just before growth arrest. The scale is 5 μm

3.8 Quantitative Analysis of Dynamic [Ca^{2+}] Changes During PT Growth

1. With ImageJ, import the original raw data-containing file, e.g., the .lif file from Leica SP8 confocal microscope. Use the import function and the Bio-Formats plug-in. By this means, the pixel-to-μm scale information will be retained so you do not have to manually set it. The raw data will be opened as a hyperstack with DIC, FRET, and CFP channels and the time course for each channel. Click on "Image", "Duplicate" to first duplicate the FRET channel. Go to the first frame.
2. Define a circular Region of Interest (ROI, 3–4 μm diameter) for the PT tip region. It will be the template for all the ROIs that will be defined hereafter. Make sure, the circle does not exceed or touch the PT boundaries and is equally spaced to all sides (1–2 μm).
3. Select the next frame and move the ROI accordingly to follow the PT path. Place it in a way that the circle always stays at the same distance (1–2 μm) from the PT tip boundary. Add the new position to the ROI manager. Repeat this process for all 40 frames subsequently and save the 40 ROIs.
4. In the ROI manager, go to "Results" and "Set Measurements", tick "stack position" (gives the frame number from 1 to 40), "mean gray value" (yields the mean fluorescence intensity) and "centroid" (reads out x and y coordinates of the circular ROI center) which are the only parameters required for our analysis. Save these FRET-related results in an Excel file.
5. Go back to **step 1** and duplicate the CFP channel. Recall the same 40 ROIs you defined in **step 3** and get the CFP-related results.
6. To determine fluorescence intensity in the shank of the PT, return to the FRET channel. Take the template ROI and place it approximately 10–12 μm away from the tip ROI in the center of the shank in the first frame.
7. Repeat **steps 3–5** for the shank region. Figure 6 illustrates the ROI set defining process.
8. From these data, different kinds of information can be retrieved. A selection is presented in the following lines:
   (a) The relative distance (Δd) covered by the PT per frame

$$\Delta d = \sqrt{\left((x(t) - x(t + \mathrm{d}t))^2 + (y(t) - y(t + \mathrm{d}t))^2\right)}.$$

   (b) The growth rate per minute (e) for each frame. The time interval (Δt) passing between the respective frames is 3 s here.

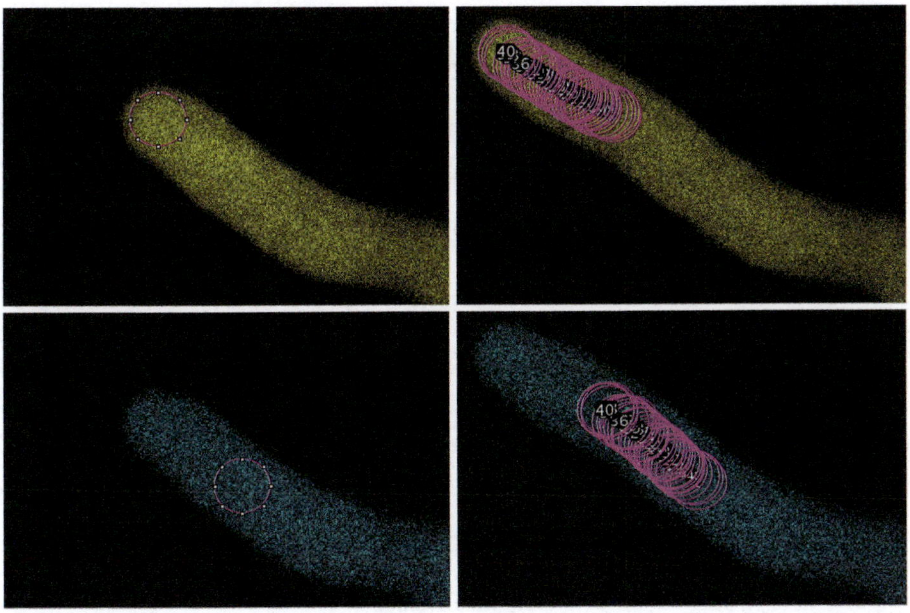

Fig. 6 Setting up your region of interests (ROI) for the analysis of [Ca^{2+}] dynamics in a growing PT. At the top, the ROI selection in the FRET channel for the tip-ratio determination is depicted; at the bottom, the ROI selection for the shank-ratio in the CFP channel is displayed. On the *left side*, $t = 0$; on the *right side*, $t = 2$ min

$$e = \frac{\Delta d}{\Delta t} \times 60 \frac{\mathrm{s}}{\mathrm{min}}.$$

(c) The cumulative distance (*d*) the PT covered over the recorded time course.

$$d = \sum_{n=1}^{n=40} \Delta d.$$

(d) The relative Ca^{2+} concentration (ratio of FRET channel signal/CFP channel signal) at the PT tip (tip-ratio) and in the shank (shank-ratio) as well as a representation of the tip-focused Ca^{2+} gradient (tip-ratio/shank-ratio). If you want an approximation of the absolute [Ca^{2+}], you need to perform with your microscope calibration of the YC3.6 ratio to known Ca^{2+} concentrations (*see* **Note** 7).

9. Figure 7 displays representative examples for a WT and an *rbohH-3 rbohJ-3* PT with their corresponding tip-ratio and cumulative distance over the 2 min-long time course (*see* **Note 8**). For WT, a steady gradual increase in cumulative distance over time and relatively mild oscillations of the [Ca^{2+}] at the tip can be observed as previously reported [8].

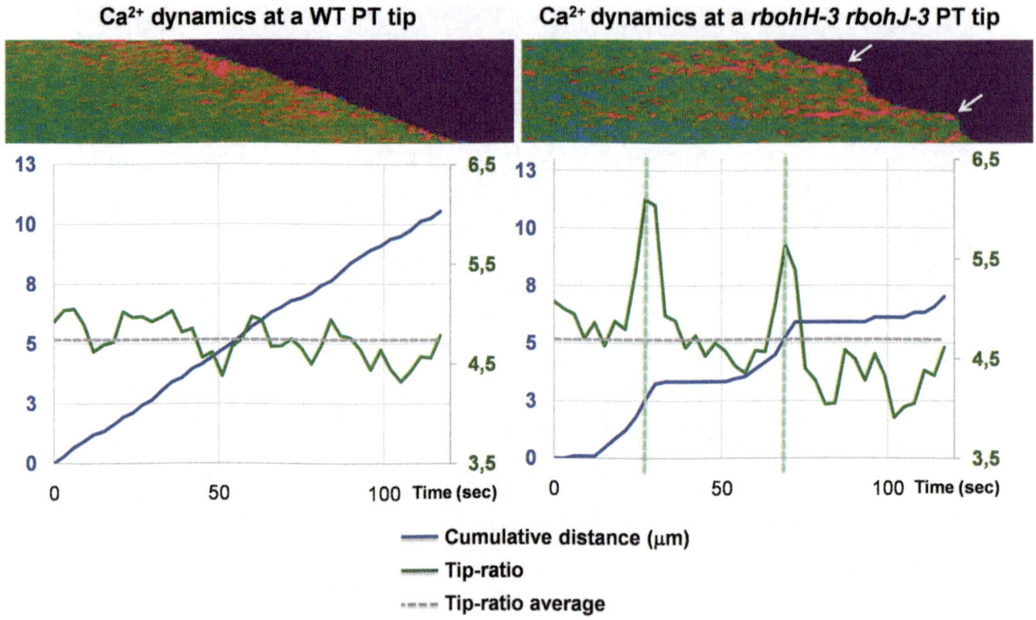

Fig. 7 Kymographs of a representative WT (*left*) and *rbohH-3 rbohJ-3* (*right*) PTs with quantitative analysis of cumulative distance (*blue line*), tip-ratio (*green line*), and average tip-ratio over time (*dashed gray line*). The *white arrows* in the *rbohH-3 rbohJ-3* kymograph indicate the [Ca^{2+}] peaks in the late phase of rapid growth and correspond to the *green* peaks in the *bottom graph*

In contrast, the *rbohH-3 rbohJ-3* PT shows major growth perturbations in cumulative distance over time, accompanied by [Ca^{2+}] peaks at the PT tip in the late phase of the rapid elongation process. This behavior is almost identical to the one reported for the allelic *rbohH-1 rbohJ-2* mutant, thereby further supporting that *RbohH* and *RbohJ*, probably through [Ca^{2+}] homeostasis regulation, are involved in dampening growth rate fluctuation [8, 9].

4 Notes

1. Preparing fresh medium for every germination assay greatly decreases the variability in pollen germination and PT growth rates.
2. Transformants that display different levels of fluorescence among pollen grains and/or in more than half of the grains are very likely to carry more than one T-DNA insertion. In the subsequent generations, combine selection on antibiotic/herbicide plates and screening for fluorescence in pollen to obtain and propagate line(s) homozygous for the YC3.6 transgene. The double mutant *rbohH-3 rbohJ-3* used here (SALK_136917 x SALK_050665) does not silence the YC3.6 transgene unlike

the *rbohH-1 rbohJ-2* (GABI_028G04 × SAIL_31_D07) independent double mutant [8, 9].

3. For cell wall integrity-related mutant PTs with bursting tendency, an incubation time of 3 h is a good compromise between germination rate and occurrence of intact PTs. After 3 h the germination rate increases along with the bursting rate, making it difficult to still find intact growing PTs. For WT or other mutants, PTs can be cultivated for a longer time period.
4. If your confocal system is equipped with normal detectors, adjust Gain and Offset so as to minimize dark noise and improve signal-to-noise ratio without photobleaching. Keep the same Gain and Offset parameters for all your samples. Moreover, during the quantification analysis phase, perform a background subtraction on your FRET and CFP channels before drawing your set of ROIs.
5. A transmission image can be acquired by using a 488 nm laser line at very low power (e.g., 3–5%).
6. Note that you can also generate a kymograph from the FRET channel only or the DIC channel for example.
7. Examples of YC3.6 calibration can be seen in [4, 21].
8. If you want to represent traces for tip-ratio data of a single PT over time, display as well the corresponding cumulative distance traveled by the PT (or its growth rates). Keep in mind that calcium dynamics data with no proof of growth are pointless.

Acknowledgments

We are grateful to Megumi Iwano (Osaka University, Japan) for gifting us seeds of the pACT1-YC3.60 line. We thank all members of Martin Hülskamp's group (University of Cologne, Germany) for sharing their facilities and CEPLAS for access to the Leica SP8 confocal microscope. This work was supported by the University of Cologne, the Deutsche Forschungsgemeinschaft Grant BO 4470/1-1, and a grant from the University of Cologne Centre of Excellence in Plant Sciences to A.B.D.

References

1. Permyakov EA, Kretsinger RH (2009) Cell signaling, beyond cytosolic calcium in eukaryotes. J Inorg Biochem 103:77–86
2. Pierson ES, Miller DD, Callaham DA, van Aken J, Hackett G, Hepler PK (1996) Tip-localized calcium entry fluctuates during pollen tube growth. Dev Biol 174:160–173
3. Michard E, Alves F, Feijó JA (2009) The role of ion fluxes in polarized cell growth and morphogenesis: the pollen tube as an experimental paradigm. Int J Dev Biol 53:1609–1622
4. Iwano M, Entani T, Shiba H, Kakita M, Nagai T, Mizuno H et al (2009) Fine-tuning of the cytoplasmic Ca^{2+} concentration is essential for

pollen tube growth. Plant Physiol 150:1322–1334

5. Steinhorst L, Kudla J (2012) Calcium - a central regulator of pollen germination and tube growth. Biochim Biophys Acta 1833:1573–1581
6. Reiss HD, Herth W (1985) Nifedipine-sensitive calcium channels are involved in polar growth of lily pollen tubes. J Cell Sci 76:247–254
7. Cárdenas L, Lovy-Wheeler A, Kunkel JGG, Hepler PKK, Cardenas L, Lovy-Wheeler A et al (2008) Pollen tube growth oscillations and intracellular calcium levels are reversibly modulated by actin polymerization. Plant Physiol 146:1611–1621
8. Boisson-Dernier A, Lituiev DS, Nestorova A, Franck CM, Thirugnanarajah S, Grossniklaus U (2013) ANXUR receptor-like kinases coordinate cell wall integrity with growth at the pollen tube tip via NADPH oxidases. PLoS Biol 11:e1001719
9. Lassig R, Gutermuth T, Bey TD, Konrad KR, Romeis T (2014) Pollen tube NAD(P)H oxidases act as a speed control to dampen growth rate oscillations during polarized cell growth. Plant J 78:94–106
10. Kong SK, Lee CY (1995) The use of fura 2 for measurement of free calcium concentration. Biochemist 23:97–98
11. Lückhoff A (1986) Measuring cytosolic free calcium concentration in endothelial cells with indo-1: the pitfall of using the ratio of two fluorescence intensities recorded at different wavelengths. Cell Calcium 7:233–248
12. Gee KR, Brown KA, Chen W-NU, Bishop-Stewart J, Gray D, Johnson I (2000) Chemical and physiological characterization of fluo-4 Ca^{2+}-indicator dyes. Cell Calcium 27:97–106
13. Lee S-K, Lee J-Y, Lee M-Y, Chung S-M, Chung J-H (1999) Advantages of calcium green-1 over other fluorescent dyes in measuring cytosolic calcium in platelets. Anal Biochem 273:186–191
14. Swanson SJ, Choi W-G, Chanoca A, Gilroy S (2011) In vivo imaging of Ca^{2+}, pH, and reactive oxygen species using fluorescent probes in plants. Annu Rev Plant Biol 62:273–297
15. Loro G, Drago I, Pozzan T, Lo SF, Zottini M, Costa A (2012) Targeting of Cameleons to various subcellular compartments reveals a strict cytoplasmic/mitochondrial Ca^{2+} handling relationship in plant cells. Plant J 71:1–13
16. Krebs M, Held K, Binder A, Hashimoto K, Den Herder G, Parniske M et al (2012) FRET-based genetically encoded sensors allow high-resolution live cell imaging of Ca^{2+} dynamics. Plant J 69:181–192
17. Nagai T, Yamada S, Tominaga T, Ichikawa M, Miyawaki A (2004) Expanded dynamic range of fluorescent indicators for Ca^{2+} by circularly permuted yellow fluorescent proteins. Proc Natl Acad Sci U S A 101:10554–10559
18. Schiøtt M, Romanowsky SMM, Bækgaard L, Jakobsen MKK, Palmgren MGG, Harper JFF (2004) A plant plasma membrane Ca^{2+} pump is required for normal pollen tube growth and fertilization. Proc Natl Acad Sci U S A 101:9502–9507
19. Twell D, Yamaguchi J, Wing RA, Ushiba J, McCormick S (1991) Promoter analysis of genes that are coordinately expressed during pollen development reveals pollen-specific enhancer sequences and shared regulatory elements. Genes Dev 5:496–507
20. Boavida LC, McCormick S (2007) Temperature as a determinant factor for increased and reproducible in vitro pollen germination in *Arabidopsis thaliana*. Plant J 52:570–582
21. Monshausen GB, Messerli MA, Gilroy S (2008) Imaging of the yellow cameleon 3.6 indicator reveals that elevations in cytosolic ca^{2+} follow oscillating increases in growth in root hairs of Arabidopsis. Plant Physiol 147:1690–1698

Chapter 11

Immunolocalization of AGPs and Pectins in *Quercus suber* Gametophytic Structures

Mário Luís da Costa, Ana Lúcia Lopes, Maria Isabel Amorim, and Sílvia Coimbra

Abstract

The arabinogalactan proteins (AGPs) are highly glycosylated proteins, ubiquitous in plants that have been linked to numerous aspects of sexual reproduction in several plant species, including the monoecious tree species *Quercus suber*. AGPs are found in cell membranes and cell walls of all types of tissues, including reproductive cells and organs. Pectins are cell wall components that also have been shown to change in composition and quantity during the maturations of the male and female gametophyte in cork oak. These findings were only possible to reveal, due to the histological study of AGP and pectins epitopes by immunolabeling. The immunofluorescence microscopy technique uses antibodies linked to fluorophores and relies on the specificity of the antibody binding to its antigen, labeling the epitope with a fluorescent dye.

In the method presented here, we explore the immunolocalization technique performed in male and female flowers of *Quercus suber*, using London Resin (LR-White) as the embedding medium, after vacuum fixation with formaldehyde/glutaraldehyde. An extensive description of all the aspects of this technique is provided, from the plant material developmental stages selection to the critical analysis of results performed, continuously supported by troubleshooting recommendations.

Key words Immunolocalization, *Quercus suber*, Arabinogalactan proteins, Pectins, LR-White, Fixation, Inclusion

1 Introduction

1.1 Quercus suber *Importance and Reproduction*

The evergreen cork oak, *Quercus suber*, is a dominating monoecious tree species in the semi-arid regions of Southern Europe, with high ecologic value and economically crucial for the sustainability of various regions, especially because of the cork and seed production. The cork oak acorn is crucial for the traditional system of Iberian pig farming and for the quality and certification of its derived products [1].

In the Iberian Peninsula cork oak forests, the importance of *Q. suber* has facilitated the development of new strategies for

Anja Schmidt (ed.), *Plant Germline Development: Methods and Protocols*, Methods in Molecular Biology, vol. 1669, DOI 10.1007/978-1-4939-7286-9_11,

reforestation programs. Nevertheless, there is still a limited conventional breeding of cork oak due to its long reproductive cycle, complex reproductive biology, and recalcitrant seeds. The existence of fruits with different morphology and physiology and the inability to predict seed yield, help to constrain the research on this plant breeding [2].

In cork oak, female flower buds appear in spring and the inflorescences emerge on the axil of new leaves, in spikes with 3–5 flowers. The flowers are inside a cupule, partially enclosed by a dome-shaped bud of imbricate scales, each flower contains three carpels, each one with two ovules inside. During spike elongation, 3–5 styles arise from the cupule and the stigma turns receptive. The ovules only become fully developed a few months later, if pollinated and can be phenologically distinguished according to stages of development, before fertilization and then after fertilization [3].

Male flowers occur in early spring and sometimes in autumn, in catkins at the base of current season branches. At anthesis, female flowers are not completely developed. 10–12 days after pollination, pollen tube growth is arrested, and its growth only resumes 2 months later. Once fertilization has occurred, only one of the six ovules develops into a monospermic fruit, which matures during autumn [4]. Such characteristics make *Q. suber* an interesting system for comparative studies of development and sexual reproduction in a non-model forest plant.

The development of the female gametophyte occurs after a delayed process of pollination, with several interactions between the pollen tube and the pistil tissues involving different players, including arabinogalactan proteins (AGPs) [5, 6].

1.2 Arabinogalactan Proteins Importance on Plant Reproduction

Arabinogalactan proteins (AGPs) are a large family of hydroxyproline-rich proteins, highly glycosylatedGlycosylation and structurally complex [7], expressed in almost all types of cells. AGPs are globally present in land plants, largely spread from Bryophyta *Physcomitrella patens* [8] to all seed plants [9], but also present in many algae, showing their ancient origin [10]. They are abundant in cell walls, plasma membranes, and extracellular secretions [11]. Their polypeptide core characteristics define the common division of AGPs into four subgroups: the classical AGPs with an N-terminal signal peptide that is removed in the mature protein, a proline/hydroxyproline-rich domain, and a C-terminal signal able to link a glycosylphophatidylinositol (GPI) anchor; the arabinogalactan (AG) peptides, with a smaller protein backbone (10–13 aminoacids) and structurally similar to the classical AGPs; the lysine-rich AGPs, with one or more lysine domains; and the fasciclin-like AGPs (FLAs) with one or more fasciclin-like domain. In AGPs, more than 90% of the molecules are sugars, they are probably the most highly modified proteins in nature undergoing several posttranslational modifications [12].

AGPs are essential molecular players involved in diverse biological processes, such as cell expansion, proliferation and differentiation, cell–cell recognition, somatic embryogenesis, pollen tube growth, programmed cell death, seed germination, and resistance to infection [13]. AGPs may act as signaling molecules due to their GPI anchor [14], they can be cleaved by specific phospholipases and released to the extracellular matrix or can activate any downstream signal transduction pathway [15, 16]. Besides, the carbohydrate moieties of AGPs might be released and the generated oligosaccharides act as signaling molecules [17].

AGPs have long been suggested to play important roles in sexual plant reproduction and during the past 20 years, several studies were deepened to unravel their biological way of action [18]. They have been found in the extracellular matrix of the transmitting tract of *Gladiolus gandavensis*, *Lilium longiflorum*, *Nicotiana alata*, and *Lycopersicon peruvianum* [19] and implicated in pollen tube growth from the stigma to the ovules in *Amaranthus hypochondriacus*, *Actinidia deliciosa*, *Catharanthus roseus*, and *Nicotiana tabacum* [20, 21]. These studies were performed either with the β-glycosyl Yariv reagent that binds specifically to the β-1,3-galactooligosaccharides AGPs, precipitating them [22, 23], or using monoclonal antibodies that identify the glycosidic epitopes of AGPs [24, 25]. These two techniques allow modeling the distribution and localization of AGPs [26] permitting inferring their possible roles [27, 28], showing that AGPs can be used as molecular markers for different stages of development and for different cell/tissues involved in sexual plant reproduction and embryogenesis [5, 9, 26, 29–32].

1.3 Immunolocalization Techniques in Flowers of Quercus suber

The study of the reproductive structures is a challenge, especially in plants with asynchronous gametic development, where female maturation spans over several months from the emergence of the flower to the maturation of the embryo sac [2–4].

The arabinogalactan proteins (AGPs) have been linked to numerous aspects of sexual reproduction in several plant species, being found in cell membranes and cell walls of reproductive cells and organs. Pectins are components of the cell wall and also have been shown to change in composition and quantity during the maturation of the male and female gametophytes in *Quercus suber* [5, 6]. Such observations would not be possible without the histological study of AGP and pectin epitopes by immunolabeling.

Several methods for histological studies are available, from free hand sectioning, to high pressure freezing followed by cryostat sectioning. However, precision of results and available resources must be taken under consideration when performing a histological study of a difficult sample. The most usual histological procedure for reproductive structures uses chemical fixation usually employing crosslinking fixatives, like aldehydes. Crosslinking fixatives create covalent links between molecules and the fixative, granting both protection against decay and rigidity to the tissue, it is the fixative of choice for protein and lipids. Formaldehyde fixation alone is mild and temporary. Glutaraldehyde is a stronger more stable fixative and usually is employed as a complement to Formaldehyde milder characteristics. Though beneficial for histological studies, the crosslinking of protein and other molecules can block the antibody access to their target epitopes. For this reason, the quantity and duration of the fixation must be carefully essayed. After fixation the tissue sample must be embedded in a solid medium, conferring rigidity and support to the sample and easing the sectioning process. The embedding medium must have the same hardness of the sample and easily penetrate the tissues. For woody tissues, like the ones found in *Quercus* flowers, medium hardness acrylic resins are the best embedding solution. London Resin (LR-White) acrylic resin is the option of choice for plant tissues inclusion and sectioning, due to its hydrophilic nature, reduced viscosity, and low toxicity. Also after polymerization, the section is permeable to aqueous solution, allowing for the penetration of stains and antibodies solution for immunocytochemical assays. An additional advantage is its low autofluorescence helping to reduce the background noise in Immunofluorescence imaging [33].

Immunofluorescence is a microscopy technique for the use with a fluorescence microscope, which employs antibodies linked fluorophores. This technique relies on the specificity of the antibody binding to its antigen, labeling the epitope with a fluorescent dye. The immunofluorescence technique can be direct or indirect. On the direct method, the antigen is targeted with an antibody chemically linked to a fluorescent dye. This method is simpler and less time consuming than the indirect approach, but it is less sensitive, more expensive, and less versatile, due to the cost and availability of specific labeled antibodies. The most common method used is the indirect approach. In this method two antibodies are used, a primary unlabeled monoclonal antibody specifically raised against the target antigen, and a secondary antibody raised against the immunoglobulin of the primary antibody. It is the secondary antibody that carries the fluorophore [34]. The indirect method is more time consuming, yields a stronger signal, and is far less expensive. The immunofluorescence technique can be used to localize epitope containing molecules distribution and evolution along treatments or development stages on cells and tissues, on all-mount or in tissue sections, making it an essential tool for studying AGPs and pectins involvement in organogenesis and tissue differentiation.

2 Materials

2.1 Equipment

1. 10 × 10 cm Square Petri dish.
2. Eight wells Teflon-printed reaction slides.
3. Cover slips.
4. Disposable glass vials.
5. Gelatin capsules sizes 1 (0.5 mL) and 2 (0.37 mL).
6. Oven.
7. Ralph glass knifes.
8. Razor blades.
9. Ultramicrotome.
10. Upright epifluorescence microscope, with DAPI (4′,6-Diamidin-2-phenylindol) and FITC (fluorescein) fluorescence filter blocks or confocal microscope.
11. Vacuum chamber.
12. Vacuum pump.

2.2 Reagents

1. Anti-Rat IgG (whole molecule)–FITC antibody produced in goat (Sigma-Aldrich).
2. $_{dd}H_2O$.
3. Ethanol absolute, dilutions of 96%, 90%, 80%, 70%, 60%, 50%, 40%, and 30% ethanol in $_{dd}H_2O$.
4. Fluorescent brightener 28 (Calcofluor).
5. 36% (w/v) Glutaraldehyde.
6. LR-white embedding resin, medium grade.
7. 16% (w/v) Paraformaldehyde, aq. soln., methanol free.
8. Rat generated Monoclonal Anti-Body (MAB).
9. Several antibodies that recognize cell wall components are available at both the Complex Carbohydrate research center (CCRC, Georgia University USA) and Plant Probes (Paul Knox Cell Wall Lab, at Leeds University UK). A short list of the most commonly used MABS and where they can be purchased is presented in Table 1.
10. Toluidine blue O.
11. Tween® 20.
12. Vectashield® Mounting Medium.
13. Vital red stain.

2.3 Solutions

1. PBS (Phosphate-buffered saline): 2.7 mM KCl, 137 mM NaCl, 10 mM Na_2HPO_4, 1.8 mM KH_2PO_4 in ddH_2O. Adjust pH to 7.4 with HCl. Filter sterilize or autoclave. Store at 4 °C.

Table 1
Commonly used AGPs and pectin antibodies

MAB	Recognizes	Availability
LM2	Arabinogalactan protein	L
LM5	Pectins	L
LM6	Neoglycoprotein	L
LM7	Homogalacturonan	L
LM8	Xylogalacturonan	L
LM16	Processed Arabinan - RG-I	L
LM19	Homogalacturonan	L
LM20	Homogalacturonan	L/G
JIM5	Low methyl-esterified homogalacturonan	L/G
JIM7	Highly methyl-esterified homogalacturonan	L/G
JIM4	Arabinogalactan protein	G
JIM8	Arabinogalactan protein	L
JIM13	Arabinogalactan protein	L
JIM20	Extensin	G
MAC207	Arabinogalactan protein	G

G CarboSource Georgia, *L* plant probes leeds

2. 0,025 M PIPES (1,4-Piperazinediethanesulfonic acid) buffer. Adjust to pH 7.2 with few drops of 1 M HCl.
3. 0.025 M Phosphate buffer (pH 7).
4. Fixative solution: 2% (w/v) Formaldehyde, 2.5% (w/v) Glutaraldehyde, 0.025 M PIPES buffer (pH 7.2), 0.001% (v/v) Tween 20. Prepare fresh.
5. Blocking solution: 5% (w/v) nonfat dried milk in onefold PBS, mix slowly until dissolved, and sift through a Whatman n°1 paper filter. The filtration may be accelerated by gentile vacuum filtration (*see* **Note 1**, Fig. 1). The final solution will look pail translucent yellow. Prepare fresh.
6. Primary antibody solution: Primary antibody fivefold dilution in blocking solution (solution 2).
7. Secondary antibody solution: Anti-Rat IgG (whole molecule)–FITC antibody 100-fold dilution in blocking solution (solution 2).
8. Calcofluor: 0.01% (w/v) Fluorescent Brightener 28 in PBS, filter sterilize and store at 4 °C. Filter again through a 20 μm syringe filter application for staining sections.

Fig. 1 Application of vacuum. The low fat dry milk powder never dissolves completely in the blocking solution leading to lumps and deposition on the slides wells; therefore, it is necessary to filter the solution (**b1**) to obtain a clear solution, (**b2**) the solution can be vacuum filtered (**a**) under low vacuum (no more than −10 kPa). The immunolocalization procedure has a great number of wash steps, a low vacuum can also help in this case (**c**). The use of a vacuum pump can accelerate the tedious task of buffer removal from the slide wells (**d**)

9. 1 M Borax.
10. 0.001% poly-L-Lysine: Dilute 10% (w/v) poly-L-Lysine with ddH_2O.
11. Cleaning solution: 70% (v/v) Ethanol and 0.1% (v/v) Triton x 100.
12. Toluidine Blue O solution: 1% (w/v) Toluidine Blue O in 1% (w/v) Borax.

3 Methods

In the following extended protocol, a tested, affordable, and fairly easy method for studying AGPs and pectins in *Quercus suber* reproductive structures is described. The major steps of this protocol are the plant material selection (including a quick development stage selection guide), fixation, dehydration, inclusion in LR-white, sectioning, and immunolabeling. Recommendations and tips will be provided along the major steps to ease the application of the protocol.

3.1 Plant Material Selection

The study of any specialized transient tissue or organ, like reproductive tissues, is a question of precise timing. Searching for a particular developmental stage is time consuming, and difficult if some external cues are not available. The following section is intended as a guide to help in identifying and isolate specific developmental stages of *Quercus suber* gametophytic development.

3.1.1 Quercus suber Male Flowers Selection

The *Quercus* flower development is notoriously asynchronous. During blooming several development stages of the anther and pollen can be observed. It is even common to find all stages of development in an individual catkin. Figure 2 is intended to ease the selection of male flowers when looking for a particular stage of pollen maturation (*see* **Note 2**).

Quercus suber anthers may be differentiated into five stages of development, according to the stage of pollen maturation.

1. Stage 1: The catkin starts to emerge and is still mostly covered by scales (Fig. 2a). At this stage the newly formed anthers present all tissue layers, epidermis, endothecium, middle layers, and tapetum. The pollen mother cells are also well developed (Fig. 2e).
2. Stage 2: The catkin starts to elongate and individual flowers enclosed in reddish bracts can be recognized (intermediary aspect between Fig. 2a, b). At this stage pollen mother cells undergo meiosis, and tetrads can be observed (Fig. 2e).

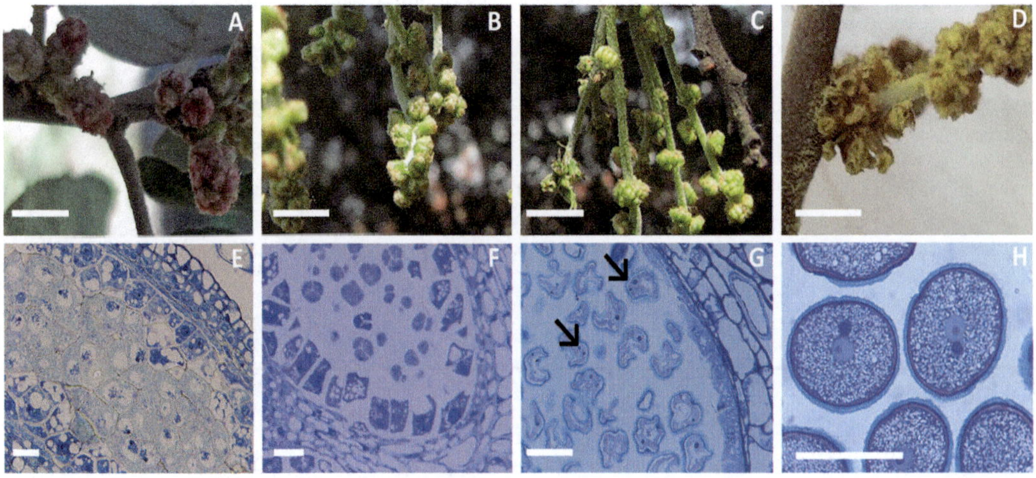

Fig. 2 Aspects of *Quercus suber* pollen development. In *Quercus suber* male flowers are arranged in catkins. At the emergence of the catkin (**a**) the anthers are very young with all anther wall tissue layers, and pollen mother cell inside the locules (**e**). During maturation, catkins elongate separating the individual flowers (**b**), at this stage meiosis occurs and tetrads are formed in the growing anthers (**f**). As the anther start peering and gaining a yellowish color (**c**) released microspores undergo mitosis (*black arrows*) (**g**). By the time the anther turns completely yellow and opens (**d**), the mature microspore is bi-cellular (**h**). Scale bars: (**a–d**) 1 cm; (**e–h**) 100 μm. (**e–h**) Sections from LR-White embedded samples with 400 nm thickness

3. Stage 3: Catkins approach their final extension (Fig. 2b). Individual flowers can be easily distinguished; green anthers can be seen emerging through the flower bracts. Tetrads break away into individual microspores.
4. Stage 4: Fully extended the catkins present flowers with significantly swollen green-yellowish anthers some clearly outgrowing the bracts (Fig. 2c). The newly released microspores start to undergo mitosis (Fig. 2g).
5. Stage 5: Full Yellow colored anthers freely swing out from the flowers (intermediary aspect between Fig. 2c, d). In the usually closed anther mature bicellular pollen (Fig. 2h) can be observed in the pollen sacs.
6. Stage 6: Anthesis is complete and most of the pollen was shedded (Fig. 2d).
7. The selected catkins or branches should be held cold and moist until processing for fixation. A good way to preserve the samples is to place them between moist paper towels in a sealable container placed on an ice box. This will prevent tissue from drying and help with preservation. Fixation should follow as soon as possible to avoid damage of the more delicate tissues.

3.1.2 *Quercus suber* Female Flower Selection

Female flowers emerge in spring, in spikes with 3–5 flowers, on the axil of new leaves, inside a cupule, partially enclosed by a dome-shaped bud of imbricate scales. Each flower usually contains three carpels, each one with two ovules. During spike elongation, 3–5 styles arise from the cupule and the stigma turns receptive. Following a similar procedure to that used with the male flower, the selection of the stages of female flower development can be helped by external visualization (Fig. 3).

Quercus suber female flowers may be differentiated into five stages of development, according to their maturation.

1. Stage 1: Female flower buds appear enclosed by protective scales (*see* **Note 3**), Then the female buds with open scales turn reddish, next buds show the elongation of the spike axis and finally, the emergency of the first pair of flowers (Fig. 3a, e).
2. Stage 2: Later, female flower appears with distinct and erect green to yellow curved stigmatas with curved brownish tips. At this time, 3–5 receptive stigmata appear from an imbricate scales dome, giving the flower full receptivity. At the time of pollination, the ovary is not yet differentiated and the transmitting tissue reaches only the base of the styles.
3. Stage 3: After 1–2 weeks, it is possible to find shining yellow and viscous stigmata in a clear divergent position. Inside the ovary, the ovules primordia start to develop. If the receptive stigmatic surface has collected enough pollen, the pollen tubes

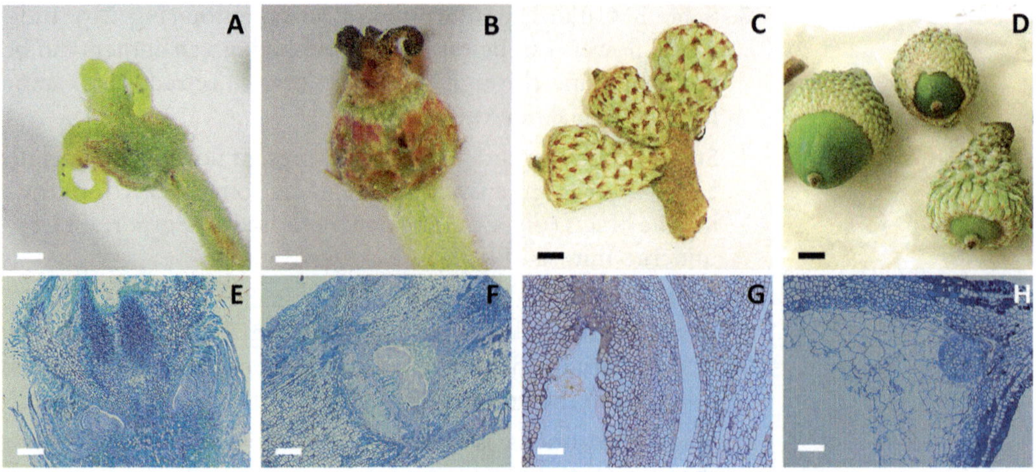

Fig. 3 *Quercus suber* female flower developmental stages before and after fertilization. Pollen receptive female flowers can be distinguished by their bright green stigmas (**a**). The ovules are not yet differentiated at this stage (**e**). After pollination the stigmas turn brown (**b**) and the ovules start to differentiate (**f**). As the ovule matures and becomes receptive (**g**), the scale covered cupule starts to grow (**c**). Post-fertilization (**h**), the nut emerges from the cupule (**d**). Scale bars: (**a**–**c**) 500 mm; (**d**) 1 cm; (**e**–**h**) 100 μm. (**e**–**h**) Sections from LR-White embedded samples with 500 nm thickness

will start to grow along the transmitting tissue. Stigmata become brown, lose receptivity, and close up (Fig. 3b). At this stage (*see* **Note 4**), different stages of ovule development are visible within a single ovary, and locules with similar size containing 2 or just 1 developing ovule are common (Fig. 3f). The more differentiated ovule shows inner and outer integuments and the embryo sac appears, not fully differentiated.

4. Stage 4: Nearly 2 months after pollination (*see* **Note 5**), the *Polygonum* type embryo sac becomes mature. At this stage, immunolocalization clearly helps to identify molecular and structural developmental changes, distinguishing basal structures like synergids, embryo sac walls, and central cell, but also highlighting their precise positions, like the filiform apparatus clearly visible at the micropylar end of the synergid cells (Fig. 3c, g) [6].

5. Stage 5: After fertilization, a large primary endosperm nucleus is generated, and subsequent division of the primary nuclei produces a free nuclear endosperm. The nuclei are peripheral with dense cytoplasm connections and the zygote remains undivided, whereas the inner integument degenerates successively. The young globular embryo appears at the micropylar end at the cellular endosperm stage, and the zygote starts to expand (Fig. 3d, h).

3.2 Sample Preparation: Fixation, Dehydration, and LR-White Inclusion

3.2.1 Fixation and Dehydration

The fixation process is critical to ensure cellular integrity, by stopping the cellular metabolism and preventing molecular diffusion. Fixative agents and concentration must be adjusted to that purpose, but also must not block accessibility for the antigens for the immunolocalization reaction to occur. The following protocol combines Paraformaldehyde, a mild fixative, with the strong fixative Glutaraldehyde. Their proportions were optimized for AGPs and cell wall components detection. The subsequent dehydration process will prepare the samples for LR-white embedding.

1. Collect the plant tissues (male and female flower organs), if using large flower organs (more than 8 mm) reduces the size by clipping around the structures of interest (*see* **Note 6**).
2. Rapidly submerge the material in a glass vial with a cold fixative solution on an ice bucket. After gathering all the samples replace the fixative solution with a fresh solution (*see* **Note 7**).
3. Apply vacuum, the floating material will start to sink to the bottom of the vial.
4. Keep under vacuum for 2 h at room temperature (*see* **Note 8**).
5. Replace the fixative solution with a fresh fixative solution, seal the glass vial and store overnight at 4 °C.
6. Remove all material that did not sink during fixation to a container for later decontamination (*see* **Note 9**).
7. Wash the sunken material with 0.025 M phosphate buffer (pH 7) for 10 min, followed by a 20 min wash in 0.025 M PIPES buffer (pH 7.2).
8. Dehydrate the samples in a series of ethanol (25%, 35%, 50%, 70%, 80%, 90%, and 3x 100% ethanol) 20 min in each concentration (*see* **Note 10**).
9. Transfer the dehydrated samples to labeled glass vials for inclusion.

3.2.2 LR-White Resin Inclusion

LR-White has a very low viscosity, which makes it ideal for penetrating many layers of cell walls; it also tolerates moderate amounts of water and offers a fairly adequate hardness for cutting plant samples.

1. Infiltrate the resin by incubating the samples for 24 h each in a crescent resin series 1:3, 2:3, 1:1, 3:2, 3:1, 1:0 LR-White in ethanol (*see* **Note 11**).
2. Substitute the last resin solution and incubate for an additional 12 h with fresh LR-white.
3. Prepare gelatin capsules and tags (*see* **Note 12**).
4. Apply one drop of LR-white resin to the bottom of each capsule.

5. Place one single sample to each capsule and fill to capacity with resin.
6. Seal the capsules by placing the tops back on and incubate for 24 h in an oven at 58 °C, or until the resin hardens (*see* **Note 13**)
7. Store samples at room temperature.

3.3 Slide Preparation and Sectioning of Samples

3.3.1 Slide Washing

Good quality glass ware is a requirement for good microscopy imaging. However, good quality is not always a synonym of expensive. With some good cleaning and a polilysine coating, Teflon-coated slides are very good to perform good quality fluorescent immunolocalization, by providing both precision and saving expensive antibody solution.

The preparation of the slides for immunolocalization will now be addressed.

1. For washing the slides (**steps 1–4**, *see* **Note 14**), place them in a staining rack and cover with the cleaning solution.
2. Maintain mild agitation for 20 min.
3. Dip the staining racks in ddH_2O with mild agitation for 10 min, repeat four times.
4. Carefully drain the racks before dipping briefly in 100% ethanol and dry in a dust-free environment.

3.3.2 Polilysine Coating (Optional)

Polilysine coating may help with the adhesion of larger sections on the slides; however, it is not required for smaller sections if the slides used were properly cleaned and the sections were left to dry completely on the slides before storage.

1. Place clean slides in a square petri dish.
2. Apply 50 μL of 0.001% poly-L-Lysine solution to each hole of the slides.
3. Dry the slides overnight by placing the closed petri dishes in a 40 °C oven.
4. Store the coated slides in a dust-free environment at room temperature.

3.3.3 Thin Sectioning of Samples and Mounting on Slides

A microtome with glass knives will be used. Attention should be given to the sections thickness. The hydrophilic resin nature of the polymerized LR-white resin will allow for the antibody penetration into the sections. The immunolabeling reaction will depend on the ability of the antibody to connect to the epitope. Thinner section will present lesser opportunities for the primary antibody to find its target, resulting in a faint labeling. Section too tick will create difficulties during image acquisition. The same holds true for the calcofluor stain that will be later applied, which will allow locating the sections and assisting with the imaging process. A good compromise would be to use sections between 200 and 500 nm, depending on the tissue characteristics.

1. With an ultra-microtome, make sections with a thickness between 200 and 500 nm (*see* **Note 15**).
2. Periodically check the section by staining with Toluidine Blue O solution.
3. Transfer one or two sections, depending on section size, to a drop of $_{dd}H_2O$ previously placed on each well of a clean and coated slide (*see* **Note 16**).
4. Place each slide in closed clean 10 cm square petri dish and let it dry in a 50 °C oven.
5. Store slides in a clean archive box until use.

3.4 Immunolocalization Procedure

The fluorescent immunolocalization relies on the binding of the primary antibody to a specific antigen. Then a secondary antibody FITC conjugated, raised specifically against the primary antibody will be used. This double immunolocalization will allow for a stronger stringency of the target detection. Controls must always be performed to insure removal of unspecific binding if present. One slot on the slide should be reserved for use as a negative control. The primary antibody treatment will be skipped for the control well and no FITC signal should be detected at the end of the immunolocalization procedure.

1. Prepare an incubation chamber (Fig. 4; *see* **Note 17**).
2. After placing the slides in the incubation chamber, incubate each well for 10 min with the blocking solution (*see* **Note 1**), about 50 μL per 8 mm well.
3. Wash all wells for 10 min with PBS.
4. Prepare the necessary volumes of primary antibody solutions (1:5 antibody in blocking solution), about 40 μL per 8 mm well.
5. Incubate each well with the appropriate antibody at room temperature for 2 h followed by an overnight incubation at 4 °C. Place blocking solution on the control well.
6. Wash all wells two times for 10 min with PBS followed by $_{dd}H_2O$ for 10 additional minutes. After this all the wells must be clear without any trace of the blocking solution or deposits.
7. Prepare the proper volume of secondary antibody solution (1:100 in blocking solution) needed, about 40 μL per well.
8. Incubate each well with the secondary antibody solution for 3–4 h at room temperature, protect from light.

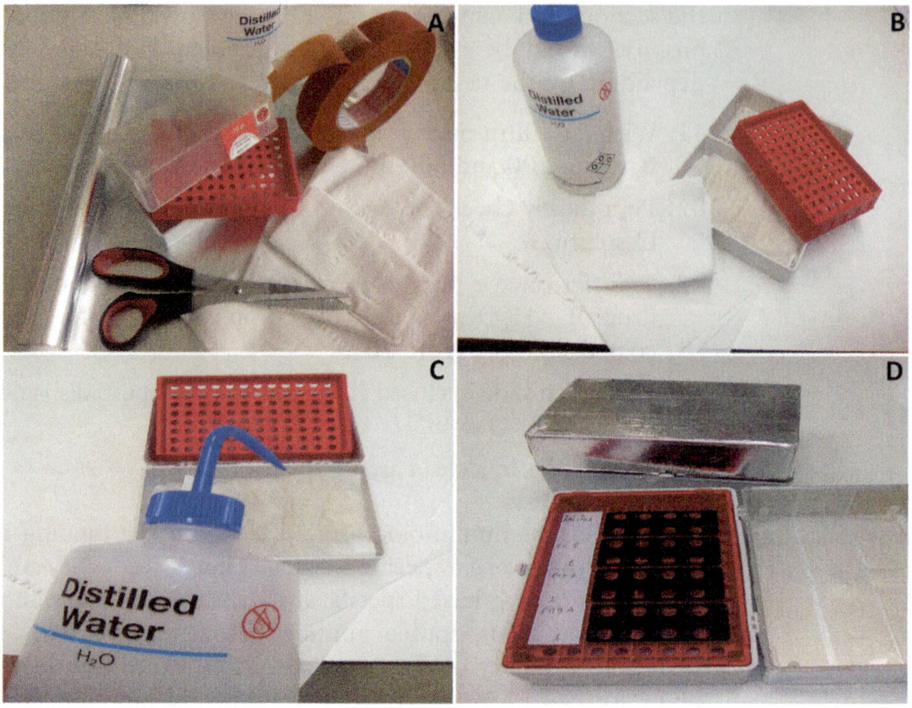

Fig. 4 Preparing an incubation chamber. An incubation chamber can be made with available laboratory supplies. All that is needed is an empty and clean pipette tip box that can be used as the base, some paper towels and aluminum film (**a**). Remove the tip tray from the box and place folded paper towels at the bottom of the box (**b**), to maintain humidity add deionized water to soak the paper towels (**c**). Use double face duct tape or glue and aluminum foil to make the incubation chamber light-proof (**d**). By placing back the tips holder tray a solid support for the slides is created (**d**). This box can be reused by rising it with water after removing the damp paper towels

9. Wash all the wells two times for 10 min with PBS followed by 10 min with $_{dd}H_2O$.
10. Apply a drop of calcofluor (1:10,000 (w/v) fluorescent brighter 28 in PBS) to each well, protect from light (*see* **Note 18**).
11. Without washing apply a drop of Vectashield® to each well (*see* **Note 19**).
12. Place cover slip and fix in place with nail varnish.
13. Observe using DAPI (for calcofluor stain) and FITC filter setting on your upright microscope, to detect cell wall and immunolocalization respectively (*see* **Note 20**).
14. *See* Fig. 5 for common problems and Fig. 6 for typical results.

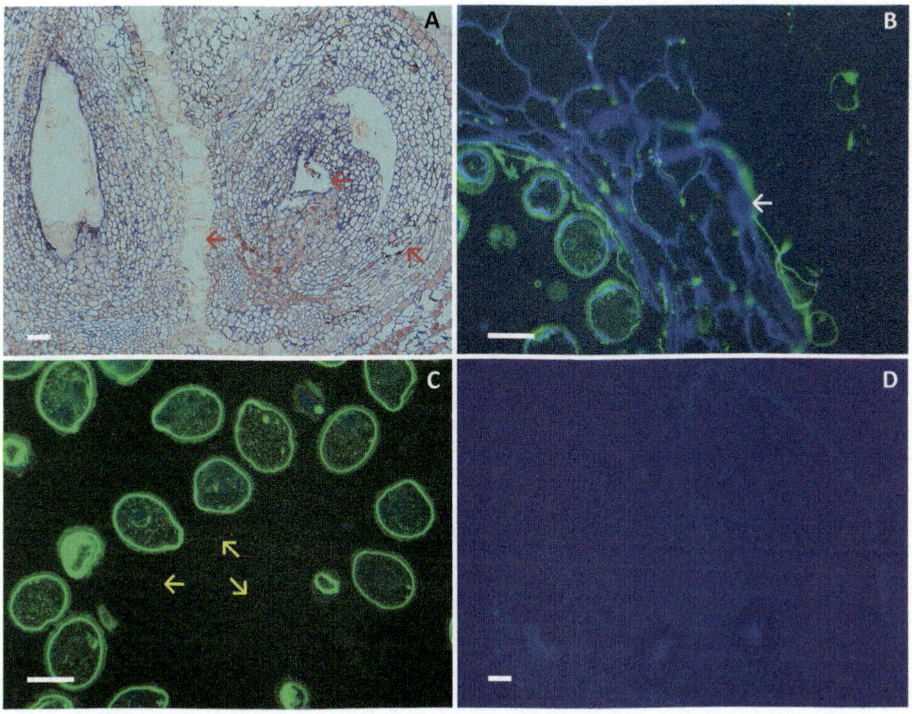

Fig. 5 Some common problems. The immunolocalization in LR-white section of plant tissues is no way a simple task. Problems that can arise are: resin inclusion (**a**), poor resin inclusion can lead to rips and resin gaps in the section (*red arrows*). The used thin plastic section can easily fold and rippling (**b**), rending impossible to obtain a clear observation (*white arrow*). For this reason, section adhesion is very important, also during the immunolocalization process the section must not be disturbed. Another common problem is the appearance of unspecific labeling (*yellow arrows*) across the section (**c**) making the observation unreliable. This situation is normally due to one of three problems: the deposition of precipitates from use of unfiltered blocking solution; inefficient or incomplete wash steps; and finally excess of antibody concentration or incubation time. Finally, and not less dramatic, it can be due to calcofluor stain failure (**d**). The excess of calcofluor solution will cause a blue blur effect, making image acquisition almost impossible. Scale bars 100 μm

4 Notes

1. Blocking solution (nonfat dry milk in 1 M PBS) must be filtered prior to use to avoid the formation of precipitates. The precipitates may retain antibodies making results unreliable (*see* Fig. 5c). This time-consuming process can be accelerated applying a very low vacuum (not more than −10 kPa) (*see* Fig. 1a, b). Vacuum can also be used to speed up the wash steps (*see* Fig. 1c, d).
2. Due to the asynchronistic male blooming of *Quercus suber*, it is highly recommended to assess the maturation stage of the anther when looking for specific pollen development stages. The use of external markers for maturation can be of great

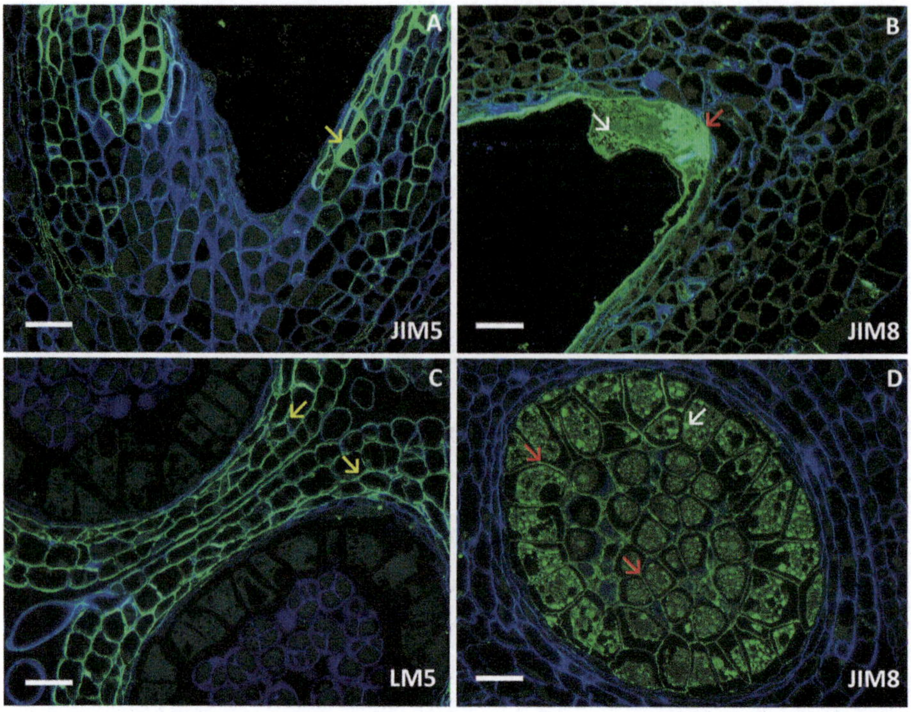

Fig. 6 Examples of typical Arabinogalactan proteins (AGPs) and pectin immunolocalization results. Jim5 and LM5 are monoclonal antibodies (MABs) raised against partially methyl esterified homogalacturonan. The fluorescent immunolocalization reveals their presence in the tegument cell walls of the mature ovule (**a**, *yellow arrow*), and in the anther walls (**c**, *yellow arrows*). Jim8 recognizes an AGP epitope. In *Quercus suber*, and other species, it has been typically found on the cell wall of gametes, along the pollen tube pathway, and it strongly labels the filiform apparatus of synergid cells of a receptive ovule (**b**). Also labeled is the anther tapetum and pollen mother cells (**d**). The epitope can generally be found on the cell membrane and cell wall (*red arrows*); in some cases it can also be found on the Golgi apparatus and transport vesicles were AGPs mature (*white arrows*). Scale bars: 100 μm

assistance to determine an approximate developmental stage. Select one anther of the flower, place it on a drop of vital red stain or DAPI, on a microscope slide and gently press it to release its contents. Observe under the microscope to determine the pollen development stage.

3. The protective scales should be removed as much as possible, so that the fixation and the resin inclusion procedures become efficient.
4. External visualization should be done with the help of a magnifying glass, which will help to remove, as much as possible, the protective scales from the flower cupule.
5. At this stage, cupule removal should be careful and meticulous. Ideally, if possible, the maximum amount of material can be withdrawn, so that the ovules become fully accessible to the fixative solution and resin.

6. *Quercus suber* has large amounts of tannins that will react with metals. Always use surgical grade steel when cutting tissues; otherwise, a dark reddish brown precipitate will start to form and will interfere with the fixation process.
7. Upon wounding most organs will start to excrete tannins and other exudates that may react with the formaldehyde reducing the fixative solution performance. Keeping the samples on ice reduces the amount of exudates. It is important to replace the fixative solution whenever it becomes cloudy.
8. All air must be evacuated of the tissue for proper fixative penetration. The presence of air in the sample will make resin infiltration and polymerization difficult. Female flowers mostly, but also the anther later stages of development, contain significant pockets of air. During the vacuum treatment air bubbles will rise from the samples, it is important to keep applying vacuum until no more bubbles are observed. If the vacuum must be extended further than 2 h, replace the fixative solution and continue for 1 additional hour or until no more air comes out of the samples.
9. In some rare cases some air may remain in samples after vacuum treatment and the overnight bath in the fixative solution will cause the sample to float. Unfortunately, in this case there is no way to guaranty that a proper fixation of this samples occurred so it is better to discard it. If this phenomenon is observed in a large number of samples, review the previous note.
10. The dehydration step is of paramount importance for the proper infiltration of the resin, as LR-White resin is almost insoluble in water. During the alcohol solutions changes the samples should not be allowed to completely dry out. After the last steps with absolute ethanol the samples are completely dehydrated and may become brittle; therefore, they should be handled with special care to avoid breaks or cracks.
11. The infiltration process must be slow and gradual for best results, also the LR-White resin should be mixed with 100% ethanol before adding to the samples.

    Small specimens (less than 3 mm diameter), like individual anthers in the first developmental stages, will be successfully infiltrated with shorter incubation times in this resin series (12 h in each concentration is effective). Nonetheless, it is not recommendable to shorten the infiltration periods without prior testing. It is also not unusual for the samples to darken during the inclusion, disturbing the preservation of the sample outer layer. Performing the infiltration at 4 °C seems to prevent this phenomenon with no negative effects on the infiltration process.

12. Heat resistant gelatin capsules are typically used as molds for the sample inclosing blocks. They provide a practical support for microtome cutting and can be hermetically closed to insure proper polymerization of LR-White resin. However, the task of tagging this small resin block is difficult and tedious. A simple way to overcome this difficulty is to tag in advance the capsules with a small strip of paper containing sample ID and preparation date. Water is a problem, not only for LR-White resin polymerization but in extreme cases it may induce deformation of the gelatin capsules when heat is applied. To avoid this inconvenience, the capsules and paper tags must be dried before, stored for at least 24 h in advance in a 58 °C oven.
13. The procedure depends on the perfect polymerization of LR-White Resin. Uncured resin blocks will give bad sections under the microtome. When cutting a not completely cured resin block with the ultra-microtome the sections will tend to lose shape, tear (*see* Fig. 5a), and/or leave an oily substance (uncured resin) that spreads across the water. The capsules should feel hard when pressed and the resin should go from almost colorless to a light golden/amber color; this usually occurs after 24 h at 58 °C for size 2 (0.37 mL) capsules. When using differently sized capsules curing time must be adjusted.
14. Depending on the section and reaction well dimensions, one or several sections may be transferred to the slides. Section must not overlap and must be perfectly flat against the slide for the best results.
15. Slides must be clean and degreased. Under the microscope any imperfection, like dust or lint, may greatly impact the specimen observation. Most glass slides suppliers use detergents or oils to prevent the slides from sticking together. If left on the slides these contaminants may both interfere with the immunolocalization reaction and adhesion of the section to the slide (*see* Fig. 5b). Also slides may be reused several times if properly cleaned and checked for imperfections or contaminants under the microscope.
16. A microtome with glass knives will be used. Attention should be given to the sections thickness. The hydrophilic resin nature of the polymerized LR-white resin will allow for the antibody penetration into the sections. The immunolabeling reaction will depend on the ability of the antibody to diffuse to and bind to the epitope. Thinner section will present few opportunities for the primary antibody to find its target, resulting in a faint labeling. Too thick sections will generate difficulties during image acquisition. The same holds true for the calcofluor stain that will be later applied, which will allow locating the

sectioned tissues and assisting with the imaging process. A good compromise would be to use sections between 200 and 500 nm, depending on the tissue characteristics.

17. During the immunolocalization process special care must be taken to avoid the evaporation of solutions in the wells of the slides, especially during the long antibody incubation times. By using an incubation chamber this problem can be avoided. An incubation chamber can be made by simply placing a few moist paper towels in the bottom of a used tip box (*see* Fig. 4) or petri dish. During secondary antibody incubation light exposure must be avoided to preserve the fluorochrome degradation. Covering the incubation box with aluminum foil is a simple way to make it light-proof.
18. The fluorescent brighter 28 solution precipitates and also gets contaminated fairly easily, both the problems will look as white dots or spikes. To avoid this problem, it is best to periodically prepare a fresh working solution from the stock and filter sterilize it prior to use. It is also important to wash off any excess of calcofluor staining solution, as it may cause a cloudiness during the slide imaging (*see* Fig. 5d).
19. Photo bleaching is the decay of fluorochromes by light. Under the intense light exposure, to which the specimens are subjected during microscope observation, photo bleaching can greatly reduce the observation time. This effect is especially dramatic with faint signals that quickly fade before image acquisition is possible. Antifade mounting medium greatly expands the fluorochromes lifespan, like Vecta-Shield® that is used. Other mounting medium with antifade properties may be used with similar results. However, it is recommendable to test the antifade mounting medium before use as, although rarely, in some cases the antifade may react with the calcofluor stain and/or FITC leading to the formation of precipitates.
20. The observation of the slides is fundamental; therefore, it is important to use the proper microscope, and the use of an adequate optic system with fluorescence grade objectives is highly recommended. Another aspect is the light filter settings on the microscope. If using a conventional upright fluorescence microscope, the filters should be set to Excitation/Emission (nm) 358/461 for DAPI and 485/530 for FITC. A confocal microscope may also be used, but due to the reduced thickness of the sections the benefits of using this equipment are limited. The excitation laser should be set to 355 nm for DAPI and 488 nm for FITC, Emission filters should be set between 450 and 460 nm for DAPI, and 515 to 525 nm for FITC. Finally, a brief reference to the image acquisition system; Refrigerated monochromatic digital cameras are an

uncontested staple for Fluorescence microscopy applications due to their high speed and sensitivity, but sacrificing true color information. On the other hand, polychromatic cameras are able to sort out signal from background fluorescence, but apart from the high-end expensive models, most are slower and far less sensitive than their monochromatic counterparts.

References

1. Charneca R, Nunes J, Le Dividich J (2010) Body composition and blood parameters of newborn piglets from Alentejano and conventional (large white × landrace) genotype. Span J Agric Res 8:317–325
2. Varela M, Valdiviesso T (1996) Phenological phases of *Quercus suber* L. flowering. Forest Genet 3:930–102
3. Boavida L, Varela M, Feijó J (1999) Sexual reproduction in the cork oak (*Quercus suber* L.). I. The progamic phase. Sex Plant Reprod 11:347–353
4. Boavida L, Silva J, Feijó S (2001) Sexual reproduction in the cork oak (*Quercus suber* L). II. Crossing intra- and interspecific barriers. Sex Plant Reprod 14(3):143–152
5. Costa M, Sobral R, Costa M et al (2015) Evaluation of the presence of arabinogalactan proteins and pectins during *Quercus suber* male gametogenesis. Ann Bot 115:81–92
6. Lopes A, Costa M, Sobral R et al (2016) Arabinogalactan proteins and pectin distribution during female gametogenesis in *Quercus suber* L. Ann Bot 117(6):949–961
7. Showalter A (2001) Arabinogalactan-proteins: structure, expression and function. Cell Mol Life Sci 58:1399–1417
8. Fu H, Yadav M, Nothnagel E (2007) *Physcomitrella patens* arabinogalactan proteins contain abundant terminal 3-*O*-methyl-lrhamnosyl residues not found in angiosperms. Planta 226:1511–1524
9. Costa M, Pereira A, Rudall P et al (2013) Immunolocalization of arabinogalactan proteins (AGPs) in reproductive structures of an early- divergent angiosperm, *Trithuria* (Hydatellaceae). Ann Bot 111:183–190
10. Popper Z, Michel G, Hervé C et al (2011) Evolution and diversity of plant cell walls: from algae to flowering plants. Annu Rev Plant Biol 62:567–590
11. Majewska-Sawka A, Nothnagel E (2000) The multiple roles of arabinogalactan proteins in plant development. Plant Physiol 122:3–9
12. Johnson K, Jones B, Bacic A et al (2003) The fasciclin-like arabinogalactan proteins of *Arabidopsis*. A multigene family of putative cell adhesion molecules. Plant Physiol 133:1911–1925
13. Seifert G, Roberts K (2007) The biology of arabinogalactan proteins. Annu Rev Plant Biol 58:137–161
14. Borner G, Sherrier D, Stevens T et al (2002) Prediction of glycosylphosphatidylinositol-anchored proteins in *Arabidopsis*. A genomic analysis. Plant Physiol 129:486–499
15. Gaspar Y, Johnson K, McKenna J et al (2001) The complex structures of arabinogalactan-proteins and the journey towards understanding function. Plant Mol Biol 47:161–176
16. Schultz C, Ferguson K, Lahnstein J et al (2004) Posttranslational modifications of arabinogalactan-peptides of *Arabidopsis thaliana*. J Biol Chem 279:45503–45511
17. Van Hengel A, Tadesse Z, Immerzeel P et al (2001) *N*-Acetylglucosamine and glucosamine-containing arabinogalactan proteins control somatic embryogenesis. Plant Physiol 125:1880–1890
18. Pereira A, Pereira L, Coimbra S (2015) Arabinogalactan proteins: rising attention from plant biologists. Plant Reprod 28:1–15
19. Gane A, Clarke A, Bacic A (1995) Localization and expression of arabinogalactan-proteins in the ovaries of *Nicotiana alata* Link and Otto. Sex Plant Reprod 8:278–282
20. Coimbra S, Salema R (1997) Immunolocalization of arabinogalactan proteins in *Amaranthus hypocondriacus* L. ovules. Protoplasma 199:75–82
21. Coimbra S, Duarte C (2003) Arabinogalactan proteins may facilitate the movement of pollen tubes from the stigma to the ovules in *Actinidia deliciosa* and *Amaranthus hypocondriacus*. Euphytica 133:171–178
22. Yariv J, Lis H, Katchalski E (1967) Precipitation of arabic acid and some seed polysaccharides by glycosylphenylazo dyes. Biochem J 105:1C–2C
23. Kitazawa K, Tryfona T, Yoshimi Y et al (2013) β-Galactosyl Yariv reagent binds to the β-1,3-

galactan of arabinogalactan proteins. Plant Physiol 161:1117–1126

24. Pennell R, Knox J, Scofield G et al (1989) A family of abundant plasma membrane-associated glycoproteins related to the arabinogalactan proteins is unique to flowering plants. J Cell Biol 108:1967–1977
25. Knox J, Linstead P, Peart J et al (1991) Developmentally regulated epitopes of cell surface arabinogalactan proteins and their relation to root tissue pattern formation. Plant J 1:317–326
26. Coimbra S, Almeida J, Junqueira V et al (2007) Arabinogalactan proteins as molecular markers in Arabidopsis thaliana sexual reproduction. J Exp Bot 58:4027–4035
27. Gao M, Showalter A (2002) Yariv reagent treatment induces programmed cell death in Arabidopsis cell cultures and implicates arabinogalactan protein involvement. Plant J 19:321–331
28. Sardar H, Yang J, Showalter A (2006) Molecular interactions of arabinogalactan proteins with cortical microtubules and F-actin in Bright Yellow-2 tobacco cultured cells. Plant Physiol 142:1469–1479
29. Pereira A, Masiero S, Nobre M et al (2014) Differential expression patterns of arabinogalactan proteins in *Arabidopsis thaliana* reproductive tissues. J Exp Bot 65:5459–5471
30. Pereira A, Nobre M, Pinto S et al (2016) Love is strong, and you're so sweet: JAGGER is essential for persistent synergid degeneration and polytubey block in *Arabidopsis thaliana*. Mol Plant 9:601–614
31. Costa M, Nobre M, Becker J et al (2013) Expression based and co-localization detection of arabinogalactan protein 6 and arabinogalactan protein 11 interactors in Arabidopsis pollen and pollen tubes. BMC Plant Biol 13:7
32. El-Tantawy A, Solís M, Costa M et al (2013) Arabinogalactan protein profiles and distribution patterns during microspore embryogenesis and pollen development in *Brassica napus*. Plant Reprod 26:231–243
33. Marttila S, Santén K (2007) Practical aspects of immunomicroscopy on plant material. Mod Res Edu Top Microsc Formatex Microsc Ser 2:1015–1021. ISBN-13:978-84-611-9418-6
34. Osborn M, Weber K (1982) Immunofluorescence and immunocytochemical procedures with affinity purified antibodies: tubulin-containing structures. Methods Cell Biol 24:97–132

Part III

Chromosome Spreading, Whole Mount In Situ, and Transmission Electron Microscopy (TEM)

Chapter 12

Optimization of Cell Spreading and Image Quality for the Study of Chromosomes in Plant Tissues

Laksana Kantama, Erik Wijnker, and Hans de Jong

Abstract

High-quality chromosome images of mitotic and meiotic cell divisions in plant tissues are inextricably connected with the technical control of cell spread preparations. Superb chromosome slides are the best for studying chromosome morphology and making karyotypes; they also are the best start for a successful fluorescent in situ hybridization experiment. In this study, we describe the essentials for fixation, enzymatic digestion, squash, spread, and dropping protocols for clean and well-differentiated nuclei and chromosome complements. In addition, we focus on the use of standard whole image processing for best sharpness, brightness and contrast adjustments, differentiation of heterochromatin/euchromatin, and high dynamic range imaging of big chromosomes. We also explain how to combine transparent layers or spot channels of different fluorescent images for making publication quality, full color photos.

Key words Chromosome technique, Cell spreading protocol, Chromosome images, Image processing microphotographs

1 Introduction

Chromosome research is one of the basic disciplines in life sciences for establishing the genetic and genomic characteristics of a eukaryotic organism. Chromosome numbers are essential features in linkage and physical mapping, whereas knowledge of chromosome morphology is a prerequisite for understanding telomeres, centromeres, heterochromatin, and repetitive sequence domains in unraveling the complexity of the genome. In most plant species chromosome analysis begins with microscopic observations of cell complements at mitotic or meiotic stages, typically obtained from dividing cell populations in fast growing root tips or flower buds, respectively. Excellent reference books with lots of background information are in [1, 2]. The production of high-quality microscopic slides with well-spread chromosomes is often regarded as an art: subtle playing with various cytological conditions to pretreat, fix, soften, and spread cells and nuclei on a slide. The use of

Anja Schmidt (ed.), *Plant Germline Development: Methods and Protocols*, Methods in Molecular Biology, vol. 1669, DOI 10.1007/978-1-4939-7286-9_12, © Springer Science+Business Media LLC 2017

sophisticated microscopes and advanced chromosome banding and Fluorescent in situ Hybridization (FISH) technologies requires full control of preparing outstanding chromosome slides. In last decades we have seen a plethora of outstanding technical papers describing the experimental conditions for chromosome morphology and banding, and FISH (e.g., [3–10]), but in most cases little attention has been paid to the experimental conditions that control cell spreads and chromosome shape quality, especially on how cell walls and cytoplasm can be removed and chromosomes be spread, without losing characteristic details of heterochromatin and euchromatin. In addition, photos of selected chromosome images were often presented in a way that the scientific message of the image was not sufficient or even entirely unclear.

Here, we discuss the different steps in making chromosome slides for high-resolution imaging and FISH technologies. We not only discuss the critical wet-lab conditions for chromosome squashing, cell spreading, and cell dropping protocols, but also explain a number of simple computer image processing tools that dramatically improve overall quality of the microphotographs of chromosomes and nuclei.

2 Materials

2.1 Equipment and Supplies

1. Binocular dissecting microscope with zoom range of about 5–40×, and a black matted stage.
2. Phase contrast microscope, with a 10× and 20×, and preferably also with 40× and 60× dry, no cover glass optics (available from most microscope companies).
3. Fluorescence microscope equipped with immersion oil Plan-Fluorite or Plan-Apochromatic objectives (40–100×, and N.A. of 1.0–1.4), a sensitive CCD camera (*see* **Notes 1** and **2** for details), epifluorescence illumination and appropriate small band filters for DAPI, and fluorescence dyes used in the FISH studies.
4. Use immersion oil specially designed for fluorescence microscopy (e.g., Zeiss Immersion oil Immersol 518F).
5. Computer system with image capturing and processing software.
6. Dissecting needles, fine forceps, glass bars, filter paper, Pasteur pipettes (or micro pipettes for more accuracy).
7. Microscopic slides, clean and grease-free, wiped dry.
8. Cover slips 24 × 24, 24 × 32, or 24 × 50 mm. For high-resolution fluorescence microscopy, the use of high-performance cover slips of 0.17 ± 0.005 mm thickness is recommended.

9. Hot plate and conical Erlenmeyer flask with boiling water (cell dropping only).
10. Liquid nitrogen or carbon dioxide ice, razor blade, and wooden laboratory peg or tube holder.

2.2 Chemicals

1. Spindle inhibitor. For most plants one of the following reagents can be used: 8-hydroxyquinoline, colchicine, α-bromonaphthalene, or paradichlorobenzene.
2. Pectolytic enzymes: Cellulase RS (Yakult), Pectolyase Y23 (Sigma), Cytohelicase (Cytohelicase from *Helix pomatia* (Sigma)).
3. DAPI (4′,6-Diamidino-2-phenylindole dihydrochloride) dry powder or 1 μg/mL DAPI in water.
4. Vectashield antifade medium (Vector laboratories) or comparable mounting media.

2.3 Solutions

1. Pretreatment of root tips. Use of the following: 0.5 mg/mL 8-hydroxyquinoline (takes overnight to dissolve), 0.1–0.5% (w/v) colchicine, 15 mg/mL para-dichlorobenzene, saturated aqueous solution of α-bromonaphthalene or ice water.
2. Carnoy's fixative (1): three parts of ethanol and 1 part of glacial acetic acid, mixed shortly before use. Use >96% absolute ethanol.
3. Carnoy's fixative (2): 1 glacial acetic acid: 3 chloroform: 6 ethanol, mixed shortly before use. Use >98% glacial acetic acid.
4. 45% (v/v) and 60% (v/v) acetic acid.
5. Citrate buffer: prepare 10 mM citric acid and 10 mM Na-citrate, autoclave both. Prepare the citric buffer by adding the Na-citrate to the citric acid solution until pH 4.5.
6. Pectolytic enzyme stock solution: 1% (w/v) Cellulase RS, 1% (w/v) Pectolyase Y23 and 1% (w/v) Cytohelicase dissolved in the citrate buffer. Store at −20 °C until use.

2.4 Root Tips and Flower Buds

1. Young root tips obtained from plants growing in soil or water (see **Note 3**).
2. Flower buds containing anthers with pollen mother cells at meiotic stages.

3 Materials

3.1 Preparation of Root Tips and Flower Buds

1. Check the appearance of fresh root tips regularly. Once the very first rootlets are seen at the surface of the soil boll, collect them in the early morning with a pair of fine forceps.

2. For metaphase accumulation, improved chromosome spreading, and clear morphology, pretreat the root tips with a spindle inhibitor solution. The most well-known spindle inhibitors are 8-hydroxyquinoline and colchicine, but some plants need saturated α-bromonaphthalene or para-dichlorobenzene (2–4 h at room temperature), or 2 days in ice water.
3. After the pretreatment, transfer rootlets to a freshly prepared Carnoy's fixative of glacial acetic acid—96% ethanol (1:3). The period of fixation is usually 15 min to a few days at 4–20 °C. For longer storage, the fixative should be replaced by ethanol 70% and the sample stored at 4 °C.
4. Flower buds containing anthers with pollen mother cells at meiotic stages are fixed directly in the Carnoy's fixative without pretreatment. Eventually, one can use a chloroform containing fixative (1 glacial acetic acid: 3 chloroform: 6 ethanol) for removing chloroplasts from the cells. Also for this fixative, replace the solution next day with ethanol 70%.

3.2 Selecting the Right Tissues for Mitotic and Meiotic Studies

1. The root tips are in general whiter and denser than the remaining part of the root. Most mitosis are in the very tip, around the quiescent center. Learn by experience how the material looks in the dissecting microscope and what cell spread preparations can be made from them.
2. As to the flower buds only use anthers that contain pollen mother cells at meiosis and tapetal cells containing well spread endomitotic metaphase complements (*see* **Note 4**). If anthers are large, open the flower bud and dissect the anthers first and treat them one by one.
3. If the anthers are too small for handling under the dissecting microscope, use the whole inflorescence instead (*see* **Note 4**).

3.3 Washing Steps and Treatment with Pectolytic Enzymes

1. On the day of making the slide preparations remove the fixative by thorough washing steps in water (three times 5 min), followed by a final washing step in 10 mM citrate buffer, pH 4.5.
2. Use a general-purpose cocktail of pectolytic enzymes containing different β-1,3 and 1,4 glucanases (e.g., pectinase RS, pectolyase Y23 and Cytohelicase; *see* **Note 5**). Dissolve the enzymes (1% of each) in the citrate buffer and keep at −20 °C until use. For the enzymatic digestion of the flower buds use the stock solution, for root tips dilute the stock two to five times.
3. The incubation takes place at 37 °C for 1–3 h until the root tips, small flower buds, or anthers are fragile and fall apart easily in small cell clumps when touched with a fine needle. The material is now ready for making a chromosome preparation.

3.4 Methods for Chromosome Spreads

1. For monitoring the different steps of one of the chromosome techniques results, it is recommended to use a dissecting microscope for transferring and spreading the tissues, and a phase contrast microscope with 10–40× dry no cover glass objectives for evaluating the quality of the chromosome spreads after cell drying.
2. In addition, one needs small petri dishes or depression slides, fine needles, forceps, glass or micro pipets and eventually a hot plate or gas flame for carefully warming the slide. Prepare small tubes with few mL of (distilled or Milli-Q) water, 45% and 60% acetic acid, and freshly prepared Carnoy's (1) fixation. Also make sure that the microscopic slides and cover slips are very clean, grease-free, and wiped dry. For high-resolution microscopy of very small chromosomes, the use of high-performance cover slips is recommended.
3. The very soft and fragile tissues are now very suitable for making chromosome preparations. We can choose between three different ways: squashing, cell spreading, and a cell dropping (*see* **Note 6**).

3.4.1 Cell Squashing Method

1. Transfer a small piece of tissue (root tip, anther, flower bud) to a grease-free microscopic slide in a small polystyrene petridish. Use a black stage or piece of plasticized carton under the slide for better contrast. In the case of a root tip, cut the tip and remove the remaining part of the root. Leave a tissue in a little amount of water and watch the following steps under the dissecting microscope.
2. Divide the tissue with a needle, forceps, or glass rod in small pieces and check under the dissecting microscope if there are no more large cell clumps. If needed repeat the treatment with the needle.
3. Add now a drop of 45–60% acetic acid and mix the cells in the solution.
4. Cover the drop with cells in acetic acid with a 24 × 24 mm or larger cover slip. Make sure that there is enough liquid to cover the complete area under the cover slip.
5. Now hold one of the corners of the cover slip firmly in place with a rubber pencil eraser and make sure that the cover slip cannot shift horizontally.
6. Next, tip with a blunt needle on the middle part of the cover slip in a way that cells spread over a large area. Eventually check the slide in the phase contrast microscope and repeat the action if needed.

7. Put now a few layers of filter paper on the slide, press gently to remove the excess of acetic acid, then press firmly to flatten the cells on the microscopic slide.
8. Freeze the slide in liquid nitrogen or carbon dioxide ice, and remove the cover slip with a razor blade.
9. Leave the slide to air-dry for at least a couple of hours.

3.4.2 Cell Spreading Method

1. Transfer a small piece of tissue (root tip, anther, flower bud) to a grease-free microscopic slide as described above. In the case of a root tip, cut the tip and remove the remaining part of the root. Leave a tissue in a little amount of water and watch the following steps under the dissecting microscope.
2. Dissect the tissue with a needle, forceps, or glass rod and check in the dissecting microscope if there are no more large pieces of tissue. If needed repeat the treatment with the needle.
3. Add now a drop of 45–60% acetic acid and mix the cells in the solution.
4. Leave the cells in the acetic acid solution for a few minutes. Warm gently on a hot plate and add more acetic acid if needed. Make sure that the cells do not dry out.
5. Now add the Carnoy's (1) fixative in a circle around the drop with cells. The drop will now retract and cells will precipitate on the glass surface. Add more Carnoy's on the top of the cells. Leave the slide in a fume hood until air-dried. Inspect the quality of the cell spreading under the phase contrast microscope and adapt eventually the spreading conditions for the next chromosome slides.

3.4.3 Cell Dropping Method

1. Dissect the material of choice (root tip, anther, etc.) in an Eppendorf tube with the blunt end of a fine needle.
2. Add freshly prepared Carnoy's fixative or, for more rigorous spreading use equal amounts of glacial acetic acid—ethanol. Dissect material further by agitating the solution using a Finnpipette. Keep on ice.
3. Before you start make sure you have ready the clean and grease free slides, wiped dry, a conical/Erlenmeyer flask with boiling water (for the hot water vapors) and a Pasteur pipet with glacial acetic acid.
4. Moisten the slide by keeping the slide above the boiling water flask.
5. Rapidly turn the slide with water droplet up and put a few droplets of cell suspension in the Carnoy on the moistened slide (fast action).
6. Wait until the drop has almost dried up.

7. Drop excess of glacial acetic acid at 20 °C, or at 45–60 °C for better spreading.
8. Moisten slide again in water vapor for 2–10 s. This is the essential step when the cells spread.
9. Put on hot plate (60 or 70 °C, but always check the difference in the final chromosome preparation!) and let air-dry.

3.5 Fine-Tuning of the Technique and Quality Assessment

The three chromosome protocols leave much room for varying conditions and optimization. Depending on the suspension at hand, it is highly recommended to check the quality of your slides! Don't make a large number of slides at once, but monitor for each slide the effect of the different steps in the dissecting microscope and the phase contrast microscope. Use the following rules as suggestions for improving your skills to make perfect slides.

1. When cells still clog together in the tissue: enhance the enzyme solution, or make a new fresh enzyme solution and start again. Material that was stored for several months in the fridge may need longer enzymatic treatment.
2. For softening the cells (in case of squashing and spreading) one may want to increase the acetic acid treatment, either by using 60% instead of 45% acetic acid, or by using longer treatment and higher temperature. However, prolonged or harsh acetic acid treatment may affect the chromosome structure leading to blurred and sticky chromosomes. In some species like rice and orchids, the effect of acetic acid may be very delicate. Also for this step, make slides under varying conditions and study the cell spreading and chromosome quality in the phase contrast microscope.
3. For the dropping method, it is good to realize that the actual spreading of the cell complements (chromosomes or nuclei) takes place in crucial step of the moistening of the slide in water vapor (*see* Subheading 3.4, **step 8**). We assume this is caused because of the sudden drop of acidity (and thus the "activation" of H+ molecules by adding water). When spreading is insufficient the cells will remain intact on the slide, whereas in a slide with good spreading, only naked chromosomes and nuclei not covered by cytoplasm will be obtained.
4. The longer the time of applying water vapors to the acidic slide, the better the cell complements spread (though the more likely it is chromosomes are washed out). Therefore, one looks for the optimal time of vapor application for every suspension. Hot vapors (90 °C) spread better, but quickly (generally after 2–3 s) may compromise chromosome morphology (depending on material). A longer treatment (5–10 s in water vapors at 70–80 °C) usually does the trick as well, or is even better.

5. Assessment of the air-dried slides under the phase contrast microscope is an essential part of producing high-quality DAPI and FISH images of the mitotic and meiotic chromosomes. The use of 40–60× no-cover glass phase contrast objectives is highly recommended. Without temporary staining and mounting, slides can now directly be examined for (a) staging, (b) occurrence of mitotic and or meiotic cell complements, (c) quality of cell and chromosome spreading, (d) presence of cytoplasm and organelles covering the chromosomes and nuclei, (e) chromosome morphology, and (f) differentiation of heterochromatin and heterochromatin.
6. Figure 1 displays a few examples of chromosome slide quality. The phase contrast microphotographs were obtained from

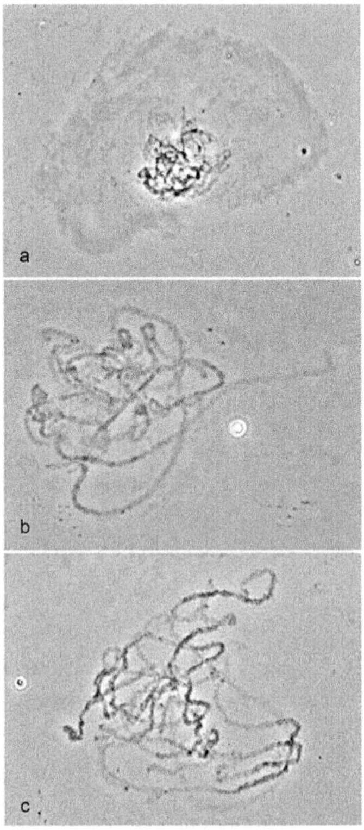

Fig. 1 Phase contrast microphotograph of prophase I cells obtained from spread tomato (*Solanum lycopersicum*) pollen mother cells. Anthers are digested with pectolytic enzymes, macerated in acetic acid 45% solution and air-dried. (**a**) The chromosomes in this early prophase I are still surrounded and covered by cytoplasm, and dark chromatin often displays halos at the borders. (**b**) Prolonged or hotter acetic acid treatment does improve spreading of the chromosomes and better removes remnants of cytoplasm, but differentiation of heterochromatin—euchromatin is poor, or chromosomes and nuclei get a fuzzy appearance. (**c**) An ideal example of a well-spread pachytene complement, free of cytoplasm and with well-differentiated heterochromatin

slides of prophase I cells obtained from spread tomato (*Solanum lycopersicum*) pollen mother cells. Prior to spreading anthers were digested with pectolytic enzymes, macerated in acetic acid 45% solution and air-dried. If spreading is poor or chromosomes still highly condensed, the chromosomes are still surrounded and covered by cytoplasm, whereas the dark chromatin displays phase halos at the borders (Fig. 1a). Such halos often point at dense chromatin structures in DAPI slides and may show up as solid un-differentiated structures that do not allow probe DNA to reach the chromosomal target in FISH experiments. Prolonged or hotter acetic acid treatment does improve spreading of the chromosomes and removes remnants of cytoplasm more effectively, but chromosomes may look fuzzy, blurred and lack heterochromatin–euchromatin differentiation (Fig. 1b). In a more idealistic situation chromosomes do spread well, have no phase contrast artifacts, are free of cytoplasm, and display well-differentiated heterochromatin (Fig. 1c).

3.6 DAPI Staining of the Chromosome Preparations

The most common fluorescent stain for visualizing DNA in mitotic and meiotic cells is DAPI. Alternatives are Hoechst 33258, Ethidium Bromide, Propidium Iodide, and Acridine Orange, but for some of them excitation/emission wavelengths are longer and hence display a slightly lower resolution power. DAPI also features a greater photostability, especially in combination with antifading mounting agents as Vectashield, Citifluor (http://citifluor.com) and Prolong antifade (https://www.thermofisher.com).

1. An easy and direct way of staining your slides is a 12 μL drop of 1 μg/mL (w/v) DAPI in Vectashield on the slides covered with a 32 × 24 mm cover slip. Make sure that the amount of DAPI/Vectashield is just enough to cover the area of the cover slip. If the DAPI fluorescence is too bright or too weak, the concentration of the fluorescence dye in the Vectashield is adapted.
2. Figure 2 shows examples of high-quality meiotic cell complements obtained from enzymatic digested pollen mother cells. Figure 2a, c, d displays well spread DAPI stained pachytene complements or cucumber, tomato and Arabidopsis with clear differentiation of heterochromatin and euchromatin; Fig. 2b is a spread pollen mother cell of banana at anaphase I. The chromosomes are well spread and differentiated, but the images except that of tomato show noticeable fluorescent particles in the cytoplasm. The origin of this cytoplasmic background can be polysaccharides (banana) or polyphenoles, and can even be caused by mitochondria (Arabidopsis). Such granular background is difficult to avoid: extended enzyme treatment has no effect, and increasing the acetic acid maceration

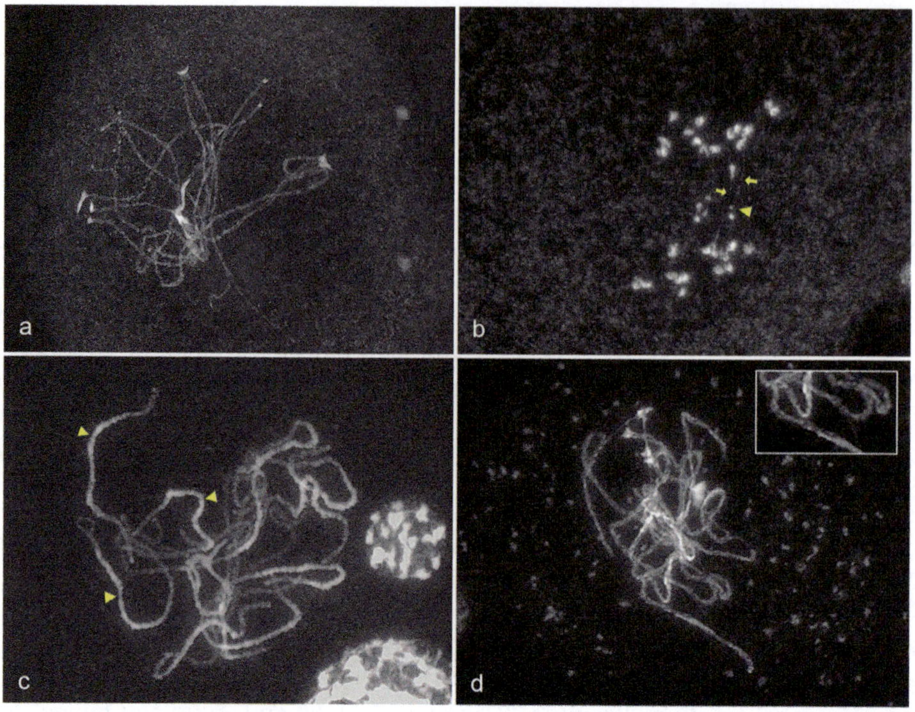

Fig. 2 Examples of fluorescence microphotographs of spread pollen mother cells obtained after enzymatic digestion, cell spreading, and staining with DAPI. (**a**) Pachytene complement of cucumber (*Cucumis sativum*). Although some cytoplasm is still present the chromosomes in the nucleoplasm are clean and well-spread. (**b**) Spread pollen mother cell at anaphase I of triploid banana (*Musa acuminata*) cultivar "Hom." Such cells have large amounts of polysaccharides or polyphenolics which makes it difficult to spread the cells and minimize the cytoplasm. Subtle sharpening of the image produce fine details allowing clear detection of anaphase bridges (*arrows*) and acentric fragment (*arrow head*). (**c**) Well-spread pachytene complement of tomato, with clear morphology in which euchromatin, heterochromatin centromeres (*arrow heads*) can be distinguished. (**d**) Pollen mother cells of a mutant *Arabidopsis thaliana* at pachytene. The 2D deconvolution filter reveals obvious details of aberrant chromosome pairing (*inset* for higher magnification). The small particles around the pachytene chromosomes are mitochondria, which do appear at later stages

may damage the chromatin structure. In addition, such substances do in general not cover the chromosomes and hence have no effect on DAPI and FISH quality.

3.7 Image Processing

1. The final phase in chromosome image acquisition is the processing and optimization of single or multiple camera captures of DAPI stained or FISH images. This part of often neglected or at least under-estimated in many publications of cytological artwork. Chromosome researchers may be insufficiently aware that the final presentation of their photomicrographs of chromosome images lose quality or are not displayed as good as they could be. Sedgewick's book [11] is one of the very few documents that deals with typical image improvements for

scientific publications. Other papers deal more with the physical properties of a good digital microscopic image (e.g., [12]).

2. In this part, we will explain a couple of simple image processing tools that all chromosome scientists may want to know about. The aim is to show how to make an image better by adjusting all pixels in the image equally in terms of color, contrast, and sharpness, and not change particular areas or details of the image, in a way that it can be interpreted as cheating practices of scientific image manipulation. In other words, no specific feature within an image may be enhanced, obscured, moved, removed, or introduced. In the case of nonlinear adjustments like changing the gamma settings, such information must be disclosed in the figure legends. More about the guidelines on image manipulation is discussed in [13, 14].
3. Companies that offer integrated fluorescence microscope, sophisticated camera, and image capturing provide software that allows capturing and processing of the fluorescent images. Those who do not have such equipment can process the image later by various image processing software, of which Adobe Photoshop CS (http://www.adobe.com) is the most common choice. Moreover, several low cost or public domain image analyses and processing packagesBioconductor package like ImageJ/Fiji (https://imagej.net/Fiji), and Gimp (https://www.gimp.org/) are available, but their profusion of tools may require a steep learning curve.
4. In all cases of image processing keep a copy of the original native camera file, preferably in a raw format and 12-bit (or more) gray levels. Original images may later be needed upon request from the editor of the scientific journal if the improvements of the digital images need further inspection. During the image processing steps always keep working with one of the lossless image file types (RAW, BMP, PSD, or TIFF). After cropping the original image to the desired part for publication, first adjust to 300–500 dpi (depending on the quality criteria of the journal) and standard 8-bit gray or 24-bit RGB color mode. Only when publication ready eventually convert the processed image into the more general lossy JPEG format at lowest compression (highest quality).
5. Images can in the next step be adjusted or improved to clarify the scientific message of the figure. In such processing steps, which mostly involve contrast, brightness, and color balance and saturation, it is demanded that all parts of the original image be treated equally.

3.7.1 Optimizing Contrast and Brightness

1. The first and most simple improvement of images is adjusting of contrast and brightness using the image processing software.

It is highly recommended to change the settings using the sliders of the input level histogram of the *levels* or *curves* dialog boxes of Photoshop or other software programs, rather than optimizing the image on the computer screen just by eye. As to the presentation of fluorescence images of chromosomes few notes come to mind. First, move the left slider that controls that dark part of the histogram palette to the right to enhance contrast, but never do so to the left end of the histogram. Most screens and even more so printed photographs in the scientific journals cannot discriminate between complete black and dark grays (low keys), and hence lead to loss of detail in the dark areas. Second, the white slider on the right of the input level now can be moved to the left to increase total brightness and contrast. This change can be done by eye, but keep in mind that no bright parts of the image (high keys) get clipped to pure white.

2. A special case of contrast management deals with image captures of chromosomes and nuclei that exhibit a greater dynamic range of exposure than is obtained through the digital camera of the microscope. A well-known example is the image capture of a DAPI stained image of large chromosomes, such as those of *Lilium*, *Alstroemeria*, *Secale*, and *Allium*. The limited dynamic range of a typical 12-bit digital camera does not allow revealing complete details of the high and low keys and leads to under- or over exposure of parts of the image. An example of a tetraploid *Allium* metaphase plate is shown in Fig. 3a, where most parts of the chromosomes and nucleus are too bright and lose essential details. The solution to this problem is high dynamic range imaging (HDR), a tool that is standard in most image processing programs. In this it is

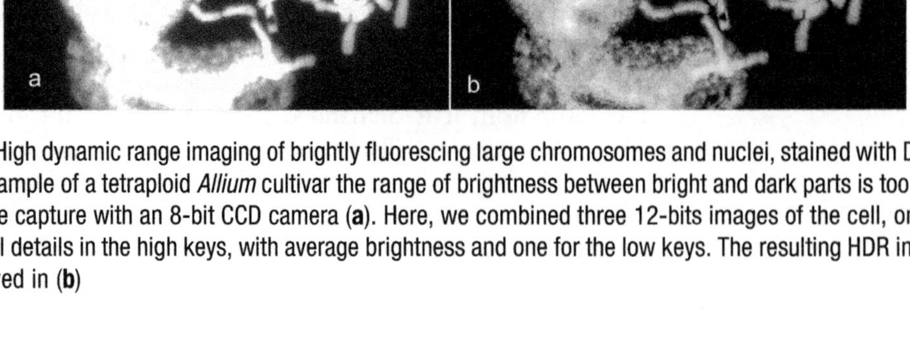

Fig. 3 High dynamic range imaging of brightly fluorescing large chromosomes and nuclei, stained with DAPI. In this example of a tetraploid *Allium* cultivar the range of brightness between bright and dark parts is too big for a single capture with an 8-bit CCD camera (**a**). Here, we combined three 12-bits images of the cell, one with optimal details in the high keys, with average brightness and one for the low keys. The resulting HDR image is displayed in (**b**)

recommended to capture three or more images, one with one stop underexposure, one with average exposure, and one with one stop overexposure. The images can then be combined by HDR, in which the dynamic range is reduced to an 8-bit tone map, thus demonstrating all details in the bright and dark areas of the chromosomes and nucleus. This method has been applied to generate Fig. 3b.

3.7.2 Sharpening Tools of Fluorescence Photomicrographs

1. Chromosome slides with DAPI counterstaining often display some level of image blurring of bright area. Blurring is about spreading some of the fluorescence at a given point over its neighbor structures. The use of a narrow band DAPI filter will decrease the blurring, and also a separate DAPI staining, followed by short rinsing in PBS buffer or distilled water and mounting in an antifading solution may have a positive effect on this image artifact.
2. Images of DAPI fluorescence can also afterward be deblurred using one of the advanced sharpening filters of the image processing software. Most of these tools are based on 2D or blind deconvolution and produce much sharper images than regular "sharp edges" or "unsharp mask" tools. Examples of such algorithms can be found in the "smart sharpen" tool of Photoshop CS, Focus Magic (http://www.focusmagic.com/), or Topaz Infocus (https://www.topazlabs.com/infocus).
3. The images in Fig. 2a, b, d are examples that were sharpened with the Focus Magic deconvolution tool. The inset in Fig. 2d displays obvious details of aberrant chromosome pairing. The small particles around the pachytene chromosomes are mitochondria, which disappear at later stages.

3.7.3 DAPI Fluorescence Displayed as Color or Gray Image?

1. One of the most common annoyances is the dark blue presentation of DAPI images whether or not as part of a multicolor FISH picture in many scientific papers. The hue of native DAPI in the fluorescence microscope is light blue, clearly visible, and well defined. But this color is very difficult displayed as a pseudo-color in a simple RGB color mode. Regrettably, many papers show such images in the blue channel of an RGB color mode, not being aware that such images are hardly visible, especially in the low-key areas of the image. An example of this phenomenon is shown in Fig. 4a–c, in which a comparison was made between the same chromosome complement of a banana tapetal cell displayed in the blue channel, full gray and pseudo-colored versions. It is obvious from these images that the gray version (b) of the image capture excels that of the blue channel image (a), but that also the pseudo-colored version (c) produced by retains enough details and image accuracy.

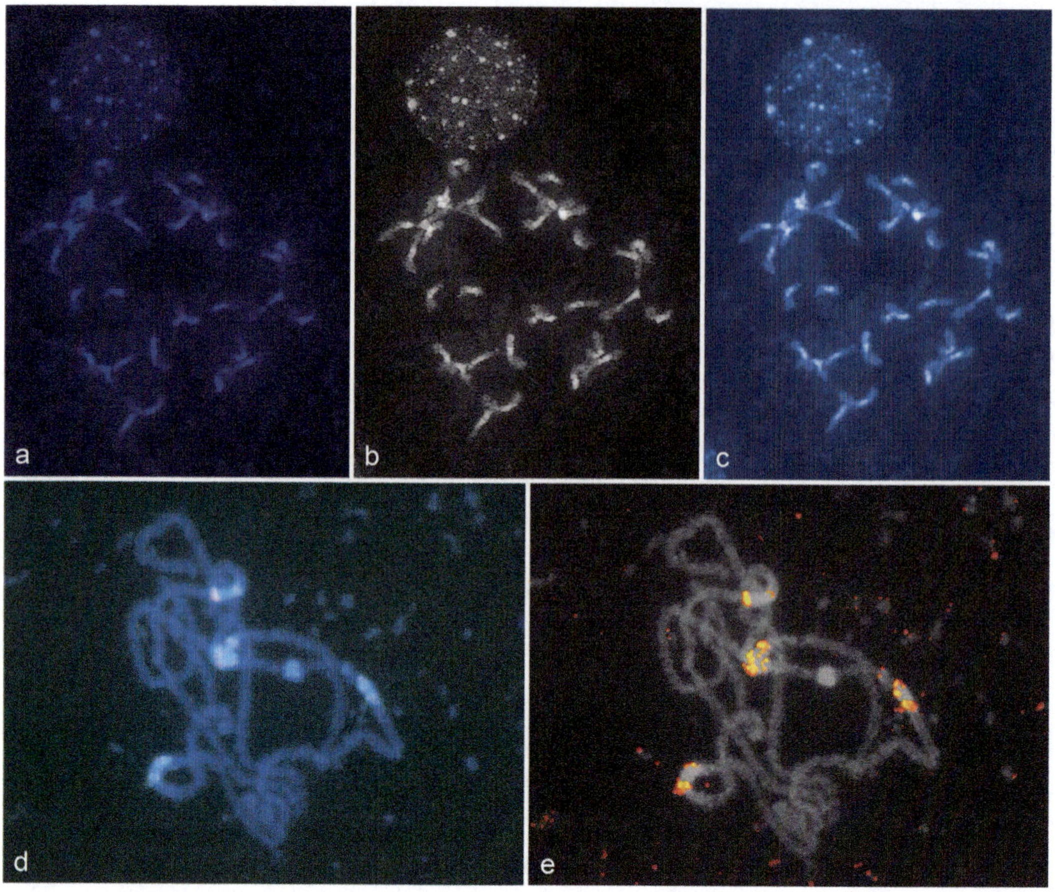

Fig. 4 Cell complement of a DAPI stained fluorescent image of a triploid banana in different color modes. (**a**) The microphotograph is displayed in the blue channel of a standard 24-bit RGB color mode. The *blue image* is simply too dark for clear distinction of the darker parts of the chromosomes. (**b**) Same cell in an 8-bit gray image. (**c**) Same monochromatic cells converted into a 24-bit RGB, followed by pseudo-coloring in a more realistic DAPI like hue, using the Hue saturation tool/colorize in Photoshop with the settings Hue: 202, Saturation: 77 and Lightness: 0. (**d**) A pachytene FISH image probes with (peri-)centromere repeats. The original image was obtained with a multiband fluorescence filter block simultaneously displaying the blue DAPI, green FITC, and red Texas Red images. (**e**) Same image, but now with color separation of DAPI in gray, green, and red fluorescent signal in separate spot-channels for green and red

2. A second example of improving color representation and separation is shown in Fig. 4d, e. The original photomicrograph was obtained with an image capture using a multicolor filter block with transmissions for blue, green, and red fluorescence (d). The separate channels in the original RGB image were copied to monochromatic channels for the three colors and merged together as a multichannel image (e).

3.7.4 Combining Colors from FISH Experiments: Multi-channels or Layers?

1. The combination of multiple captures from the same image is one of the major applications of image processing in a FISH experiment using the simultaneous hybridization with fluorescent label probes of various pooled single copy DNA sequences and repetitive sequences. In the epifluorescence microscope that we use, we have six narrow band filters, for the UV-blue DAPI (375 excitation/510 emission), blue DEAC (436/480), green FITC (495/519), orange Cy3 (512/570), red Cy3.5 (581/594), and far-red Cy5 (625/650) fluorescent probes, which allow simultaneous study of five probes on a chromosome target counterstained with DAPI. Most commercial FISH microscopes come up with a multichannel mode that allows merging of the six images into one. If an alternative image processing tool (e.g., in Photoshop) is required, then the following steps can be followed: (a) each monochrome image is converted into an RGB image and then pseudo-colored using a New Adjustment Layer | gradient map; (b) the pseudo-colored images are then copied in a multiple layer file using a screen blending overlay mode; and (c) the layers are finally merged into a simple RGB image.
2. A second mode to combine various fluorescent images involves the creation of a multichannel photo, consisting of a RGB (gray) image of the DAPI fluorescence with on top the multiple spotchannels, each of them containing one of the five monochrome probe images. The advantage of this method is that each monochrome image can be optimized individually for contrast, brightness, transparency, and spot color. A Photoshop example in Fig. 5 shows a pachytene complement of *Arabidopsis thaliana* that is hybridized with probes for different DNA pools and repetitive sequences. In the channel dialog box different combinations of fluorescent signals can be activated to visualize the spatial relationship of the probe sequences under study.

4 Notes

1. Microscopy, optical system, and camera: The fluorescence microscope has epifluorescence illumination and is capable of transmitting emission and excitation light covering a spectral range of 350–800 nm. Most chromosome studies require high-resolution optics, preferably 60× and 100× Plan Fluorite and Plan Apochromatic objectives with Numerical Aperture (N.A.) of at least 1.35. For highest quality make use of the high-performance cover slips of 0.17 ± 0.005 mm, and immersion oil optimized for fluorescence microscopy with a refractory index of 1.5.

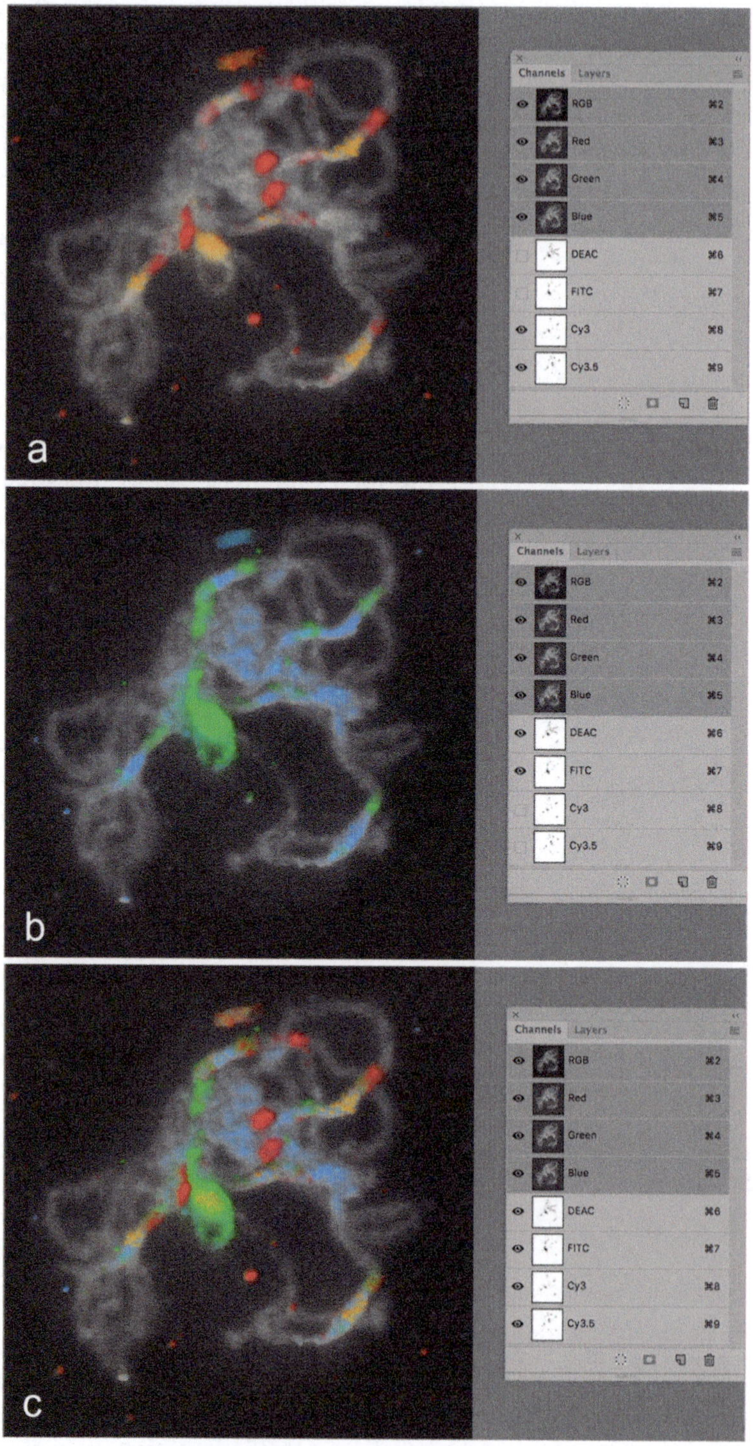

Fig. 5 Fluorescence *in situ* Hybridization of an Arabidopsis pachytene complement. Probes of repeats and BAC pools were labeled and visualized in DEAC (*blue*), FITC (*green*), Cy3 (*orange*), Cy3.5 (*red*), and Cy5 (*purple*). Chromosomes

2. The great advantage of digital imaging is that the result of the microscopic image capture can be seen directly, an advantage that is especially important for the short life time of some of the fluorescent chromosome preparations. In addition, scientific-grade CCD cameras feature outstanding dynamic range that is sufficient for even weak fluorescent signals, have a spatial resolution more than enough to meet the properties of the superior microscopic optics and a spectral bandwidth that covers near UV to near infrared images. A typical high-quality digital fluorescence microscope camera features a cooled monochrome chip, of at least 1.4 Mega-pixels, which is enough for 60–100× high N.A. objectives. In addition, the camera should cover spectral bandwidth of 400–900 nm, show very low noise values, and is capable of producing images that support 12 or more bit gray levels.
3. There is no general rule for the best medium, but some pilot tests with sandy soil, coconut fiber peat, or other soil type and watering can help to find the best conditions for root development.
4. Rules of thumbs for staging anther development: if pollen mother cells are at premeiotic or early meiotic, the pollen sac will still be intact and not easy to spread; if the pollen mother cells are at different meiotic stages they can be released easily from the supporting anther tissues and tapetum cells. Post-meiotic anthers contain tetrads and pollen grains which can be identified easily under the dissecting microscope.
5. β-1,3-glucanase breaks down β-1,3-glucans such as callose in the cell wall of the pollen mother cells; β-1,4-beta-D-glucosidic or cellulase breaks down cellulose. The snail gut cytohelicase also contains other glucanase/laminarinase activity and is essential for the enzymatic digestion of the callose walls of pollen mother cells.
6. The squashing method works often better with large chromosomes (*Vicia*, *Allium*, *Alstroemeria*, *Secale*, etc.), the dropping method is better than the cell spreading, but gives more loss of material and requires more practicing. It is above all a matter of personal taste which method works best.

Fig. 5 (Continued) were counterstained in DAPI (*gray*). The fluorescent images were captured with a monochromatic CCD camera using epifluorescence equipment and narrow band filters for the six fluorescent dyes. The multicolor image thus achieved contains the RGB image of the DAPI image, and spot color channels for the five probe images. The panels of the channel dialog box in Photoshop on the *right* show how different fluorescent signals can be excluded (**a**, **b**), or combined (**c**), giving a better impression on how signals are overlapping and located on the chromosomes

Acknowledgments

We thank Xianwen Ji for providing the images of Fig. 5.

References

1. Sharma AK, Sharma A (2014) Chromosome techniques. Butterworth-Heinemann, Oxford
2. Singh RJ (2016) Plant cytogenetics, 3rd edn. CRC Press, Boca Raton, FL
3. Guzzo F, Campagnari E, Levi M (2000) A new FISH protocol with increased sensitivity for physical mapping with short probes in plants. J Exp Bot 51:965–970
4. Devi J, Ko JM, Seo BB (2005) FISH and GISH: modern cytogenetic techniques. Ind J Biotechnol 4:304–315
5. Li J, Zhao X, Zhang X, Li M, Li J, Yuan J, Zhao X, Zhang X, Li M (2014) A simple and rapid method for visualization of male meiotic chromosomes in Arabidopsis thaliana. Biosci Biotechnol Biochem 78:1026–1028
6. Andres RJ, Kuraparthy V (2013) Development of an improved method of mitotic metaphase chromosome preparation compatible for fluorescence in situ hybridization in cotton. J Cotton Sci 17:149–156
7. Kirov I, Divashuk M, Van Laere K, Soloviev A, Khrustaleva L (2014) An easy "SteamDrop" method for high quality plant chromosome preparation. Mol Cytogenet 7:21
8. Aliyeva-Schnorr L, Ma L, Houben A (2015) A fast air-dry dropping chromosome preparation method suitable for FISH in plants. J Vis Exp 106:e53470
9. Younis A, Ramzan F, Hwang Y-J, Lim K-B (2015) FISH and GISH: molecular cytogenetic tools and their applications in ornamental plants. Plant Cell Rep 34:1477–1488
10. Dang J, Zhao Q, Yang X, Chen Z, Xiang S, Liang G (2015) A modified method for preparing meiotic chromosomes based on digesting pollen mother cells in suspension. Mol Cytogenet 8:80
11. Sedgewick G (2007) Quick Photoshop for Research. Springer, New York, NY
12. Kervrann C, Sanchez Sorzano CO, Acton ST, Olivo-Marin J-C, Unser M (2016) A guided tour of selected image processing and analysis methods for fluorescence and electron microscopy. IEEE J Sel Top Signal Process 10:6–30
13. Rossner M (2004) What's in a picture? The temptation of image manipulation. J Cell Biol 166:11–15
14. Frow EK (2012) Drawing a line: setting guidelines for digital image processing in scientific journal articles. Soc Stud Sci 42:369–392

Chapter 13

Whole Mount RNA-FISH on Ovules and Developing Seeds

Andrea Bleckmann and Thomas Dresselhaus

Abstract

A key element to understand developmental and reproductive processes like germline development, double fertilization, and embryogenesis is the study of cell-specific gene expression patterns which is best analyzed by RNA in situ hybridization. Different visualization techniques have been established to mark either the region of mRNA production (using the classical chromogenic detection system) or the specific localization of mRNAs (using fluorescent labeled probes). In this chapter, we describe and compare whole mount RNA in situ hybridization techniques on ovules and young developing seeds from *Arabidopsis thaliana* using three different detection systems. The alkaline phosphatase (AP) coupled antibody detecting the antigen labeled probe facilitates the production of a precipitating dye indicating mRNA presence: (1) using BCIP/NBT as substrates, it is converted to a blue staining that can be visualized using differential interference contrast (DIC) microscopy. Alternatively, (2) using Fast-Red as a substrate it is converted to a purple fluorescent staining that can be visualized either by light microscopy or, for a higher cellular resolution, by confocal microscope. To analyze mRNA distribution with subcellular resolution we (3) describe a third, highly sensitive fluorescent detection system, which is based on the enzymatic activity of a peroxidase. In combination with a tyramide signal amplification (TSA) system, it leads to multi-fluorescent labeled antibodies marking the mRNA bound probe locally.

Key words Fluorescent RNA in situ hybridization, Whole mount, F-WISH, TSA, Ovule, Subcellular RNA detection, Arabidopsis

1 Introduction

The spatial-temporal regulation of mRNA translation is a key mechanism to control germline and embryo development in animals [1, 2]. An increasing number of recent publications showed that short and long-range mRNA transport exist also in plants, but their visualization still lacks mainly due to technical difficulties [3–5]. In this chapter, we describe and compare three whole mount in situ hybridization methods to visualize mRNA in the female gametophyte and during the first steps of embryogenesis using *Arabidopsis thaliana* as a model. Two methods are based on the enzymatic activity of the alkaline phosphatase, which converts substrates to a precipitating dye. The third method involves

Anja Schmidt (ed.), *Plant Germline Development: Methods and Protocols*, Methods in Molecular Biology, vol. 1669, DOI 10.1007/978-1-4939-7286-9_13, © Springer Science+Business Media LLC 2017

enzymatic peroxidase activity, which converts tyramides to radicals leading to multi-fluorescent labeling of the probe bound antibody and thereby permits analyzing mRNA distribution with high sensitivity and resolution at the subcellular level. The describe whole mount RNA-FISH or F-WISH is a modified protocol of formerly described procedures [6–8].

2 Materials

All the liquids should be prepared in RNase-free bottles or tubes (*see* Subheading 3.1). If possible, solutions are treated with diethylenepyrocarbonate (DEPC) (*see* Subheading 3.1). Compounds with primary amine groups (e.g., Tris) will react with DEPC and therefore cannot be DEPC-treated. All other solutions are prepared in RNase-free water (either DEPC-treated water or from a water purification system equipped with a special ultrafilter, e.g., Biopak®-Polisher from Milli-Q®). If not otherwise stated, solutions and buffers can be stored at room temperature.

2.1 Synthesis of Labeled RNA Probes

1. 10× DIG RNA-labeling mix: 10 mM ATP, 10 mM CTP, 10 mM GTP, 6.5 mM UTP, and 3.5 mM Digoxigenin-11-UTP (stored at −20 °C).
2. 10× Fluorescein RNA-labeling mix: 10 mM ATP, 10 mM CTP, 10 mM GTP, 6.5 mM UTP, 3.5 mM Fluorescein-12-UTP, pH 7.5; stored at −20 °C; Roche).
3. 10 U/μl T7 or SP6 RNA polymerase (stored at −20 °C).
4. 40 U/μl RNase inhibitor (stored at −20 °C).
5. 20 mg/ml tRNA from *Escherichia coli* (in RNase-free water; store in aliquots at −20 °C).
6. 1 U/μl DNase I (RNase-free; stored at −20 °C).
7. 0.5 M Ethylene-diamine-tetraacetic acid (EDTA), pH 8.0 (RNase-free).
8. 10 M Lithium chloride (LiCl), DEPC-treated (*see* Subheading 3.1, **step 4**).
9. 2× Alkaline hydrolysis buffer: mix freshly 600 μl RNase-free water, 240 μl 0.5 M Na_2CO_3 and 160 μl 0.5 M $NaHCO_3$.
10. 10× 3-Morpholinopropane-1-sulfonic acid (MOPS) buffer: 0.4 M MOPS, 0.1 M sodium acetate, and 0.01 M EDTA, pH 7.0. Prepare in RNase-free water and store in the dark, discard when buffer turns yellow.
11. 2× RNA loading dye: 95% (v/v) formamide; 0.5 mM EDTA; 0.025% (w/v) SDS; 0.025% (w/v) bromophenol blue; 0.025% (w/v) xylene cyanol FF; 0.025% (w/v) ethidium bromide.

2.2 Whole Mount In Situ Hybridization (WISH)

1. 10% (w/v) formaldehyde stock solution from paraformaldehyde in RNase-free water: Paraformaldehyde will not resolve directly. Heat the solution (≥60 °C) carefully and add a few drops of 1 N NaOH until the solution becomes clear (pH ~8); check the pH using a pH paper indicator and store the solution in single use aliquots at −20 °C. (Cautions: vapors are toxic; work under a fume hood).
2. 10× Phosphate-buffered saline (PBS): 1.3 M NaCl, 70 mM Na_2HPO_4, 30 mM NaH_2PO_4, pH 7.4 treated with DEPC. All dilutions should be prepared with RNase-free water.
3. Fixative: 4% (w/v) formaldehyde (from stock solution), 15% (v/v) dimethyl sulfoxide (DMSO), 0.1% (v/v) Tween®20, in 1 × PBS.
4. n-Heptan.
5. PBS-T: 1× PBS + 0.1% (v/v) Tween®20.
6. HistoClear.
7. 30% H_2O_2.
8. Freshly prepared 3% (v/v) H_2O_2 in ethanol.
9. 20× Saline sodium citrate buffer (SSC): 3 M NaCl, 0.3 M sodium citrate, adjust the pH to 7.0 with HCl; treated with DEPC. All dilutions should be prepared with RNase-free water.
10. 75% and 80% (v/v) ethanol in RNase-free water.
11. 50% and 25% (v/v) ethanol in PBS.
12. Proteinase K stock solution: 25 mg/ml Proteinase K in RNase-free water. Store in single use aliquots at −20 °C.
13. 10× Glycine solution: 200 mg/ml glycine in RNase-free 10× PBS, stored in aliquots at −20 °C. Dilutions should be prepared with RNase-free water.
14. Hybridization solution: 50% (v/v) formamide, 5× SSC, 0.1% (v/v) Tween®20, and 0.1 mg/ml heparin in RNase-free water. Add freshly denatured fragmentized salmon sperm DNA (10 min at 95 °C) to a final concentration of 1 mg/ml.
15. 50 mg/ml heparin stock solution.
16. 12 mg/ml RNase-free fragmentized sperm DNA: purify a DNA solution by phenol-chloroform extraction followed by an ethanol precipitation and resuspension in RNase-free water. Adjust DNA concentration to 12 mg/ml using RNase-free water.
17. Wash solution 1: 50% (v/v) formamide, 2× SSC, and 0.1% (v/v) Tween®20 in RNase free-water.
18. Wash solution 2: 2× SSC and 0.1% (v/v) Tween®20 in RNase-free water.

19. Wash solution 3: 0.2× SSC and 0.1% (v/v) Tween®20 in RNase-free water.
20. Blocking solution: 3% (w/v) bovine serum albumin (BSA) in PBS-T. Prepare freshly before use.
21. 1 M $MgCl_2$ (DEPC-treated).
22. 2 M Levamisol in RNase-free water. Store in aliquots at −20 °C.
23. Alkaline phosphatase (AP) buffer: 0.1 M Tris–HCl (pH 9.5), 0.1 M NaCl, 50 mM $MgCl_2$ (freshly added), 0.1% (v/v) Tween®20.
24. 50 mg/ml Nitroblue tetrazolim chloride (NBT) stock solution (Roche) store at −20 °C.
25. 50 mg/ml 5-Bromo-4-chloro-3-indoxylphosphate (BCIP) stock solution (Roche), store at −20 °C.
26. AP staining solution: AP Buffer +2 mM Levamisole, 110 μg/ml NBT and 90 μg/ml BCIP.
27. TNT buffer: 0.1 M TRIS–HCl (pH 7.5), 0.15 M NaCl, 0.05% Tween®20.
28. TSA staining solution: always prepare fresh following the manufacturer's instructions (e.g., TSA-Plus-Cyanine3-Kit; PerkinElmer).
29. Fast-red staining solution: always prepare fresh following the manufacturer's instructions (e.g., SIGMAFAST Fast Red TR/Naphthol AS-MX tablets).
30. Clearing solution: 70% (w/v) chloral hydrate and 10% (v/v) glycerol.
31. Baskets for holding back samples (e.g., from Intavis, available with different mesh sizes (*see* **Note 1**)).
32. DAPI-counterstaining solution: 77% (v/v) glycerol, 300 nM 4′,6-diamidine-2′-phenylindole dihydrochloride (DAPI) in 1× PBS.
33. Multi-well plates or InsituPro VS liquid handling robot (Intavis AG).
34. Glass petri dishes.

3 Methods

3.1 RNAse Decontamination of Glassware, Plastic Ware, and Solutions

1. Use sterile, disposable plastic ware whenever possible—these should be RNase-free.
2. Decontaminate other plastic ware by soaking in 3% (v/v) hydrogen peroxide for 10 min followed by extensively rinsing with RNase-free water.

3. Glassware can be decontaminated by baking at 180 °C for several hours (caution: use bottles with heat resistant lids and pour ring).
4. Treat all solutions that are not containing compounds with primary amine groups with DEPC. Prepare 1 ml/l DEPC, stir for 2 h, and autoclave for DEPC inactivation.

3.2 Synthesis of Labeled Probes

1. We commonly use linearized plasmid or purified PCR-product as template for labeled RNA probe synthesis. The selected sequence is cloned into a plasmid that contains T7-(5′-TAATACGACTCACTATAG-3′) or/and SP6-(5′-ATTTAGGTGACACTATAG-3′) RNA polymerase promoter sequences for in vitro transcription as well as several unique recognition sites for restriction enzymes creating blunt end or 5′-overhangs for plasmid linearization in their flanking regions (*see* **Note 2**). Alternatively, the T7- or SP6-specific promoter sequence is added to a gene-specific mRNA region by PCR-mediated ligation (at the 5′ for sense probe and at the 3′ for antisense probe).
2. Clean up template DNA by a DNA-purification kit (e.g., MicroSpin Cycle-Pure Kit; PeqLab) according to the manufacturer's instructions and elute in RNase-free water.
3. Set up a DIG- or Fluorescein RNA labeling reaction according to the manufacturer's instructions (e.g., DIG RNA Labeling Kit (SP6/T7)).
4. Add 2 μl of tRNA stock solution and 1 μl of DNase I and incubate for 15 min at 37 °C.
5. Transfer on ice and add 0.8 μl EDTA, 1 μl LiCl and 75 μl ethanol.
6. Precipitate RNA for ≥30 min at −80 °C.
7. Centrifuge for 30 min at ≥16,000 rcf, 4 °C.
8. Wash pellet in 80% (v/v) ethanol.
9. Centrifuge for 10 min at ≥16,000 rcf, 4 °C.
10. Dissolve pellet after drying in 100 μl RNase-free water.
11. Probe quality should be checked by electrophoresis on a denaturing agarose gel (*see* **Note 3**).
12. Store probe in aliquots at −80 °C and avoid repeated freeze/thaw cycles.
13. Long probes (>500 nt) can have negative effects on hybridization efficiencies or tissue penetration depth. The size can be reduced by hydrolysis to generate an optimal probe size of ~300–500 nt. For that propose, add 1 volume of freshly prepared alkaline hydrolysis buffer and incubate at 60 °C for a calculated time (t).

((t[min] = (probe length [kb] – desired probe length [kb])/(0.11 × probe length [kb] × desired probe length [kb])) [6].

14. Stop hydrolysis by adding acetic acid to a final concentration of 0.5% (v/v) and sodium acetate to a final concentration of 0.1 M and mix gently.
15. Add 1 volume of isopropanol and precipitate for ≥30 min at −80 °C.
16. Centrifuge for 30 min at ≥16,000 rcf, 4 °C.
17. Wash pellet in 80% (v/v) ethanol.
18. Centrifuge for 10 min at ≥16,000 rcf, 4 °C.
19. Dissolve pellet after drying in 100 μl RNase-free water.

3.3 Tissue Preparation and Fixation

1. Prepare fresh fixative:n-Heptane (1:1) emulsion by strong vortexing and shaking.
2. Add 1 ml of the emulsion to a 1.5 reaction tube or fill the bottom of a glass petri dish containing sieves and store them on ice (*see* **Note 4**).
3. A flower of the desired stage is collected on a glass slide and flower organs like sepals, petals, and stamen are removed using forceps. The pistil is glued on a double-sided adhesive tape in a way that the fusion line of the two carpels (replum) is pointing upward (Fig. 1). Carpels are removed by cutting with a cannula along the valve margin on the upper side. Then carpels are pushed down and glued to the tape and removed by cutting underneath the visible ovules. Ovules connected to the transmitting track are transferred to the prepared sieves or tubes using forceps (*see* **Note 5**).
4. Penetration depth of the fixative can be increased by vacuum infiltration for 10–30 min (*see* **Note 6**).
5. Incubate the tissue for additional 30-60 min on a shaker (*see* **Note** 7).
6. Wash the tissue at least three times for 5 min with methanol or until the tissue is white and all chlorophyll is removed.
7. Wash the tissue two times for 5 min with ethanol.
8. The fixed/dehydrated tissues can be stored at −20 °C for weeks.

3.4 Degradation of Endogenous Peroxidase

This step is only required for the HRP/TSA-detection system.

1. Incubate the tissue for 1 h in freshly prepared 3% (v/v) H_2O_2/ethanol.
2. Wash the tissue twice for 5 min with ethanol.

Steps 3–10 can also be performed by a liquid handling robot.

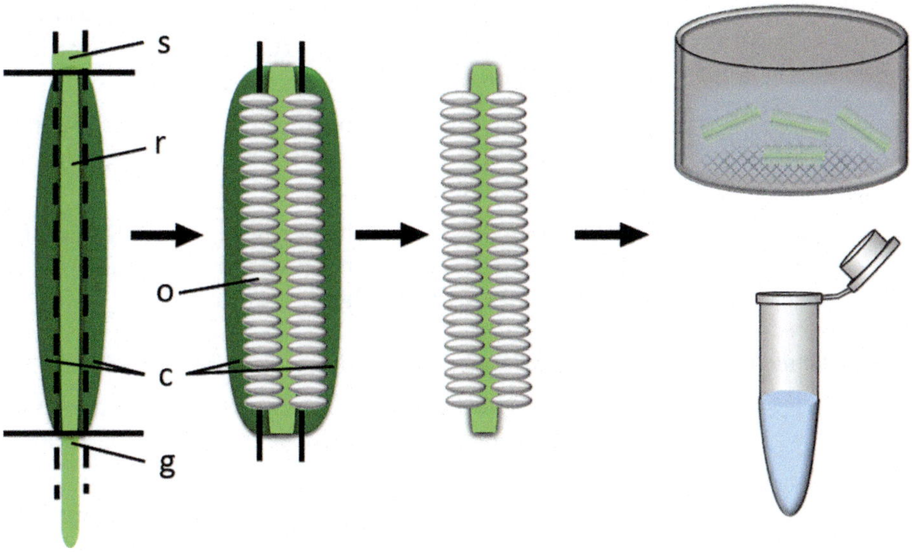

Fig. 1 Concept of an ovule/developing seed preparation procedure. A flower of the desired stage is collected on a glass slide and floral organs like sepals, petals, and stamen are removed using forceps. The pistil is glued on a double-sided adhesive tape in a way that the fusion line of the two carpels (replum, r) is pointing upward. Style (s) and gynophore (g) of a silique are removed by cutting along the *black lines* using a cannula. Ovaries are opened by cutting carpels (c) along the valve margin (*dashed line*) with a cannula without touching the ovules. Next maternal carpel tissue is glued to the tape by pushing it down and removed by cutting underneath the visible ovules (o) (along the *lines* indicated in the *middle*). Ovules connected to the transmitting tract are then transferred to a sieve or 1.5 ml reaction tube filled with fixative

3.5 Tissue Permeabilization and Probe Hybridization

1. Incubate the tissue in HistoClear/ethanol (1/1) solution for 30 min (*see* **Note 8**).
2. Wash the tissue three times each for 5 min with ethanol.
3. Incubate the tissue in 75% and 50% (v/v) ethanol followed by 50% and 25% (v/v) ethanol/PBS for 20 min each.
4. Re-fixate the tissue for 20 min in fixative.
5. Wash the samples twice in PBS-T for 10 min each.
6. To remove RNA bound proteins the tissue has to be incubated in freshly prepared pre-warmed 150–200 μg/ml Proteinase K solution (company recommended buffer or water) for 15 min at 37 °C (*see* **Note 9**).
7. The reaction is stopped by incubation with glycine solution for 5 min.
8. Wash the samples twice in PBS-T 10 min each.
9. Re-fixate the tissue for 20 min in fixative.
10. Wash the samples twice in PBS-T 10 min each.

3.6 Probe Hybridization

1. Incubate the tissue in pre-hybridization mix for 10 and 60 min at the desired hybridization temperature (*see* **Note 10**).

2. Incubate the tissue for 12–16 h with hybridization mix containing the denaturated probe (10 min at 65 °C) at the desired hybridization temperature (*see* **Note 11**).
3. To remove any excess probe and hybridization mixture, wash the samples three times in pre-warmed washing solution 1 for 10, 60, and 20 min, respectively.
4. Wash the samples twice with washing solution 2 and twice in washing solution 3 for 20 min each to remove nonspecific and/or repetitive DNA/RNA hybridization.

3.7 Antibody Binding

1. Samples are washed three times in PBS-T for 10 min.
2. To block unspecific protein binding sites, the tissue is saturated with BSA by incubation in blocking solution for 90 min.
3. Add the primary labeled antibody in blocking solution onto the samples and incubate for 4–15 h at 21 °C (*see* **Note 12**).
4. Wash the samples eight times with PBS-T (AP labeled antibody) or TNT (HRP labeled antibody) each for 10 min to remove excess antibody.

3.8 Visualization of RNA Probes and Mounting of Samples

3.8.1 Chromogenic AP-Based Detection Using the Substrate NBT/BCIP

1. Wash samples twice with AP buffer.
2. Incubate samples in a pre-warmed (37 °C) AP staining solution to initiate colorimetric detection.
3. Incubate samples in the dark at 37 °C until a dark blue indigo staining is visible (initially check every 10 min; afterward the periods can be extended).
4. Stop the reaction with 100% (v/v) ethanol.
5. Wash tissues with 50% (v/v) ethanol for 10 min.
6. Wash tissues with PBS-T for 10 min.
7. For DIC microscopy, ovules are dissected with cannulas and cleared by mounting the samples in clearing solution on a glass slide (*see* **Note 13**).

3.8.2 Fluorescent AP-Based Detection Using Fast-Red as a Substrate

1. Wash samples twice with AP buffer.
2. Incubate samples in pre-warmed (37 °C) Fast-Red staining solution to initiate colorimetric detection.
3. Incubate samples in the dark at 37 °C until a red staining is visible (initially check every 10 min; afterward the periods can be extended).
4. Stop the reaction by washing the samples four times with PBS-T (*see* **Note 14**).
5. For light microscopy, ovules are dissected with cannulas and mounted in 77% (v/v) glycerol/PBS on a glass slide.

6. For fluorescent microscopy samples are mounted in 77% (v/v) glycerol/PBS; 300 nM DAPI for DNA counterstaining (*see* **Note 13**).
7. Samples are analyzed using a confocal laser scanning microscope (CLSM) (Fast-Red: excitation wavelength 561 nm and emission bandpass of 570–640 nm; DAPI: excitation wavelength 405 nm and emission bandpass of 420–470 nm).

3.8.3 HRP-Based Detection Using TSA Detection System

1. Incubate samples for 5–15 min in fresh TSA staining solution.
2. Wash samples four times for 5 min followed by five times for 10 min in TNT-buffer to remove unbound dye.
3. Ovules are dissected with cannulas and mounted in 77% (v/v) glycerol/PBS; 300 nM DAPI for DNA counterstaining (*see* **Note 13**).
4. Samples are then analyzed using a CLSM. Microscope setting depends on the fluorescent dye.

3.9 Comparison of the Three Different Detection Methods

Classical RNA in situ detection using an AP labeled antibody in combination with the substrate NBT/BCIP can be easily analyzed by simple light microscopy. Using clearing solutions, the best cellular resolution can be achieved when ovules/developing seeds are oriented on the flat side if it is not covered by leftovers from other tissues. An example of this WISH detection system using *SUBTILASE 4.13 (SBT4.13; AT5G59120.1)* antisense DIG labeled probe can be seen in Fig. 2a, b.

The substrate Fast-Red provides the possibility of optical sectioning of the tissue using a CLSM. Moreover, very faint signal can still be visualized by CLSM. In combination with DAPI counterstaining, a good cellular resolution can be achieved independent of ovule orientation and tissue leftovers. However, it is more difficult to obtain good standard light microscopic images because the above-mentioned clearing procedure, which is essential for a good cellular resolution, bleaches the dye. An example of this WISH detection system using *SBT4.13* antisense DIG labeled probe can be seen in Fig. 2c–f.

The staining process of these two detection systems can be monitored over time and be stopped when an optimal coloration of the tissue is visible. Moreover, the AP enzyme is active over a longer period of time, allowing continuous amplification of the signal only limited by background production due to unspecific probe or antibody binding as well as substrate instability. Very strong signals can therefore also over-stain the tissue leading to signal spreading or dye diffusion into neighboring cells.

The third TSA method is suitable to study mRNA localization also at a subcellular resolution. An example of this WISH detection system using *SBT4.13* antisense DIG labeled probe can be seen in

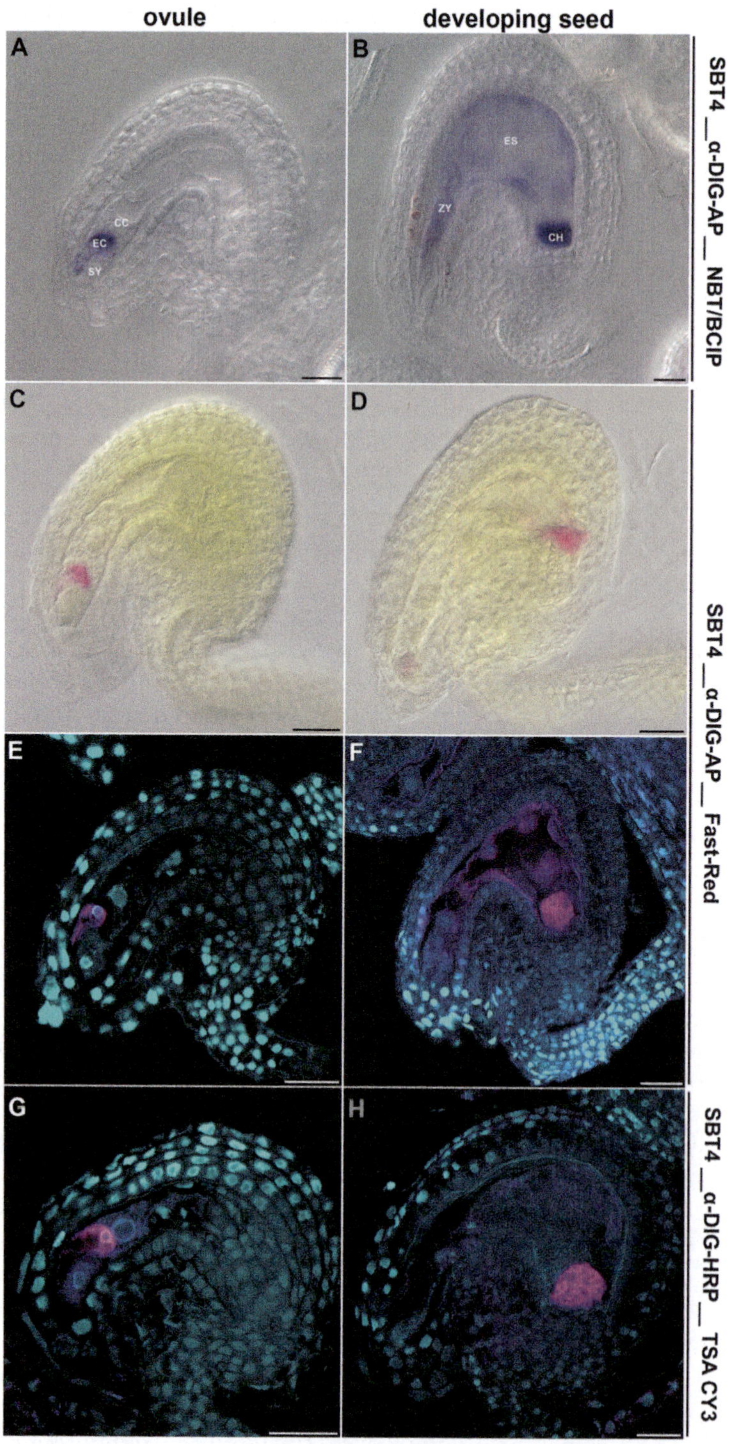

Fig. 2 Expression of the subtilase *SBT4.13* in the ovule (*left row*) and developing seed (*right row*). (**a**, **b**) DIC image of a WISH using *SBT4.13* antisense probe and a phosphatase linked DIG antibody with NBT/BCIP as a substrate. (**c**, **d**) DIC and (**e**, **f**) CLSM images of a (F)-WISH using *SBT4.13* antisense probe and a phosphatase

Fig. 2g, h. This method has a very high sensitivity due to the enzyme catalyzed multi-labeling process of the antibody. In this system the antibody linked peroxidase oxidized the substrate, a fluorophore-linked tyramide, leading to the formation of radicals. The activated fluorophore will react with tyrosine in close proximity, thus with tyrosine residues of the peroxidase enzyme itself and those of the antibody. This multi-labeling of the peroxidase leads also to its inactivation. Therefore, an extended labeling reaction as discussed for the other methods is not suitable for an increase of specific signals and instead results in an increase of background signals. Unbound activated tyramides will react with each other and form dimers, which can be washed out. A disadvantage of this method is that the staining reaction cannot be monitored by eye; thus, different conditions have to be tested for an optimal result.

In conclusion, all three methods have their advantages and disadvantages. Their application depends on the equipment available as well as on the individual research question and experimental goals.

4 Notes

1. Alternatively, a stainless steel mesh (mesh width 25–100 μm) glued to reaction tubes slices by heat can be used.
2. In vitro transcription creates labeled "run off" transcripts; the selected restriction site causes transcript termination and therefore should be located as close to the insert site as possible.
3. The added tRNA distorts photometric concentration measurements of the transcribed probe, but simultaneously increases precipitation efficiencies and serves as a blocking reagent in the probe hybridization step.
4. For vacuum infiltration or longer storage of the tissue incubation in reaction tubes is recommended although liquid exchange is slightly more time-consuming.

Fig. 2 (Continued) linked DIG antibody with Fast-Red as a substrate (*magenta*). *SBT4.13* is expressed in the egg cell until fertilization. Afterward, it is downregulated in the zygote but strongly expressed from the chalazal pole nucleus of the endosperm (ES). (**g**, **h**) F-WISH using *SBT4.13* antisense probe and a peroxidase linked DIG antibody with TSA Plus Cyanine 3 substrate (PerkinElmer) for detection. The mRNA of *SBT4.13* is evenly distributed in the cytosol of the egg cell. In the syncytium of the developing endosperm *SBT4.13* mRNA is detected in the cytosol, which surrounds the nucleus at the chalazal pole. (**e–h**) The DNA is counterstained with DAPI (*cyan*). *EC* egg cell, *CC* central cell, *SY* synergid, *ZY* zygote, *ES* endosperm, *CH* chalazal region. Scale-bars are 20 μm

5. The tissue should be as clean as possible to achieve high sensitivity and low background. Tissue preparation has to be conducted as fast as possible without touching ovules directly, to avoid damaging the tissue. Any kind of sample drying, lesion, or squeezing will lead to a strong background and unspecific staining.
6. For vacuum infiltration tissue should be collected in a reaction tube, which can be placed in a vacuum concentrator. Centrifugation will prevent tissue loss due to air evaporation from the emulsion and the tissue.
7. We have extended this step up to 5 h and never saw any difference, but it is described elsewhere that over-fixation can lead to signal loss.
8. Use glass ware or plastic tubes resistant against HistoClear.
9. In our hands this is one of the most critical steps of the whole procedure. Higher Proteinase K concentration can have negative effect on tissue stability, but still will result in a good signal. Too low Proteinase K concentration will result in signal loss. So we recommend starting with high concentration.
10. We usually start with a hybridization temperature of 55 °C, which works in most cases.
11. Compared to tissue sections, higher probe concentrations are required for a strong hybridization signal due to the thickness of the tissue. However, a very high probe concentration will lead to elevated background due to unspecific binding. We usually start with a series of 0.1–2% of the original probe reaction mix in 1 ml hybridization mix.
12. Best results are achieved by diluting the antibody in blocking solution shortly before usage (also when using an automated system).
13. Tissue is transferred into a drop of the desired mounting solution and dissected with cannulas under the dissecting microscope. To prevent tissue crushing, we use a drop of clear nail polish as a cover glass spacer.
14. The precipitated Fast-Red is washed out by ethanol and bleached by clearing solution. Thus, it is not possible to clear the tissue for an optimal DIC image.

Acknowledgment

This study was funded by the Collaborate Research Center SFB960 of the German Research Foundation (DFG—Germany) to T.D.

References

1. Medioni C, Mowry K, Besse F (2012) Principles and roles of mRNA localization in animal development. Development 139:3263–3276. doi:10.1242/dev.078626
2. Buchan JR (2014) mRNP granules. RNA Biol 11:1019–1030. doi:10.4161/15476286.2014.972208
3. Notaguchi M, Higashiyama T, Suzuki T (2015) Identification of mRNAs that move over long distances using an RNA-seq analysis of Arabidopsis/Nicotiana benthamiana heterografts. Plant Cell Physiol 56:311–321. doi:10.1093/pcp/pcu210
4. Thieme CJ, Rojas-Triana M, Stecyk E et al (2015) Endogenous Arabidopsis messenger RNAs transported to distant tissues. Nat Plants 1:15025. doi:10.1038/nplants.2015.25
5. Zhang W, Thieme C, Kollwig G et al (2016) tRNA-related sequences trigger systemic mRNA transport in plants. Plant Cell 2016: tpc.01056.2015. doi:10.1105/tpc.15.01056
6. Hejátko J, Blilou I, Brewer PB et al (2006) In situ hybridization technique for mRNA detection in whole mount Arabidopsis samples. Nat Protoc 1:1939–1946. doi:10.1038/nprot.2006.333
7. Jackson D (1991) In situ hybridization in plants. In: Bowles DJ, Gurr SJ, McPherson M (eds) Molecular plant pathology: a pract approach, series 1. Oxford University Press, Oxford, pp 163–174
8. Bleckmann A, Dresselhaus T (2016) Fluorescent whole-mount RNA in situ hybridization (F-WISH) in plant germ cells and the fertilized ovule. Methods 98:66–73. doi:10.1016/j.ymeth.2015.10.019

Chapter 14

Analysis of Peroxisome Biogenesis in Pollen by Confocal Microscopy and Transmission Electron Microscopy

Peng-Fei Jia, Hong-Ju Li, and Wei-Cai Yang

Abstract

Peroxisome is an essential single-membrane bound organelle in most eukaryotic cells and functions in diverse cellular processes. De novo formation, division, and turnover of peroxisomes contribute to its biogenesis, morphology, and population regulation. In plants, peroxisome plays multiple roles, including metabolism, development, and stress response. Defective peroxisome biogenesis and development retard plant growth, adaption, and reproduction. Through tracing the subcellular localization of fluorescent reporter tagged matrix protein of peroxisome, fluorescence microscopy is a reliable and fast way to detect peroxisome biogenesis. Further fine-structural observation of peroxisome by TEM enables researchers to observe the detailed ultrastructure of its morphology and spatial contact with other organelles. Pollen grain is a specialized structure where two small sperm cells are enclosed in the cytoplasm of a large vegetative cell. Two features make pollen grain a good system to study peroxisome biogenesis: indispensable requirement of peroxisome for germination on the stigma and homogeneity. Here, we describe the methods of studying peroxisome biogenesis in Arabidopsis pollen grains by fluorescent live-imaging with confocal laser scanning microscopy (CLSM) and by DAB-staining based transmission electron microscopy (TEM).

Key words Pollen, Peroxisome biogenesis, Live-imaging, CLSM, TEM

1 Introduction

Peroxisome is a kind of round or oval-shaped organelle that is surrounded by a single lipid bilayer. Diverse metabolic activities are compartmentalized in this organelle, including fatty acid β-oxidation, hormone production, and photorespiration in plants [1]. Peroxisome is the major organelle where catalase resides, thus catalase is often used as a molecular marker for peroxisome in eukaryotes. Catalase is an enzyme that catalyzes the decomposition of hydrogen peroxide to water and oxygen. Catalase can also catalyze the oxidation, by H_2O_2, of various metabolites and 3,3′-diaminobenzidine (DAB) [2]. Therefore, peroxisome can be visualized by DAB staining for its catalase activity.

Anja Schmidt (ed.), *Plant Germline Development: Methods and Protocols*, Methods in Molecular Biology, vol. 1669, DOI 10.1007/978-1-4939-7286-9_14, © Springer Science+Business Media LLC 2017

De novo biogenesis and proliferation by division contribute to peroxisome biogenesis which includes peroxisome membrane protein transport, matrix protein import, and peroxisome division [3]. The peroxisome matrix proteins usually contain the importing signals, known as PTS1 and PTS2. Therefore, PTS1- or PTS2-tagged fluorescent proteins are often used to label peroxisomes for confocal microscopy. In mutants defective in peroxisome biogenesis, the PTS1- and PTS2-tagged fluorescent reporters are distributed in the cytosol instead of inside the peroxisome, and furthermore, no intact peroxisomes are detected by DAB-staining-based TEM studies. Thus combining CLSM and TEM strategies, the peroxisome biogenesis, morphology, turnover, and its contact with other organelles can be studied [4–6]. Here, we described the methods of CLSM and TEM to observe the peroxisomes using heterozygous *dayu/DAYU* mutant (*dau/DAU* for short) [6] as an example, in which half pollen grains are defective in germination due to defective peroxisome biogenesis. Lack of peroxisome leads to pollen sterility and embryo lethality; thus lack-of-peroxisome mutations are maintained as heterozygotes. For easier statistical analysis, *dau* mutation is introduced into the *quartet 1* (*qrt*) background which generates four attached functional tetrad microspores (2 *dau* and 2 wild-type pollen) [7].

2 Materials

2.1 Plant Materials, Reagents, and Material for Microscopy

1. *Arabidopsis thaliana* ecotype Landsberg *erecta* (L*er*) with either *DAU/DAU qrt/qrt* or *dau/DAU qrt/qrt* genotype [6, 7].
2. Zeiss LSM 780 laser scanning microscope with 10×, 20×, 40×, and 100× objectives.
3. Leica UC6 microtome.
4. JEM-1400 transmission electron microscope (JEOL, 80KV).
5. 0.2 M cacodylate buffer (pH 7.2, adjusted with NaOH).
6. Fixative solution: 2.5% (w/v) glutaraldehyde in 0.1 M cacodylate buffer (pH 7.2) (*see* **Note 1**).
7. 1% (w/v) osmium tetroxide (OsO_4) (*see* **Note 2**).
8. 0.2% DAB (3,3′-diaminobenzidine tetrahydrochloride) (Sigma-Aldrich) (*see* **Note 3**): 0.2 g DAB and 57 μL H_2O_2 in 100 mL 50 mM Tris–HCl, pH 9 (pH adjusted with NaOH).
9. EMbed 812 resin (Electron Microscopy Sciences, Hatfield) (*see* **Note 4**): 12 mL EMbed 812, 8 mL DDSA (Dodecenyl succinic anhydride), 7 mL NMA (Methyl nadic anhydride), 1.5–2% DMP-30 [2,4,6-tri (dimethylaminomethyl) phenol] (*see* **Note 5**).
10. Ethanol.

11. Propylene oxide.
12. 1% (w/v) uranyl acetate.
13. 0.1–0.4% (w/v) lead citrate.
14. Coverslip (no. 1, or thinner than 0.17 μm).
15. 100-mesh copper grids.
16. Formvar membrane.

3 Methods

3.1 Plant Transformation

1. Grow Arabidopsis of *DAU/DAU qrt/qrt* or *dau/DAU qrt/qrt* genotypes, respectively, grown in green house under a 16 h–light/8 h–dark cycle at 22 °C.
2. The pollen-specific expression construct *LAT52:mCherry-PTS1* can be transformed with either vacuum infiltration or floral dip method into *dau/DAU qrt/qrt* and *DAU/DAU qrt/qrt* plants, respectively. PTS1, in frame with the three amino-acid-residue sequence SKL, is fused C-terminally to the fluorescent reporter mCherry.
3. Transgenic plants are selected by growing on the MS media supplemented with the corresponding antibiotics. The antibiotic resistant seedlings are transferred to a new MS media-containing plate for further antibiotic-resistance confirmation and let grown for two more days before being transferred to the soil in the green house.
4. Use transgenic plants with healthy flowers for further analysis.

3.2 CLSM Assay for Peroxisome Matrix Protein Import

1. Choose open flowers at 10–12 o'clock in the morning.
2. Pick up dehiscent anthers with tweezers.
3. Dip the anthers several times on a drop of water on the slide to release pollen grains under a stereoscopic microscope.
4. Cover the water drop containing pollen grains with a coverslip (*see* **Note 6**).
5. The pollen grains are first visualized with a Zeiss LSM 780 laser scanning microscope with 10× objective to locate the grains. Laser scanning microscopes from other manufacturers are also applicable.
6. Then, view the pollen grains at either 63× or 100× oil immersion objectives (1.4 NA). The mCherry fluorescent signals can be detected with filter set (excitation at 561 nm and emission between 570 and 650 nm). Pinhole is adjusted to 1 Airy unit for each image. Image acquisition and processing are performed according to the manufacturer's instructions.

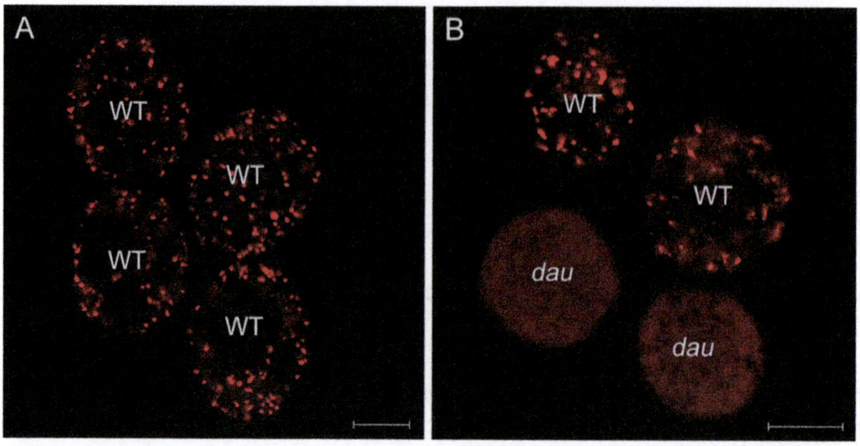

Fig. 1 Tetrad pollen expressing mCherry-PTS1. (**a**) Tetrad pollen from *DAU/DAU qrt/qrt* (WT) plants expressing *mCherry-PTS1*. The fluorescent puncta are peroxisomes. (**b**) Tetrad pollen from *dau/DAU qrt/qrt* plants expressing *mCherry-PTS1*. The pollen grains with punctate fluorescence are WT (*DAU qrt*) and the other two grains with dispersed mCherry fluorescence are *dau* pollen defective in peroxisome biogenesis or peroxisome import. Bar, 10 μm

7. mCherry-PST1 signal in *DAU qrt* pollen appears as puncta that are indicative of peroxisomes (Fig. 1a), while mCherry-PST1 signal in *dau qrt* mutant pollen appears dispersed in the cytosol (Fig. 1b). Similarly, in other mutant pollen lacking peroxisomes, mCherry-PST1 displays similar distribution pattern in the cytosol [4–6].
8. Based on this localization pattern of mCherry-PTS1 by CLSM, it can be concluded that either mCherry-PTS1 is not imported into peroxisome in mutant pollen grains as that of the wild-type, or peroxisome biogenesis is impaired in the mutant. Thus, a further conclusion on peroxisome biogenesis can be evidenced by the ultrastructural imaging by TEM (*see* **Note 7**).

3.3 TEM Assay on the Ultrastructure of Peroxisomes

1. Collect 20 anthers at stage 12 or 13 (Fig. 2) with tweezers and immerse them immediately into the fixative solution [2.5% glutaraldehyde in 0.1 M cacodylate buffer (pH 7.2)] and vacuum-infiltrated for 15–30 min at room temperature. Then the samples are incubated in fresh fixative for 2 h or overnight at 4 °C.
2. Remove the fixative with a pipette, and wash the samples with 1 mL 0.1 M cacodylate buffer for four times, 30 min each.
3. Replace the buffer with 0.2% 3, 3′-diaminobenzidine tetrahydrochloride (DAB) solution (containing H_2O_2) and incubate for 1–2 h in the dark at room temperature. Here, the DAB was oxidized in peroxisomes by the catalase in the presence of H_2O_2 into a water-insoluble brown-colored substance [2, 8–10].

Fig. 2 Inflorescence and flower of Arabidopsis. (**a**) *Arrow*, flower at stage 12; *arrowhead*, flower at stage 13. (**b**) Flower at stage 13. *Arrow*, anther

4. Wash the samples with 1 mL 0.1 M cacodylate buffer (pH 7.2) for four times (30 min each).
5. Discard the supernatant and add 0.5 mL 1% osmium tetroxide to the Eppendorf tubes to fix the samples for 2 h in ventilation at room temperature. (OsO_4 is extremely volatile and toxic. **Steps 5–6** should be strictly handled with care in a ventilated hood).
6. Wash the samples with 1 mL distilled water four times, 30 min each.
7. Dehydrate the samples by passing through a graded ethanol series as follows: distilled water, 10, 20, 30, 40, 50, 60, 70, 80, 90, 95%, with 15–20 min each. Percentages are given as (v/v).
8. The samples are further dehydrated three times in 100% ethanol (make sure that the ethanol used is 100%) with 20 min each.
9. Substitute 100% ethanol with propylene oxide by incubating the samples in 100% propylene oxide twice with 30 min each.
10. Substitute the propylene oxide by incubating the samples sequentially in a mixture of propylene oxide and EMbed 812 resin, namely 3:1 (V/V) for 2 h, 1:1 for 2 h, 1:3 for 2 h at room temperature.
11. Infiltrate the sample four more times in EMbed 812 resin, each step for 8 h at room temperature. Vacuum-infiltration is not necessary here.
12. Transfer the samples to a plastic mold and orientate them as required or into a capsule. Fill up the mold with EMbed 812 resin and leave it to polymerize sequentially at 35 °C for 12 h, at 45 °C for 12 h, and at 60 °C for 24 h.
13. Cut the samples with Leica UC6 microtome mounted with a diamond knife or glass knife, into about 60-nm thick sections

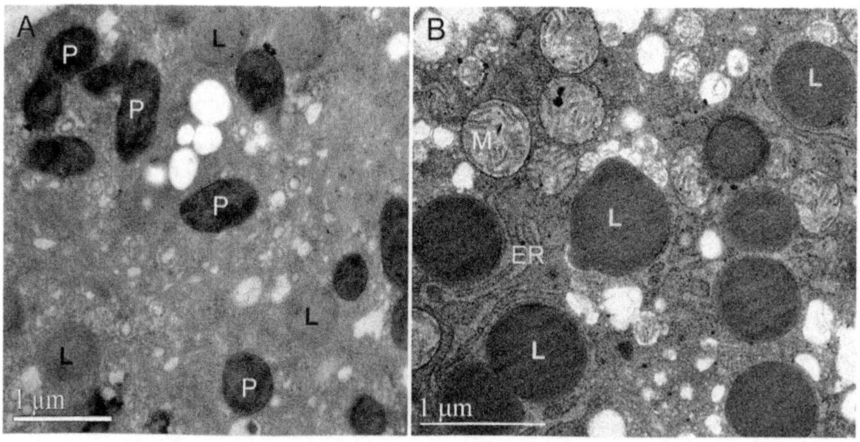

Fig. 3 TEM of wild-type and *dau* pollen. (**a**) TEM micrograph showing the peroxisomes (P) stained by DAB. Note the lipid bodies are often wrapped by ER cisternae; (**b**) TEM showing the heavily stained lipid body, while no visible peroxisomes are detected. *L* lipid bodies, *ER* endoplasmic reticulum, *M* mitochondrion. Bar, 1 μm

(silver to gray color), and collect the sections on the 100-mesh copper grids covered by formvar membrane. Microtomes from other manufacturers are also applicable.

14. Stain the sections with uranyl acetate by putting the copper grid with the section side down onto a drop of 1% uranyl acetate in a petri dish, and keep in the dark for 30 min.
15. Gently rinse the grids three times with distilled water and handle them as in **step 14**.
16. Stain the grids with 0.1–0.4% lead citrate for 5 min. CO_2 content should be kept minimum by placing some NaOH powder in the petri dish (*see* **Note 8**).
17. The copper grids are air dried and examined using a JEM-1400 transmission electron microscope (JEOL, 80KV).
18. The images of TEM show that the *DAU qrt* pollen grains contain heavily DAB-stained peroxisomes, while the *dau qrt* mutant pollen grains lack peroxisomes (Fig. 3). It should be noticed that the lipid bodies are heavily stained with osmium tetroxide in cells lacking peroxisomes due to the disrupted lipid metabolism. Through this DAB-staining-based TEM assay, the ontogeny, development, and morphology of peroxisomes can be observed [6] (*see* **Note 9**).

4 Notes

1. The fixative must be freshly prepared just before use.
2. Osmium tetraoxide can be prepared by putting the OsO_4-containing ampoule into a bottle with distilled water a week

prior to use as it is hard to dissolve in water. Osmium solution is highly toxic and volatile and must be stored in the dark and at a safe place. During samples fixation, all the operations must be carried out in a ventilated hood and care needs to be taken to avoid any contact to your skin.

3. DAB is hard to solubilize in alkaline buffer, thus a stock solution in Tris–HCl (pH 9) should be prepared in a capped tube before use. It is helpful for its solubilization to shake the tube at 37 °C for at least 1 h. DAB should be kept in the dark by wrapping the whole tube with aluminum foil during preparation. The solubilized DAB is colorless or slight pink and should be freshly prepared.
4. Use only freshly prepared embedding medium. The components of EMbed 812 resin are mixed thoroughly except for DMP-30 and stored at 4 °C for more than 24 h in the dark to release possible air bubbles. The proportion of resin components should be adjusted according to the local environmental conditions (temperature and humidity) for best ultrathin sectioning according to the manufacturer's recommendation (http://www.emsdiasum.com).
5. All the components of EMbed 812 resin (EMbed 812 resin kit, Electron Microscopy Sciences) can be mixed at a time. Usually, we add accelerator, 1.5–2% DMP-30 (2,4,6-tri-(dimethylaminemethyl)-phenol) to the 100% resin for the last two changes in Subheading 3.3, **steps 11** and **12**.
6. Care should be taken when covering the slip on the pollen, because pollen burst easily under pressure.
7. In tetrad pollen from *DAU/dau qrt/qrt* plants, 50% pollen grains harbor peroxisomes, while the remaining do not. Thus, the two kinds of pollen grains are observed in TEM sections.
8. NaOH powder is placed in the periphery of the petri dish, and kept at some distance from the drops of lead citrate and grids.
9. This described CLSM and TEM assay of peroxisome can also be used to analyze or identify mutants with defective matrix protein turnover, peroxisome division, or morphological defect or contact with other organelles. In the mutants with defective matrix protein turnover, it is characterized by enlarged mCherry-PTS1 labeled peroxisomes and leak of mCherry-PTS1 to the cytosol by CLSM. The cells with morphological defect, excessive, or abolished contact with organelles such as ER or chloroplasts can be identified by both CLSM and TEM.

Acknowledgment

This work is supported by the grant from the National Natural Science Foundation of China (31571385, 2013CB945103 and 31330053).

References

1. Reumann S, Bartel B (2016) Plant peroxisomes: recent discoveries in functional complexity, organelle homeostasis, and morphological dynamics. Curr Opin Plant Biol 34:17–26
2. Fahimi HD (1969) Cytochemical localization of peroxidatic activity of catalase in rat hepatic microbodies (peroxisomes). J Cell Biol 43:275–288
3. Baker A, Paudyal R (2014) The life of the peroxisome: from birth to death. Curr Opin Plant Biol 22:39–47
4. Farmer LM, Rinaldi MA, Young PG, Danan CH, Burkhart SE, Bartel B (2013) Disrupting autophagy restores peroxisome function to an Arabidopsis *lon2* mutant and reveals a role for the LON2 protease in peroxisomal matrix protein degradation. Plant Cell 25:4085–4100
5. Schumann U, Prestele J, O'Geen H, Brueggeman R, Wanner G, Gietl C (2007) Requirement of the C_3HC_4 zinc RING finger of the Arabidopsis *PEX10* for photorespiration and leaf peroxisome contact with chloroplasts. Proc Natl Acad Sci U S A 104:1069–1074
6. Li XR, Li HJ, Yuan L, Liu M, Shi DQ, Liu J, Yang WC (2014) Arabidopsis DAYU/ABERRANT PEROXISOME MORPHOLOGY9 is a key regulator of peroxisome biogenesis and plays critical roles during pollen maturation and germination *in planta*. Plant Cell 26:619–635
7. Copenhaver GP, Keith KC, Preuss D (2000) Tetrad analysis in higher plants. A budding technology. Plant Physiol 124:7–16
8. Vigil EL (1969) Intracellular localization of catalase (peroxidatic) activity in plant microbodies. J Histochem Cytochem 17:425–428
9. Vigil EL (1970) Cytochemical and developmental changes in microbodies (glyoxysomes) and related organelles of castor bean endosperm. J Cell Biol 46:435–454
10. Wanner G, Vigil EL, Theimer RR (1982) Ontogeny of microbodies (glyoxysomes) in cotyledons of dark-grown watermelon (*Citrullus vulgaris* Schrad.) seedlings. Planta 156:314–325

Chapter 15

Transmission Electron Microscopy (TEM) to Study Histology of Pollen and Pollen Tubes

Peng-Fei Jia, Hong-Ju Li, and Wei-Cai Yang

Abstract

Here, we describe methods of transmission electron microscopy (TEM) based on conventional chemical fixation and high-pressure freezing (HPF) and freeze-substitution (FS) to examine the ultrastructure of *Arabidopsis* pollen grains and pollen tubes. Compared to other plant samples, such as root or leaf, pollen grains have thick pollen coat and cell wall which limit the permeation of fixative. Thus, it is difficult to obtain high-quality ultrastructural images of pollen. Moreover, pollen tube is very soft and the traditional procedure is too harsh to get an intact pollen tube sample. Up to now, there is no available mature protocol for TEM sample preparation of *Arabidopsis* pollen tube. Here, we describe the details and step-by-step procedures of chemical fixation, HPF, FS, and resin-embedding protocols for *Arabidopsis* pollen and pollen tube. In addition, we also provide a method on how to get longitudinal sections of pollen tubes.

Key words Pollen, Pollen tube, Transmission electron microscopy

1 Introduction

In the past few decades, *Arabidopsis thaliana*, a small species of the Cruciferae, has been extensively studied as one of the primary model organisms in diverse fields of plant science. In flowering plants, sexual reproduction requires delivery of immobile sperms to the female gametophyte through growing pollen tubes. So to study *Arabidopsis* pollen and pollen tube will contribute to understanding of sexual reproduction. Strategies including mutant screening, genetic analysis, molecular and cellular biology are commonly used to study pollen and pollen tubes. Transmission electron microscopy (TEM) is a microscopy technique capable of imaging at a higher resolution than light microscopy based on ultra-thin specimen preparation. With TEM, ultrastructural characteristics of pollen and pollen tubes can be visualized. *Arabidopsis* pollen possess a hard thick pollen coat outside of the cell wall, while the germinated pollen tube is relatively soft. And the small size of pollen and pollen tubes (diameter: 10–20 μm for pollen and 3–5 μm for pollen tube)

Anja Schmidt (ed.), *Plant Germline Development: Methods and Protocols*, Methods in Molecular Biology, vol. 1669, DOI 10.1007/978-1-4939-7286-9_15, © Springer Science+Business Media LLC 2017

make them difficult to manipulate. These features also obstruct obtaining high-quality ultrastructural samples. The main difficulties are sample fixation and infiltration of pollen as well as positioning for longitudinal pollen tube section. In this chapter, we analyze the advantages and disadvantages between chemical fixation and HFP/FS [1], and provide a reliable method for TEM on *Arabidopsis* pollen and pollen tube.

2 Materials

2.1 Plant Material and Pollen Germination Medium

1. Mature *Arabidopsis* pollens can be isolated from freshly opened flowers.
2. Developing anthers can be obtained from unopened flower buds with protruded white petals.
3. Pollen tubes can be obtained from cut stigmas on which the pollen tubes grow out and penetrate the solid pollen germination medium.
4. Pollen germination medium: 1 mM $CaCl_2$, 5 mM $Ca(NO_3)_2$, 1 mM $MgSO_4$, 5 mM KCl, 0.01% (v/v) boric acid, 10% (w/v) sucrose, pH 7.5 adjusted with 0.1 M NaOH, 1% (w/v) agarose (melted in >80 °C hot water).

2.2 Reagents

1. Agarose, low melting point.
2. Acetone.
3. Cacodylate buffer: 0.1 M sodium cacodylate in water (pH 7.2, adjusted with NaOH).
4. EMbed 812 resin kit.
5. Glutaraldehyde, 50% in water.
6. Lead citrate, 0.1–0.4% (w/v) in distilled water (*see* **Note 1**).
7. 4% (w/v) Osmium tetroxide.
8. Uranyl acetate.
9. 0.1–0.15 M sucrose solution as cryoprotectant.

2.3 Fixatives and Resin

1. Prefixation fixative: 2.5% (v/v) glutaraldehyde, 4% (w/v) paraformaldehyde in 0.1 M cacodylate buffer (pH 7.2) (*see* **Note 2**).
2. Postfixation fixative: 1% (w/v) osmium tetroxide (*see* **Note 3**).
3. EMbed 812 resin (*see* **Note 4**): Components: EMbed 812, Dodecenyl succinic anhydride (DDSA), Methyl nadic anhydride (MNA), 2,4,6-tri (dimethylaminomethyl) phenol (DMP-30); 12 mL EMbed 812, 8 mL DDSA, 7 mL NMA, 1.5–2% (v/v) DMP-30 (*see* **Note 5**).

2.4 Embedding and Sectioning

1. Flat embedding mold.
2. Planchelette.
3. High pressure freezer, e.g., Leica EM ICE.
4. Cryovials.
5. Diamond knife or glass knife.
6. Leica UC7 microtome.
7. 100 mesh copper grids.
8. Formvar membrane.
9. JEM1400 TEM (80 kV) (*see* **Note 6**).

3 Methods

All procedures are carried out at room temperature unless otherwise specified.

3.1 Sample Preparation of Pollen and Pollen Tubes

Steps 1–6 describe the prefixation and pre-embedding of pollen, subsequent **steps 7–11** the prefixation and pre-embedding of pollen tubes. **Step 13** onward applies for pollen and pollen tubes.

3.1.1 Chemical Fixation Method

1. Prefixation for pollen: Mature pollen grains are immersed in the Prefixation fixative in a 1.5 mL microcentrifuge tube for 1 h.
2. For pollen pre-embedding in 1% agarose, centrifuge the pollen grains at 4000 × *g* for 5 min and discard the supernatant.
3. Wash the pellet four times by resuspending the pollen grains in 1 mL of 0.1 M cacodylate buffer and centrifuge again.
4. Discard the supernatant (about 50 μL pollen grains left), and then add 50 mL 2% agarose at 37 °C and mix gently at room temperature.
5. Place the microcentrifuge tube immediately on ice to solidify the agarose.
6. Cut the solidified agarose in which the pollen grains embedded to 1 mm × 1 mm blocks with a double-edge blade for subsequent fixation.
7. Prefixation for pollen tube: Cut stigmas after pollination and put them vertically on solid pollen germination medium plate to germinate for 1 h until pollen tubes penetrate into the style. Add the Prefixation fixative to the plates to immerse the stigma.
8. Wash the stigmas four times with 0.1 M cacodylate buffer. Add cacodylate buffer to the plates to immerse the stigma for 10 min, and then change for fresh cacodylate buffer.
9. For pollen tube pre-embedding: put the stigmas with pollen tubes into 1% agarose at 37 °C.

10. Solidify immediately on ice.
11. Shape the solidified agarose into small blocks, ensure stigma localized in the center of sample block.
12. Transfer the agarose block containing pollen tubes to 2 mL microcentrifuge tubes, and add 1 mL the prefixation fixative and fix for 2 h or overnight at 4 °C.
13. Wash the blocks four times with 0.1 M cacodylate buffer, each time for 30 min.
14. Postfix the blocks in 1% osmium tetraoxide for 2 h.
15. Wash the blocks four times in distilled water, each time for 30 min.
16. The whole blocks are stained in 1% uranyl acetate solution for 1 h, and subsequently washed four times in distilled water, each time for 30 min.
17. Dehydrate the blocks by sequentially passing through 30, 50, 70, 90, 95%, and three times 100% acetone, each time for 15 min (for pollen) (*see* **Note 7**) or 8–10 min (for pollen tube) (*see* **Note 8**).
18. Infiltrate the blocks by passing through a progressive series of EMbed 812 resin to acetone dilutions, acetone to resin (v/v) 3:1 for 2 h, 1:1 for 2 h, 1:3 for 2 h, pure resin for three times (8 h each time).
19. Transfer the blocks to flat embedding mold and fill in the mold with EMbed 812 resin, then orientate the block as required (*see* **Note 9**).
20. Polymerize at 35 °C for 12 h, at 45 °C for 12 h, at 60 °C for 24 h (*see* **Note 10**).

3.1.2 High-Pressure Freezing (HPF) Method

1. Use a double-edge blade to incise a slit on the surface of anthers under stereomicroscope before freezing to facilitate subsequent freeze substitution (FS).
2. Place the anthers into the planchette filled with cryoprotectant/filler (*see* **Note 11**). For anther samples, we commonly use 0.1–0.15 M sucrose solution as cryoprotectant. We use the 0.3 mm-deep well of planchette as a carrier for anthers.
3. Cover the loaded planchette as shown in Fig. 1.
4. Insert the planchette into the high pressure freezer (Leica EM ICE) to perform the HPF process.
5. Transfer the planchettes with the frozen samples to cryovials containing 1% osmium tetraoxide in anhydrous acetone immersed in liquid nitrogen (*see* **Note 12**).

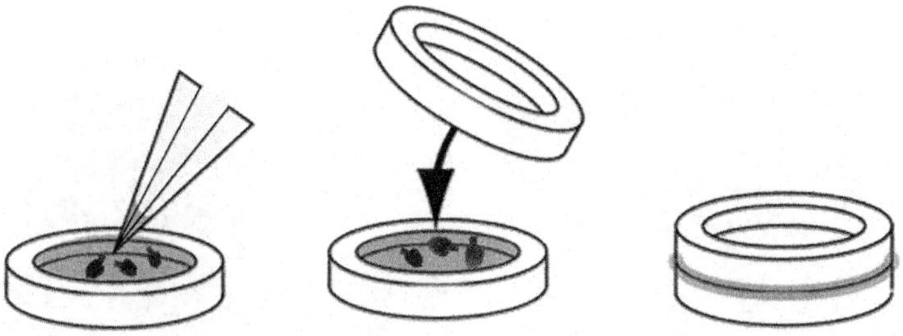

Fig. 1 Loading anther samples into HPF planchette

3.1.3 Freeze Substitution (FS) Method

FS is the process that frozen water in the samples is replaced with acetone plus fixatives. The FS protocol for anther samples is described in the following steps (*see* **Note 13**).

1. Incubate the fixed samples at −90 °C for 72 h.
2. Warm the samples gradually from −90 to −60 °C for 6 h, 5 °C rise per hour.
3. Incubate samples at −60 °C for 24 h.
4. Warm samples gradually from −60 to −30 °C for 6 h, 5 °C rise per 1 h.
5. Incubate samples at −30 °C for 12 h.
6. Gradually warm samples from −30 to 0 °C for 6 h, 5 °C rise per 1 h.
7. Incubate samples at 0 °C for 2 h.
8. Gradually warm samples from 0 to 25 °C for 5 h, 5 °C rise per 1 h.
9. Separate the samples from the planchettes if they are still attached to the planchettes, and then wash four times with anhydrous acetone at room temperature, each time for 30 min.

3.1.4 Infiltration, Embedding, and Polymerization

1. Infiltrate the samples by passing through a progressive series of EMbed 812 resin to acetone dilutions, namely acetone to resin (v/v) 3:1 for 4 h, 1:1 for 4 h, 1:3 for 4 h, and 100% resin for four times with each time for 8 h.
2. For embedding, transfer the samples into a flat embedding mold and orientate them as required, fill in the mold with 100% EMbed 812 resin.
3. Transfer the mold to an incubator and let polymerize at 35 °C for 12 h, at 45 °C for 12 h, at 60 °C for 24 h.

3.1.5 Ultrathin Section

1. Semi-thin section positioning: Confirm the position of longitudinal section of pollen tube or cross-section of pollen grains through a 2 μm thick semi-thin section, as in Fig. 2.

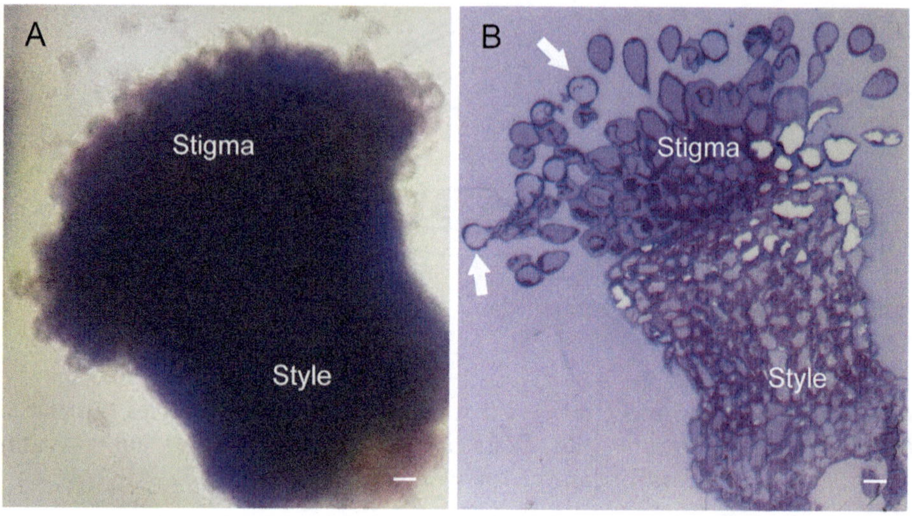

Fig. 2 Semi-thin section. (**a**) Positioning the longitudinal section of stigma in the sample block. (**b**) 2 μm semi-thin longitudinal section of stigma. *Arrow*, pollen. Bar, 20 μm

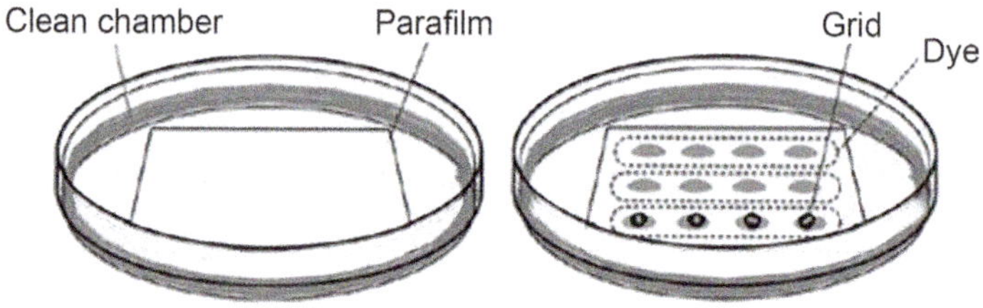

Fig. 3 A chamber for staining grids

2. Once reaching the right cutting plane, cut the sample with diamond knife or glass knife mounted on a Leica UC7 microtome into 60 nm-thick ultrathin sections (gray or white color).
3. Pick up the sections with 100-mesh copper grids covered formvar membrane.

3.1.6 Staining

Grids are routinely stained using the following methods described in **steps 1–6**.

1. Place a piece of clean parafilm into a petri dish as shown in Fig. 3, and put a drop of 1% uranyl acetate solution in distilled water on the parafilm.
2. Place the grid with the section side facing the dye and stain for 30 min in the dark.
3. Dip the grid vertically up and down in three beakers filling with distilled water to remove the dye.
4. Subsequently, the grid is stained in 0.1–0.4% lead citrate in distilled water for 5 min as in **step 2** (*see* **Note 14**).

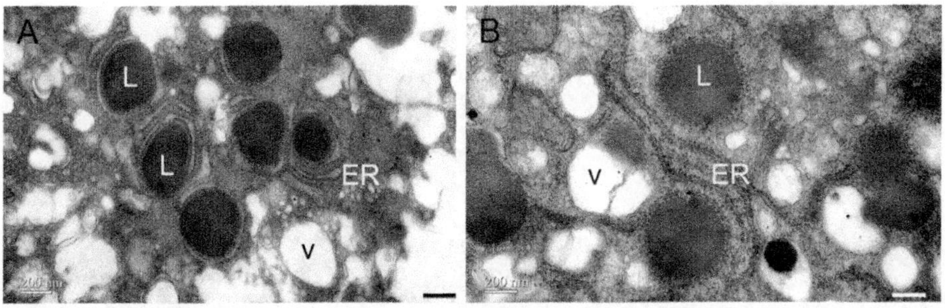

Fig. 4 Pollen ultrastructure by chemical fixation and HPF/FS. (**a**) Pollen ultrastructural image by chemical fixation. (**b**) Pollen ultrastructural image by HPF/FS. *L* lipid body, *ER* endoplasmic reticulum, *v* vesicles. Bar, 200 nm

5. Repeat **step 3** to remove the dye.
6. Remove the water left with a piece of filter paper, and air dry the grid. The stained grids can be stored in a dry container.

3.2 Imaging of TEM

1. The grids are examined using JEM1400 TEM (80 kV) and the ultra-structural images are obtained. For pollen grains, our experience shows that the ultrastructure of organelles is preserved better by the HPF/FS procedure than the chemical fixation (Fig. 4).
2. However, several problems occur when using HPF/FS. First, ice crystals easily form and damage the ultrastructure of plant cells with high water content. Second, high pressure in HPF leads to dramatic deforming of some soft samples such as pollen tubes. So for the preparation of TEM sample of pollen tubes, the HPF/FS methods are not appropriate, because high pressure will cause pollen tube burst. We prepare the pollen tube TEM samples by conventional chemical fixation. Typical results are shown in Fig. 5. The key steps of this method include placing of sample during embedding and accurate positioning of longitudinal pollen tube section during sectioning.

4 Notes

1. Lead citrate solution must be freshly prepared and is tightly sealed when stored at 4 °C.
2. The fixative must be freshly prepared just before use.
3. Osmium tetroxide is often provided with 4% solution or powder sealed in an ampoule. In case of powder, it must be prepared a week prior to use because it dissolves very slow in water. Osmium solution is highly toxic and volatile and must be stored in the dark and at a safe place. During sample fixation, all operations must be carried out in ventilation. Be careful not to

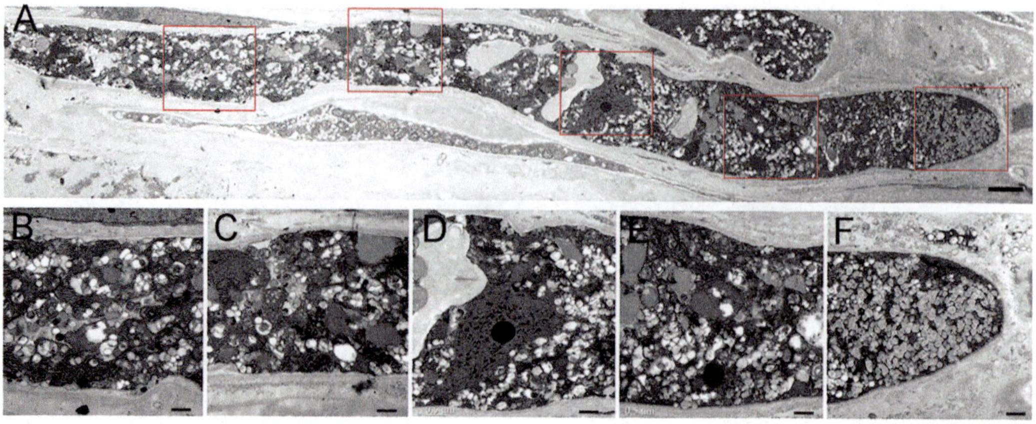

Fig. 5 TEM images of a pollen tube growing in the style. (**a**) A pollen tube growing in the style. From *left* to *right* corresponds to the basal to apical region. Bar, 2 μm. (**b–f**) Enlarged images of corresponding *boxes* shown in **a**. Note the abundant vesicles in the apical region (**f**). Bar, 0.5 μm

contact your skin. In case of powder supplied, it is advisable to cut the ampoule and immediately put it into a brown bottle with required amount of water and let it dissolve.

4. A freshly prepared embedding medium is used. The components of EMbed 812 resin are mixed thoroughly except for DMP-30, stored at 4 °C in the dark for more than 24 h to release the air bubbles in the resin. The ratio of resin components should be adjusted for best sectioning operation according to the manufacturer's recommendation ([2]; http://www.emsdiasum.com).
5. It is a simple method to mix EMbed 812 resin at once. Usually, we add accelerator, 1.5–2% DMP-30 [2,4,6-tri-(dimethylaminemethyl) phenol] to the 100% resin for the last two resin changes before embedding.
6. Other types of TEM from other manufacturers are also applicable, and the voltage of TEM is usually from 60 to 100 kV. Higher voltage will improve the resolution of TEM images, but will possibly damage the sample on grid.
7. Total time in pure acetone should be less than 60 min for membrane integrity.
8. Compared to pollen, pollen tubes take less time to dehydrate.
9. Samples are placed with appropriate position to get the longitudinal section of pollen tube.
10. The polymerization of EMbed 812 resin is an aerobic process. We use plastic silica molds to embed the samples. Desiccative environment should be maintained during polymerization process (humidity less than 50%).

11. It is important to fill the carrier planchette completely with cryoprotectant/filler. Trapped air within the planchette will interfere with the freezing and collapse during pressurization which leads to sample deformation.
12. The cryovials containing samples are always immersed in liquid nitrogen to avoid formation of ice crystals.
13. This protocol is commonly used for anther samples, the incubation time for each temperature can be modified for different samples.
14. During staining, set up an environment without CO_2 through placing some NaOH powder. NaOH powder is placed on periphery of the petri dish, and should not touch the drops of lead citrate and grids.

Acknowledgment

This work is supported by the grant from National Natural Science Foundation of China (31571385 and 31330053).

References

1. Kang BH (2010) Electron microscopy and high-pressure freezing of Arabidopsis. Methods Cell Biol 96:259–283
2. Luft JH (1961) Improvements in epoxy resin embedding methods. J Biophys Biochem Cytol 9:409–414

Part IV

Isolation of Specific Cells of the Male and Female Germlines

Chapter 16

Isolation of Arabidopsis Pollen, Sperm Cells, and Vegetative Nuclei by Fluorescence-Activated Cell Sorting (FACS)

Mário R. Santos, Cláudia Bispo, and Jörg D. Becker

Abstract

Efficient methods to isolate highly purified *Arabidopsis thaliana* pollen and the subcellular components of the male gametophyte (the vegetative nucleus and two sperm cells) have enabled genome-scale studies revealing a highly dynamic reprogramming of the transcriptome and epigenome during pollen development. Here, we describe the isolation of uninucleate microspores, mature pollen, as well as sperm cells and vegetative nuclei by Fluorescence-Activated Cell Sorting.

Key words Pollen, Microspore, Sperm cell, Vegetative nucleus, Arabidopsis, FACS, Transcriptomics, Epigenetics

1 Introduction

Male gametophyte development in flowering plants is characterized by two consecutive cell divisions, transforming the product of meiosis, the uninucleate microspore, into a mature pollen grain harboring the vegetative nucleus and a pair of sperm cells. Pollen development is among the best-characterized processes at the transcriptomic level [1], which is partly due to the fact that mature pollen grains and even preceding stages of male gametophyte development are relatively easy to collect and purify, without the need to digest complex tissues [2, 3]. But when analyzing RNA isolated from mature pollen or growing pollen tubes, the relative contributions of the vegetative cell and of the male gametes cannot be distinguished. Separation of these subcellular components of the male gametophyte (vegetative nuclei (VN) and sperm cells (SC)) is therefore relevant not only for transcriptomeTranscriptomics analyses, but also for studies of epigenetic inheritance [4–6]. Methods range from isolation through micromanipulation [7], over centrifugation-based separation methods [8, 9] to purification based on Fluorescence-Activated Cell Sorting (FACS) [10–12].

Anja Schmidt (ed.), *Plant Germline Development: Methods and Protocols*, Methods in Molecular Biology, vol. 1669, DOI 10.1007/978-1-4939-7286-9_16, © Springer Science+Business Media LLC 2017

Here, we provide detailed protocols for isolation of highly purified uninucleate microspores, mature pollen, as well as VN and SC. Protocols start with collection of *Arabidopsis thaliana* flower buds and open flowers, respectively. Samples are grinded (for microspore isolation) or washed in buffer (for mature pollen), and subjected to purification by FACS. When isolating SC and VN, mature pollen grains are broken using glass beads. Separation by FACS can then be achieved using a fluorescent marker line with GFP and RFP labels in the SC and VN, respectively [10], or by using nucleic acid stains on non-marker lines. With modern library preparation protocols for Next-Generation Sequencing (NGS), requiring relatively low starting quantities of RNA or DNA, even down to the single cell level [13], our protocols provide sufficient yield for a range of NGS applications.

2 Materials

Use ultrapure or MiliQ grade water (sensitivity of 18.2 MΩ-cm at 25 °C) and analytical grade reagents stored at room temperature to prepare all buffers. Buffers are prepared at room temperature, filter sterilized using a vacuum filtration unit (250 ml, 0.22 μm pore size) and stored at 4 °C, unless otherwise mentioned (*see* **Note 1**). Follow waste disposal rules strictly.

2.1 Plant Material

1. *Arabidopsis thaliana*, Columbia (Col-0) ecotype.
2. Double homozygous transgenic marker line harboring *MGH3p::MGH3-eGFP* and *ACT11p::H2B-mRFP* constructs in Col-0 background (available from ABRC under stock number CS67829) [10].

2.2 Sample Preparation for FACS

1. Pollen Buffer (PB) (*see* **Note 2**): 5 mM $CaCl_2$, 1 mM $MgSO_4$, 5 mM KCl, 0.01% (w/v) H_3BO_3, 10% (w/v) Sucrose. Adjust pH to 7.5. The starting pH should be about 6.5. To adjust use approximately 50 μl of 0.1 M NaOH when preparing 20 ml of PB.
2. Sperm Cell buffer (SC buffer): 1.3 mM H_3BO_3, 3.6 mM $CaCl_2.2H_2O$, 0.74 mM KH_2PO_4, 438 mM Sucrose, 7 mM MOPS, 0.83 mM $MgSO_4.7H_2O$. Adjust pH to 6.0. The starting pH should be about 4.7. To adjust use approximately 300 μl of 1 M NaOH when preparing 250 ml of SC buffer.
3. Galbraith buffer (here called Sperm Nuclei buffer (SN buffer)): 45 mM $MgCl_2$, 30 mM $C_6H_5Na_3O_7.2H_2O$ (Trisodium citrate dihydrate), 20 mM MOPS, 0.1% (v/v) Triton X-100. Adjust pH to 7.0. The starting pH should be about 5.7. Adjust with about 500 μl of 5 M NaOH when preparing 250 ml of SN buffer.

4. PBS buffer: 150 mM NaCl, 2 mM KH_2PO_4, 8 mM Na_2HPO_4. Adjust pH to 7.2–7.4.
5. Dyes: 250 μM SYTOX® Orange Dead Cell Stain (in DMSO), for flow cytometry (Thermo Fisher Scientific). 10,000× SYBR Green I nucleic acid stain stock solution.
6. DAPI (4′,6-diamidino-2-phenylindole) for nucleic acid staining. DAPI working solution: 0.1 M sodium phosphate (pH 7), 1 mM EDTA, 0.1% (v/v) Triton X-100, 0.4 μg/ml DAPI.
7. Mesh: Miracloth; 41 μm, 28 μm, and 10 μm.
8. Beads: Glass beads, acid-washed, 425–600 μm.
9. Filters: CellTrics®, 10 μm mesh. These can be re-used and autoclaved (*see* **Note 3**).
10. Vortex: Lab minishaker for mixing.
11. Centrifuges: Centrifuge with swing-bucket rotor (for 50 ml conical tubes), microcentrifuge (for 1.5–2 ml microcentrifuge tubes), mini-centrifuge.

2.3 Fluorescence-Activated Cell Sorting

1. Instrumentation: Cell sorter equipped with a 488 nm laser used for scatter, green autofluorescence and GFP detection, a 561 nm laser for RFP detection and a 405 nm laser for DAPI detection in purity checks. For example, a MoFlo (Beckman Coulter) and a FACSAria I (Becton Dickinson) can be used.
2. Optical configuration: autofluorescence/GFP and RFP; detected using bandpass filters in the range 530/40 nm and 630/75 nm, respectively. For the combination of SYBR Green and SYTOX Orange dyes use 530/40 nm and 630/75 nm bandpass filters, respectively. Use the 450/40 nm bandpass filter for DAPI detection.
3. Sheath fluid: Use Phosphate Buffer Saline (PBS) as sheath fluid in all sorting experiments.

2.4 RNA Extraction

1. RNeasy Plant Mini Kit (QIAGEN).
2. DNaseI.

3 Methods

This chapter focuses largely on sample preparation prior to FACS. For FACS, only the gating strategy is briefly indicated (*see* **Note 4**). All procedures are carried out at room temperature and on the bench, unless otherwise specified. Buffers are kept on ice throughout the protocol. All centrifugation steps should be carried out at room temperature, unless indicated otherwise.

3.1 Isolation of Mature Pollen

Mature pollen grains can be isolated by FACS based on their strong autofluorescence caused by pollen coat phenolic components, and based on their shape. The method described here yields sufficient material for standard NGS approaches. Alternative pollen collection methods can be used [2], if higher yields are required, e.g., for proteomics analyses.

3.1.1 Preparation of Sample for FACS

1. Sow and grow plants on soil.
2. Collect flowers only from healthy plants that are bolting for at least 1.5 weeks (first elongated siliques formed) (*see* **Note 5**). Sample yields will decrease with plant age [14], with the best yields being obtained in weeks 2–4 after bolting.
3. Collect a volume of 5–15 ml volume fresh, day-open Arabidopsis flowers into a 50 ml conical centrifuge tube (*see* Fig. 1a) (*see* **Notes 6** and 7).
4. Add up to 2–3 times more volume of PB than flower volume collected (*see* Fig. 1b). The total volume of PB and flowers should not exceed 20 ml.
5. Vortex the flowers at 2500 rpm for 3 min at room temperature (*see* Fig. 1c).
6. Filter the crude fraction through a Miracloth mesh into a 50 ml conical centrifuge tube (*see* Fig. 2) (*see* **Note 8**).
7. Add an extra 0.5–1 ml of PB to the crude fraction on top of the Miracloth, ensuring that all the pollen goes through and is not retained in the Miracloth.

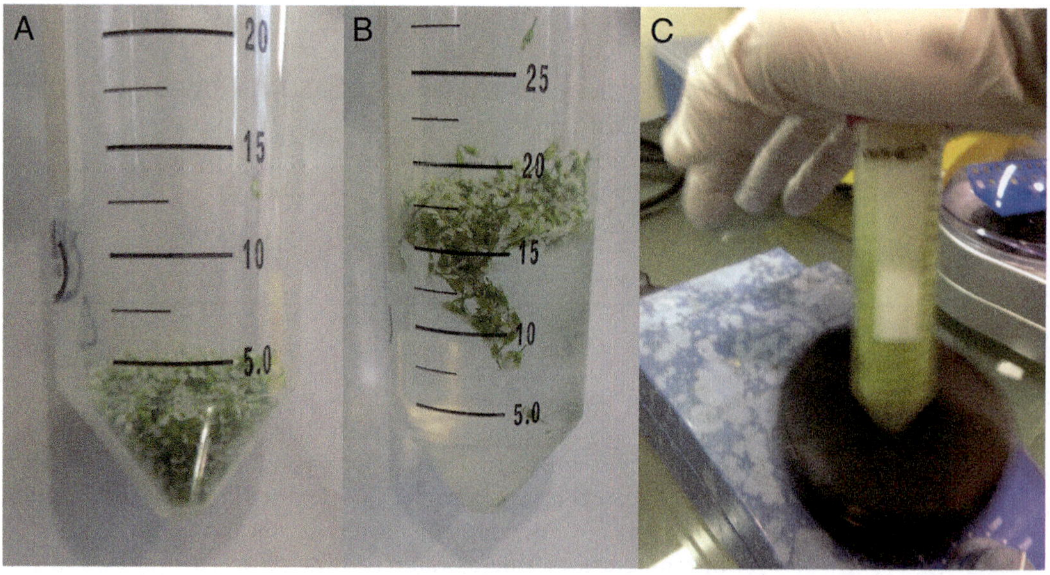

Fig. 1 (**a**) Collected flowers and (**b**) The corresponding volume of buffer added. (**c**) Flower vortex using a IKA MS2 minishaker mounted with a One-hand insert

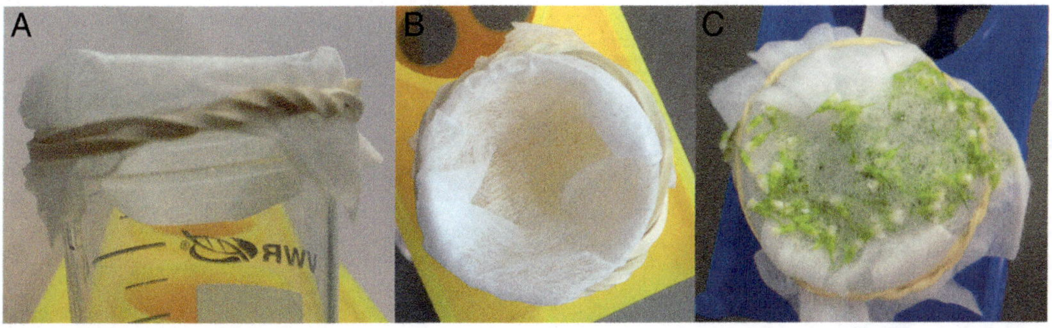

Fig. 2 (**a**, **b**) Miracloth mesh assembly on conical centrifuge tube and (**c**) Filtering of the crude fraction of vortexed flowers through the Miracloth mesh

8. Collect the flowers from the Miracloth mesh using tweezers and transfer them to the conical centrifuge tube used in **step 3**.
9. Repeat **steps 4–7**.
10. Remove the Miracloth mesh from the conical centrifuge tube.
11. Centrifuge the filtered solution from the two pollen extractions at 1620 × *g* for 3 min to pellet the pollen.
12. Carefully aspirate the supernatant by pipetting without disturbing the pellet. Leave about 5–7.5 ml of supernatant in the tube.
13. Wash the conical centrifuge tube walls using the leftover of the supernatant. This step ensures that all pollen is collected, and that it is concentrated (*see* **Note 9**).
14. Repeat **steps 11** and **12**, this time leaving only approximately 1 ml of supernatant in the tube.
15. Resuspend the pellet in the remaining volume and transfer to a 2 ml microcentrifuge tube.
16. Add 0.5–1 ml of fresh PB to the conical centrifuge tube, washing the walls and removing additional pollen; transfer to the 2 ml microcentrifuge tube (from **step 15**) containing the resuspended pollen.
17. Centrifuge at 5,900 × *g* for 1 min. Remove excess supernatant without disturbing the pellet.
18. Add 1 ml of fresh PB and resuspend the pellet (*see* **Note 9**).
19. Filter the sample through a 41 μm mesh mounted in a CellTrics® filtering system to a 2 ml microcentrifuge tube. Use a Mini-centrifuge (or a regular microcentrifuge at full speed) to spin down the pollen through the mesh excluding extra plant debris (*see* **Note 10**).
20. Add 200 μl of fresh PB to the mesh and centrifuge (*see* **Note 10**) to assure all pollen went through. The sample is ready for FACS (keep on ice until sorting).

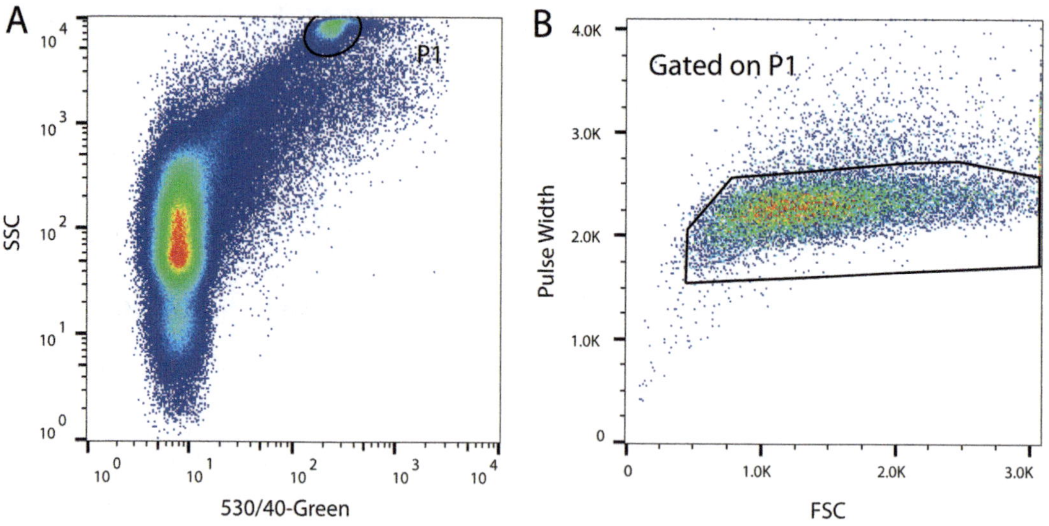

Fig. 3 Example of FACS gating for *Arabidopsis thaliana* pollen: (**a**) Pollen (population P1) selected by gating events with increased *green* autofluorescence. (**b**) Additional gating on population P1 will increase purity levels by removing pollen aggregates

3.1.2 Fluorescence-Activated Cell Sorting

1. Pollen can be purified from sample obtained in the previous step in PB running the sorter at a constant pressure of 207 kPa (30 psi) with a 100 μm nozzle and frequency of drop formation of between 30 and 40 kHz (*see* **Note 11**).
2. Determine the optimal excitation voltages using the prepared pollen sample, adjusting the scatter and fluorescence parameters to achieve a similar separation of pollen grains (population P1) from debris and other particles as seen in Fig. 3a (*see* **Note 4**).
3. Visualize the pollen population on an autofluorescence (530/40 nm) vs forward (FSC) or side scatter graph (SSC). Pollen grains (P1) are identified based on their strong green autofluorescence signal (*see* Fig. 3a).
4. Gate and sort for single cell pollen population based on FSC and Pulse Width (Fig. 3b).
5. Cells can be collected either in PB or directly into RNA or DNA extraction buffer, depending on downstream applications (adjust initial volume in collection tube, so that final ratio of sorted volume to buffer does not exceed 1:10 of the final volume). Maintain sorted cells at 4 °C for immediate processing (preferred) or snap freeze in liquid nitrogen for long-term storage at −80 °C.

3.1.3 RNA Extraction

For RNA extraction from sorted pollen (and microspores) an RNeasy Plant Mini Kit can be used, following the kit's "RNeasy Mini Protocol for Isolation of Total RNA from Plant Cells and

Tissues and Filamentous Fungi" and using the RLC buffer provided (*see* **Note 12**). All steps of the RNeasy protocol should be performed at room temperature and as quickly as possible. All centrifugation steps are performed at 20–25 °C in a standard microcentrifuge. For downstream applications requiring removal of genomic DNA (e.g., RT-PCR), on-column DNase I digestion is recommended.

1. Pre-chill a mortar and pestle using liquid nitrogen in a fume hood.
2. Centrifuge the sorted pollen sample at 5900 × *g* for 1 min using a microcentrifuge. Remove excess supernatant without disturbing the pellet.
3. Add 350 μl of RLC buffer and resuspend the pellet.
4. Add your sample to the cooled mortar (the buffer will freeze).
5. Carefully crush the frozen sample with the pestle and grind the resulting white powder for several minutes (*see* **Note 13**).
6. Allow buffer mixture to thaw and continue grinding.
7. Proceed to **step 4** of the kit's manual and finalize the preparation using the manufacturer's instructions.

3.2 Isolation of Uninucleate Microspores

Uninucleate microspores can be isolated based on their autofluorescence and size. The final sorted samples will contain a small fraction of early bicellular pollen (up to 3%), which are indistinguishable from uninucleate microspores with our method [10].

3.2.1 Preparation of Sample for FACS

1. Collect 8–10 inflorescences containing only closed flower buds, discard open flowers or developing siliques (*see* **Note 14**).
2. Gently grind the closed buds using mortar and pestle (*see* **Note 15**).
3. Add 1–2 ml of PB to the grinded tissue and gently grind it a bit more (*see* **Note 16**).
4. Add PB to a final volume of about 10 ml.
5. Filter the crude fraction through a Miracloth mesh into a 50 ml conical centrifuge tube (similar to Fig. 2) (*see* **Note 8**).
6. Add an extra 0.5–1 ml of PB buffer to the crude fraction on top of the Miracloth mesh, ensuring that all the microspores go through and are not retained in the Miracloth mesh.
7. Remove the Miracloth mesh from the conical centrifuge tube.
8. Centrifuge at 1 620 × *g* for 3 min. You will obtain a pellet, with a greenish upper layer and a yellowish bottom layer (enriched in microspores).

9. Remove most of the supernatant without disturbing the pellet. Leave about 1–2 ml of supernatant in the tube.
10. Filter through a 20 μm CellTrics® filtering system using a mini-centrifuge (or a regular microcentrifuge) to spin down microspores through the mesh and exclude extra plant debris and mature pollen (*see* **Note 10**).
11. Add 200 μl of fresh buffer to the mesh and centrifuge (*see* **Note 10**), ensuring all cells went through. The sample is ready for FACS (keep on ice until sorting).

3.2.2 Fluorescence-Activated Cell Sorting

1. Microspores can be purified in PB using the same setup and configuration as used for sorting pollen (constant pressure 207 kPa (30 psi) and 100 μm nozzle).
2. Determine the optimal excitation voltages using the prepared sample.
3. Visualize the "Microspores" on an autofluorescence (530/40 nm) vs forward (FSC) or side scatter graph (SSC). Gate for microspores (population P1) based on increased green autofluorescence of these cells (*see* Fig. 4a).
4. Within this population, gate and sort microspores based upon their characteristic smaller size captured by a diminished low FSC and time-of-flight (Pulse Width) (*see* Fig. 4b). This allows their differentiation from later stages of pollen development, like late bicellular pollen and tricellular pollen [10].

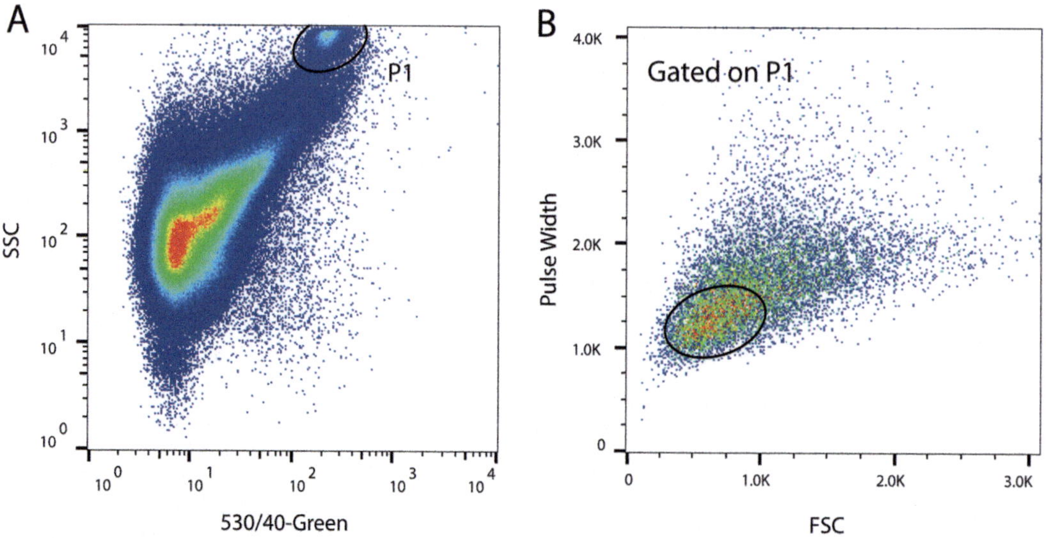

Fig. 4 Purification of microspores by FACS. (**a**) Microspores (population P1) roughly gated by events with brightest *green* autofluorescence. (**b**) Microspores are then selected by their characteristic smaller size, captured by a diminished low angle scatter (FSC) and time-of-flight (Pulse Width)

5. Similarly to when sorting pollen grains, microspores can be collected either in PB (maintaining sorted volume to 1:10 of PB in collection tube) or directly into RNA or other extraction buffer depending on experimental objective. Cells collected are maintained at 4 °C.

3.3 Isolation of Viable Sperm Cells and Vegetative Nuclei Using a Marker Line

While sorting of viable SC and VN is possible from wild-type or mutant plants (*see* Subheading 3.4), it is more efficient when using a line expressing GFP and RFP fluorescent markers in the SN and VN, respectively.

3.3.1 Preparation of Sample for FACS

Follow **steps 1–16** of Subheading 3.1.1, with the exception that SC buffer is used instead of PB.

1. Transfer the sample to a 1.5 ml microcentrifuge tube containing 125 μl (*see* **Note 17**) of glass beads (425–600 μm) and vortex continuously at 2200 rpm for 2 min (*see* Fig. 5a, b).
2. Filter the sample through a 10 μm mesh in a CellTrics® filtering system to a 2 ml microcentrifuge tube, using a mini-centrifuge (or a regular microcentrifuge) to centrifuge (*see* Fig. 5c) (*see* **Note 10**). If no more rounds of breaking are needed proceed to **step 6** (*see* **Note 18**).
3. Carefully disassemble the CellTrics® filtering system (*see* **Note 19**) and transfer the mesh into a new 2 ml microcentrifuge tube containing glass beads.
4. Wash the mesh with 1 ml of fresh SC buffer. Make sure that the yellowish powder is washed off the mesh.

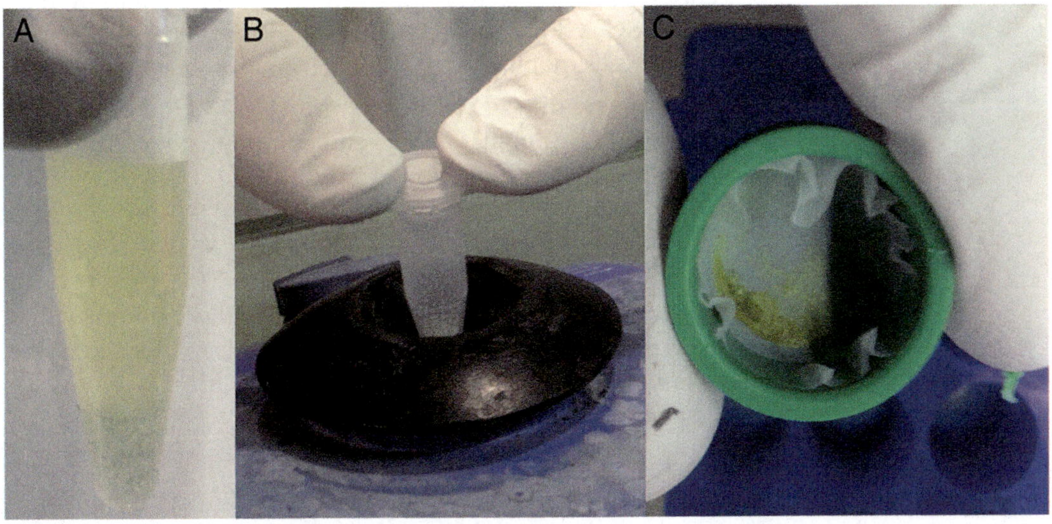

Fig. 5 (**a**) 1.5 ml microcentrifuge tube containing glass beads and pollen sample. (**b**) Breaking of pollen grains with beads using a Vortex. (**c**) CellTrics® filtering system used to filter the broken pollen sample through the mesh

5. Repeat **steps 1–4** as many times as needed (usually 2 times) until the amount of yellowish material retained in the mesh is minimal.
6. The sample is ready for FACS (keep on ice until sorting).

3.3.2 Staining of Sperm Cells and Vegetative Nuclei

To assure the sorting of only viable SC with intact plasma membranes (e.g., for transcriptome analysis), SYTOX Orange can be added to the sample (*see* **Note 20**). In this way, GFP-positive cells representing viable SC can be distinguished from nuclei released from bursted SC, exhibiting GFP/SYTOX double-positive labeling.

1. Prepare a 1:100 dilution of SYTOX Orange in SC buffer (2.5 μM final concentration) (*see* **Note 21**).
2. Measure the volume of sample to be stained (usually 1.5–2 ml).
3. Add 1:50 of the SYTOX Orange solution (1:100 dilution) to your sample, to reach a final concentration of 50 nM.
4. Wrap your sample in aluminum foil and incubate on ice for 10 min. The sample is ready for FACS (*see* **Note 22**).

3.3.3 Fluorescence-Activated Cell Sorting

1. Viable SC and VN can be sorted with high degrees of purity and integrity running the sorter at a constant pressure of 414 kPa (60 psi) with a 70 μm nozzle and frequency of drop formation of approximately 90 kHz (*see* **Note 23**).
2. Start by determining the optimal excitation voltages using the prepared sample in SC buffer.
3. Visualize both SC and VN population in the respective fluorescence channels. Use RFP (630/75 nm) vs GFP (530/40 nm) to gate SC (GFP-positive cells, population P1) and VN (RFP-positive cells, population P2). An example is shown in Fig. 6a.
4. When sorting the sample stained with SYTOX Orange solution (Fig. 6b), VN (population P2) and SN (population Nonviable P1) exhibit RFP/SYTOX and GFP/SYTOX double-positive labeling, respectively. Proceed to gate only viable SC by selecting SYTOX-negative SC (population viable P1, Fig. 6b) and/or population with increased FSC (Fig. 6e, *see* **Note 24**).
5. Gate also for both SC (P1) and VN (P2) those events with a uniform SSC profile (Fig. 6c, d, respectively). This allows the exclusion of debris and cells with altered integrity (*see* **Note 25**).
6. To avoid sorting aggregates, the auxiliary graph FSC vs Pulse width (or FSC-W) should be used (dot plot not shown).
7. SC and VN can be collected either in SC buffer (maintaining sorted volume to 1:10 of SC buffer in collection tube) or directly into an extraction buffer depending on the

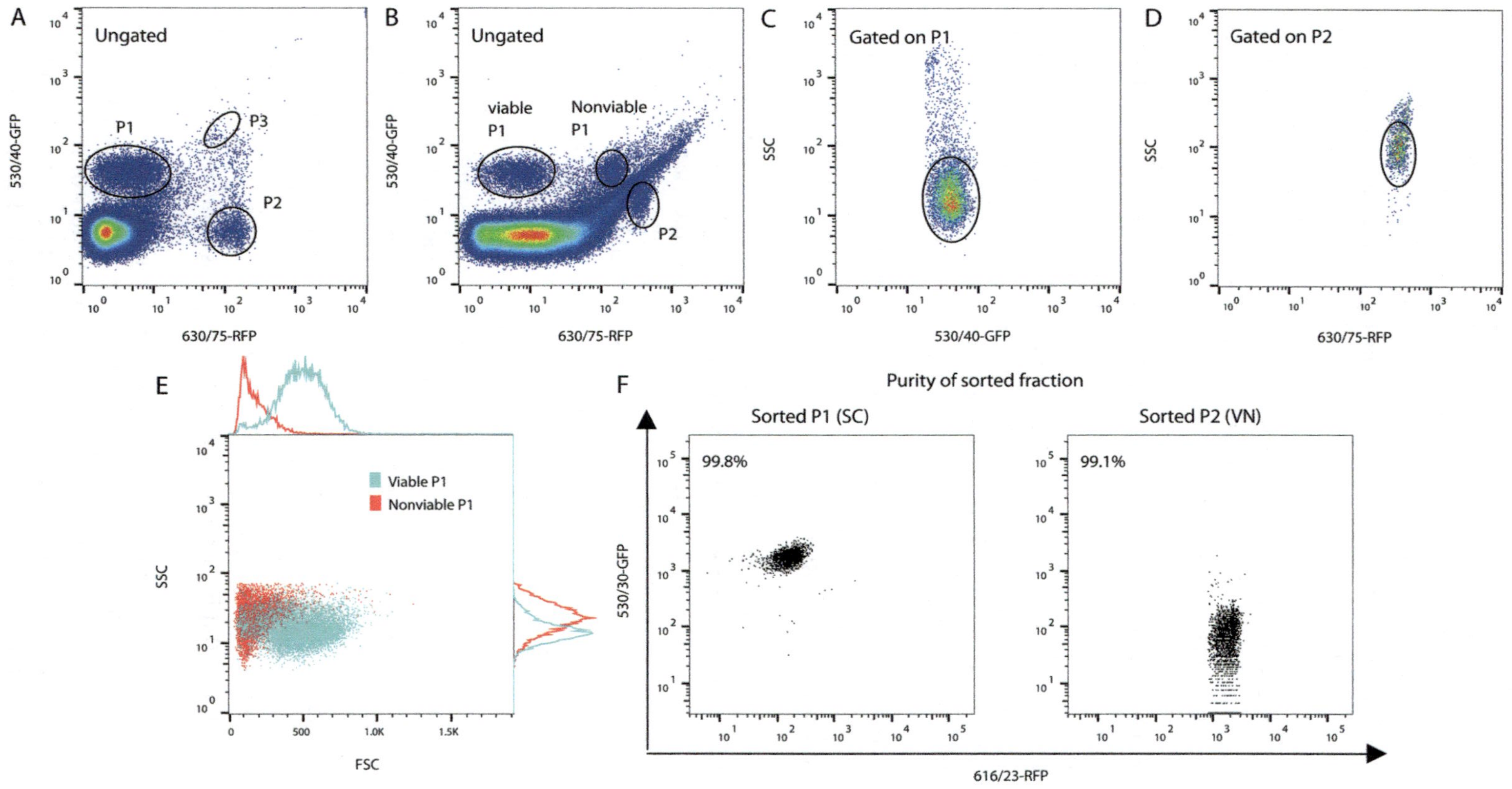

Fig. 6 Purification of viable SC and VN by FACS. (**a**) Distinct populations can be seen in the unstained, pre-sorting sample. Viable and nonviable SC are in this case identified based on their strong GFP signal (population P1), whereas VN separate toward the opposite axis based on strong RFP signal (population P2). In some sample preparations, intact pollen grains (population P3) may still be identified through their high autofluorescence. (**b**) In samples stained with *SYTOX Orange* GFP-positive only cells represent the viable SC (population viable P1), while GFP/SYTOX double-positive events identify the SN of bursted SC (population nonviable P1). Gating of viable SC (**c**) and VN (**d**) can be further improved by selecting only those cells presenting a homogeneous SSC profile. (**e**) Dot plot of FSC vs SSC of the two SC populations, in *red* nonviable SC (SYTOX-positive) and in *blue* viable SC (population P5). (**f**) Sorted SC and sorted VN fractions were stained with DAPI and run through a flow cytometer to check for purity. Respective purity is indicated in the top left corner of each graph and was determined by measuring the percentage of SC and VN present within the total number of DAPI-positive events, corresponding to all DNA-containing particles present in the sorted sample

experimental objective. Make sure to follow particular recommendations for ratio of sample/extraction buffer.

8. Maintain cells collected at 4 °C for immediate processing (preferred) or snap freeze in liquid nitrogen for long-term storage at −80 °C.
9. Purity of sorted fractions can be assessed using DAPI staining to measure the percentage of SC and VN sorted into SC buffer [10]. Add 1 μl of DAPI (0.4 μg/ml) to each 100 μl of sample. Run the stained sample on a flow cytometer to check purity (Fig. 6f).

3.4 Isolation of Viable Sperm Cells and Vegetative Nuclei from Non-marker Lines

For the analysis of mutant lines the fluorescent markers would have to be crossed in, constituting a time-consuming process. Alternatively, nucleic acid dyes can be used on lines missing the fluorescent markers. FACS of SN and VN from wild-type and mutant lines based on such a method have been described elsewhere [11, 12]. Therefore, we focus here on the isolation of intact SC and VN using our sample preparation protocol and a nucleic acid dye.

3.4.1 Preparation of Sample for FACS

Sample preparation follows the same steps as described in Subheading 3.3.1 for the fluorescent marker line, but with the following alterations from **step 6** on:

3.4.2 SYBR Green and SYTOX Orange Staining

For sorting of viable SC and VN from wild-type plants or non-fluorescent markers mutants, SYBR Green is used to distinguish all DNA-containing particles present in the sample (*see* **Note 26**). This allows the discrimination of these small cells from the debris. For sorting of viable SC, cells should be stained with SYTOX Orange (50 nM final concentration) as previously described in Subheading 3.3.2.

1. Measure the volume of your sample (usually 1.5–2 ml).
2. Add 2.5 μl of 10,000× SYBR Green stock solution to each 250 μl of sample (1:100 dilution, final concentration 100×).
3. Wrap your sample in aluminum foil and incubate on ice for 10 min.
4. Add 1:50 of the SYTOX Orange Solution (50 nM final concentration).
5. Wrap your sample in aluminum foil and incubate on ice for at least 10 min (*see* **Note 22**). The sample is ready for FACS.

3.4.3 Fluorescence-Activated Cell Sorting

1. Purification of viable SC and VN from plants without the fluorescent SC/VN markers can be achieved with similar levels of purity by using the described combination of nucleic acid dyes while running the sorter at a constant pressure of 414 kPa (60 psi) with a 70 μm nozzle and frequency of drop formation of approximately 90 kHz (*see* **Note 27**).

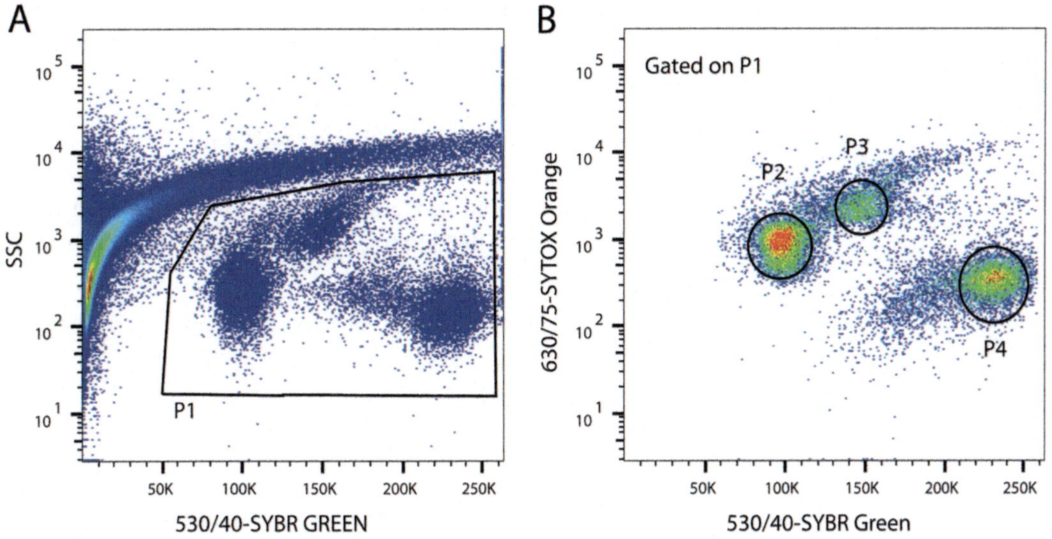

Fig. 7 Purification of wild-type SC and VN by FACS. (**a**) When staining the original pre-sort sample with *SYBR Green dye*, several distinct populations can be seen, and all DNA-containing particles with exception of pollen are gated by the P1 population. (**b**) Viable SC are identified based on their exclusion of *SYTOX Orange dye* (on the y-axis, population P4), while VN population (population P3) stains both with *SYBR Green* and *SYTOX Orange*, as do the SN (population P2) as previously described [11]

2. Start by defining the optimal excitation voltages with the single stained sample of SYBR Green (SYBR Green channel set to linear scale, *see* **Note 28**).
3. Visualize and gate all DNA-containing particles as shown in the SYBR Green (530/40) vs SSC (Fig. 7a) plot, excluding the continuum of debris and pollen seen above population P1.
4. Within P1 several populations can be identified, the SYTOX-positive/SYBR green positive SN (Fig. 7b population P2) and VN (population P3) and the SYTOX-negative/SYBR green positive viable SC (population P4).
5. To avoid sorting aggregates, the auxiliary graph FSC vs Pulse width (or FSC-W) should be used (dot plot not shown).
6. SC and VN can be collected either in SC buffer (maintaining sorted volume to 1:10 of SC buffer in collection tube) or directly into an extraction buffer depending on the experimental objective. Make sure to follow particular recommendations for ratio of sample/extraction buffer.
7. Maintain cells collected at 4 °C for immediate processing (preferred) or snap freeze in liquid nitrogen for long-term storage at −80 °C.

3.5 Isolation of Sperm Nuclei and Vegetative Nuclei Using a Marker Line

Sample preparation follows the same steps as described in Subheading 3.3.1 for the fluorescent marker line, but SN buffer is used instead of SC buffer.

3.5.1 Preparation of Sample for FACS

Instead of sorting from the fresh material, it might be desirable to collect and store pollen before cracking and sorting of nuclei. In such a case, follow sample preparation steps as described in Subheading 3.1.1 for pollen until **steps 3–17**, using SN buffer instead of SC buffer. Then add these steps:

1. Seal the microcentrifuge tube with parafilm and snap freeze it in liquid nitrogen.
2. Store the pollen sample at −80 °C.
3. Thaw the frozen pollen sample and resuspend it in 1 ml of SN buffer.
4. Continue from **step 1** in Subheading 3.3.1 for the fluorescent marker line, but using SN buffer instead of SC buffer.

3.5.2 Fluorescence-Activated Cell Sorting

1. Similarly to SC, SN and VN can be purified with high degrees of purity and integrity using the previous sorting conditions (constant pressure of 414 kPa (60 psi) and 70 μm nozzle) and same optimal excitation voltages.
2. Visualize both SN and VN population in the respective fluorescence channels. In this example, RFP (616/23) vs GFP (530/30) was used to gate SN (GFP-positive nuclei, population P1) and VN (RFP-positive nuclei, population P2) (*see* Fig. 8a).

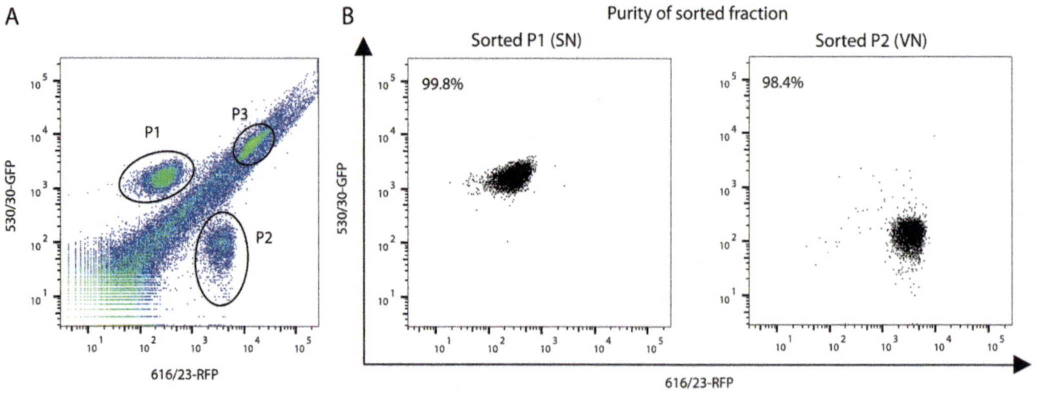

Fig. 8 Purification of SN and VN of fluorescent marker line by FACS. (**a**) Similarly to the SC, the respective SN are identified based on their strong GFP signal (population P1), whereas VN separate toward the opposite axis based on strong RFP signal (population P2). As before, some pollen grains may still be observable (population P3). (**b**) Purity, as indicated in the top left corner of each graph, can be determined by measuring the percentage of P1 (SC) and P2 (VN) detected in relation to the total number of DAPI-positive events

3. Similarly to SC/VN sorting, gating of SN and VN can be further improved by selecting only those events which present a homogenous SSC profile (dot plot not shown).
4. To avoid sorting aggregates, the auxiliary graph FSC vs Pulse width (or FSC-W) should also be used (dot plot not shown).
5. SN and VN can be collected either in SN buffer (maintaining sorted volume to 1:10 of SC buffer in collection tube) or directly into an extraction buffer depending on the experimental objective. Make sure to follow particular recommendations for the ratio of sample/extraction buffer.
6. Maintain nuclei collected at 4 °C for immediate processing or snap freeze in liquid nitrogen for long-term storage at −80 °C.
7. For checking the purity of the sorted populations, DAPI staining can be used in cells sorted into SC buffer by adding 1 μl of DAPI (0.4 μg/ml) to each 100 μl of sample. Run the stained sample on a flow cytometer to check purity (Fig. 8b).

4 Notes

1. Buffers can be stored at 4 °C for up to 1 month without having any impact on FACS sorting, unless otherwise mentioned.
2. The Pollen Buffer (PB) should be prepared freshly, on the day, and stored at 4 °C or on ice until being used. Always use fresh MilliQ water. This buffer does not need to be filtered.
3. CellTrics® filters can be acquired in different mesh sizes (5, 10, 20, 30, 50, 100, and 150 μm) and can come sterile (individually wrapped) or non-sterile (bulk packagesBioconductor package). Once used, disassemble the unit and remove the mesh. The unit can then be washed, fitted with a new mesh (cut to size ≈ 2.5 cm of circumference diameter), and autoclaved again (*see* Fig. 9).

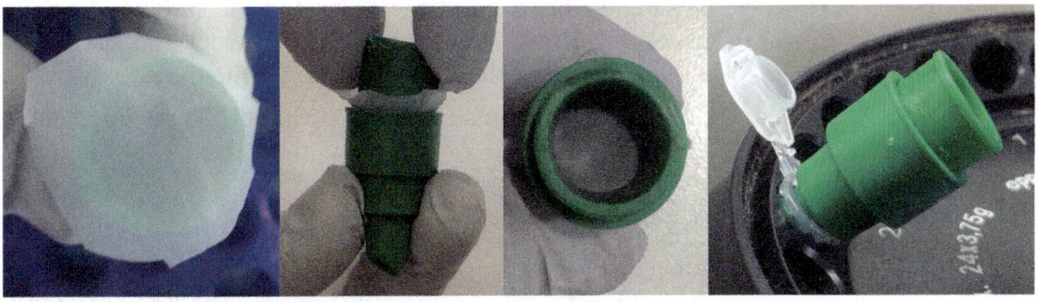

Fig. 9 Assembly of CellTrics® filtering system

4. It falls beyond the scope of this chapter to review FACS or flow cytometry in detail. For further information, please consult reference literature about this technique, such as [15].
5. Avoid using trimmed plants, since this affects pollen quality [14].
6. Flowers can easily be collected using tweezers. If no buffer is added at this stage, they can be kept at room temperature for up to 4 h without affecting the final yield.
7. This is the maximum volume of flowers that can be processed in one batch. In case smaller amounts are desired, the protocol can be scaled down to a 2 ml microcentrifuge tube. The ratio of sample volume to tube size should be 1:5 at maximum (e.g., 2 ml microcentrifuge tube filled with approximately 150 mg of flowers (about 400 μl volume)). This will allow for a proper vortex of the sample.
8. Make a funnel inside the tube using the Miracloth mesh to avoid the solution spilling over the rim of the conical centrifuge tube. This filtering step removes the flowers and bigger debris.
9. Pollen gets stuck to the walls of the centrifuge tube, so be sure to extensively wash the wall of the conical centrifuge/microcentrifuge tube to collect all the pollen.
10. This is just a quick spin, thus keep it between 3 and 6 s and maximum of 2320 × g (if using a microcentrifuge).
11. There are no significant differences in viability of pollen and sperm cells between fractions sorted in a cuvette-based FACSAria I and jet-in-air MoFlo high-speed cell sorter.
12. Other RNA extraction protocols can be used, but an efficient breaking of the pollen grains has to be assured.
13. Be careful when working with liquid nitrogen. Direct contact of skin with liquid nitrogen and long-time exposure to low temperatures (when gripping the cooled mortar and pestle during grinding) can cause cold burns and frostbite.
14. When collecting the inflorescences, use tweezers to remove open flowers and siliques.
15. Be careful not to grind the tissue too much. The idea is to gently squash the inflorescence and disrupt the buds, just enough to release the microspores from the tissue.
16. Grinding with a small volume of buffer allows for a more effective disruption of the buds than if all the volume is added at once.
17. This volume is the equivalent to a volume of 125 μl of H_2O.
18. The repetition of the process is only needed if aiming for a high yield of sperm cells or vegetative nuclei. The first breaking is

usually enough to obtain ≈ 100,000 sperm cells starting from ≈ 5 ml of flowers.

19. There should be a yellowish powder containing intact pollen grains on the mesh. Be careful not to disturb it when disassembling the filter unit. If necessary, use tweezers to handle the mesh.
20. SYTOX Orange is a membrane impermeant nucleic acid dye that is used in this protocol to discriminate viable, intact SC from SN during FACS.
21. Calculate the final volume of SYTOX Orange (1:100) that you are going to use in the experiment. This dilution should be prepared freshly and not stored for longer than 1 week at −20 °C. An "old" solution may compromise the dye stability resulting in the low viability of SC.
22. A 10 min incubation assures the proper staining of the sample, minimizing signal variations during the sorting. Proceed to FACS without long delays.
23. When sorting SC, SN, and VN and using FSC as a trigger, set the threshold level as low as possible. On BD machines values of 200–800 are recommended.
24. Coarse discrimination of viable SC is achievable using FSC parameter, although the use of a fluorescent dye (such as SYTOX Orange) is recommended to accomplish a more accurate selection of viable cells.
25. It is expected that in freshly prepared buffer over 85% of cells will present a homogeneous SSC profile after sample preparation.
26. SYBR Green is a membrane permeant nucleic acid dye that is used in this protocol to stain all DNA-containing particles in the sample, allowing their discrimination from other types of debris.
27. For SC and VN sorting from non-marker lines, it is recommended to stain the sample in 500 μl to 1 ml aliquots. Having a freshly stained sample will avoid fluctuations in SYBR Green staining.
28. When sorting SC (or SN) and VN from non-marker lines with a combination of SYBR Green and SYTOX Orange dyes, keep the instrument flow rate as slow as possible to allow a better resolution of the populations.

Acknowledgement

We thank Marta Monteiro, Mariana Fernandes (Flow Cytometry Unit), and Vera Nunes (Plant Facility) for technical support. Leonor Boavida is acknowledged for providing invaluable insight and comments. This work was supported by project EVOREPRO within the framework of the 2nd call ERA-NET for Coordinating Plant Sciences, with funding from Fundacao para a Ciencia e a Tecnologia-FCT (Portugal) through grant ERA-CAPS/0001/2014. J.D.B. received salary support from FCT through an "Investigador FCT" position.

References

1. Rutley N, Twell D (2015) A decade of pollen transcriptomics. Plant Reprod 28:73–89
2. Johnson-Brousseau SA, McCormick S (2004) A compendium of methods useful for characterizing Arabidopsis pollen mutants and gametophytically-expressed genes. Plant J 39:761–775
3. Dupl'áková N, Dobrev PI, Reňák D, Honys D (2016) Rapid separation of Arabidopsis male gametophyte developmental stages using a Percoll gradient. Nat Protoc 11:1817–1832
4. Calarco JP, Borges F, Donoghue MTA et al (2012) Reprogramming of DNA methylation in pollen guides epigenetic inheritance via small RNA. Cell 151:194–205
5. Ibarra CA, Feng X, Schoft VK et al (2012) Active DNA demethylation in plant companion cells reinforces transposon methylation in gametes. Science 337:1360–1364
6. Wibowo A, Becker C, Marconi G et al (2016) Hyperosmotic stress memory in Arabidopsis is mediated by distinct epigenetically labile sites in the genome and is restricted in the male germline by DNA glycosylase activity. Elife 5:e13546
7. Zhang Z, Xu H, Singh MB, Russell SD (1998) Isolation and collection of two populations of viable sperm cells from the pollen of *Plumbago zeylanica*. Zygote 6:295–298
8. Russell SD, Gou X, Wong CE et al (2012) Genomic profiling of rice sperm cell transcripts reveals conserved and distinct elements in the flowering plant male germ lineage. New Phytol 195:560–573
9. Lu Y, Wei L, Wang T (2015) Methods to isolate a large amount of generative cells, sperm cells and vegetative nuclei from tomato pollen for "omics" analysis. Front Plant Sci 6:391
10. Borges F, Gardner R, Lopes T et al (2012) FACS-based purification of Arabidopsis microspores, sperm cells and vegetative nuclei. Plant Methods 8:44
11. Chumak N, Mosiolek M, Schoft VK (2015) Sample preparation and fractionation of *Arabidopsis thaliana* sperm and vegetative cell nuclei by FACS. Bio-Protocol 5:e1664
12. Schoft VK, Chumak N, Bindics J et al (2015) SYBR green-activated sorting of Arabidopsis pollen nuclei based on different DNA/RNA content. Plant Reprod 28:61–72
13. Macaulay IC, Teng MJ, Haerty W et al (2016) Separation and parallel sequencing of the genomes and transcriptomes of single cells using G&T-seq. Nat Protoc 11:2081–2103
14. Boavida LC, McCormick S (2007) Temperature as a determinant factor for increased and reproducible in vitro pollen germination in *Arabidopsis thaliana*. Plant J 52:570–582
15. Wood JCS (2001) Principles of gating. Curr Protoc Cytom Chapter 1:Unit 1.8

Chapter 17

Isolation of Rice Sperm Cells for Transcriptional Profiling

Scott D. Russell, Daniel S. Jones, Sarah Anderson, Xinkun Wang, Venkatesan Sundaresan, and Xiaoping Gou

Abstract

The male germline of flowering plants displays unexpectedly divergent transcriptional profiles compared to other cell types and tissues of plants. As these are among the smallest cells, and are harbored within pollen, isolating a pure collection of germline RNA presents unusual challenges. The sperm cells of rice represent a particularly challenging subject for study as the pollen are unusually short lived upon release from the anther, and the marker gene sequences that make FACS possible in *Arabidopsis* have not yet been introduced into rice. The purity of the germline samples requires careful collection because of the limited amount of material available and potential contamination by other nearby tissues, pollen, and RNases. A discontinuous Percoll density gradient centrifuge was developed to isolate and obtain enough rice sperm cells for RNA-seq or microarray analysis.

Key words Male gamete isolation (angiosperm), *Oryza sativa*, Rice, Pollen isolation, Sperm transcriptome

1 Introduction

The germline of angiosperms originates late in the angiosperm life cycle, following spore formation, early in the life of the gametophyte. In angiosperms, the male and female gametes are initiated from an archesporial cell lineage formed in two distinct organs of the flower, the anther and ovule, respectively. Within the male gametophyte the germ lineage forms after the first mitotic cycle forming a male germ unit and within the female gametophyte after two or, more frequently three, mitotic cycles [1, 2]. In pollen of about two-thirds of angiosperms, sperm cells have yet to be formed at floral anthesis and pollen is released with the precursor generative cell, which appears to be a basal condition [3]. Sperm cells are always present at the conclusion of pollen tube elongation, but different stable cell cycle conditions have arisen such that fertilization is quite variable among the angiosperms [4]. In some plants, S-phase ensues in the pollen or pollen tube prior to fertilization,

Anja Schmidt (ed.), *Plant Germline Development: Methods and Protocols*, Methods in Molecular Biology, vol. 1669, DOI 10.1007/978-1-4939-7286-9_17, © Springer Science+Business Media LLC 2017

whereas in others, like rice, S-phase occurs after fertilization. Sperm cells of rice (*Oryza sativa*) are normally formed within 2 days prior to anthesis [5], and remain in G1 phase throughout pollen tube elongation and sperm cell delivery. Sperm cells correspondingly fuse at G1 within the egg and central cells, in an event that is characteristic of angiosperms, namely double fertilization. Variability in reproduction, however, is high and isolation of male gametes requires modification of protocols to align with their reproductive and cell biology [2].

Rice displays an unusual life history in which an exceedingly brief interval ensues between pollination, germination, pollen tube arrival, and gamete fusion. In fact, fertilization in rice may occur in as short an interval as 30–45 min, and unlike some other species, the cell cycle of sperm and egg cell are immediately receptive. Also, the pollen grain is exceedingly short lived, remaining viable for minutes after release from the anther [6] and the marker gene sequences that make FACS possible in *Arabidopsis* [7] have not yet been introduced into rice. Here, we describe a Percoll-based method to isolate rice sperm cells for transcriptional analysis.

2 Materials

Prepare all solutions using deionized water.

1. Flowering rice plants.
2. 45% (w/v) sucrose: Add 45 g sucrose to ~80 ml of water, mix into a homogeneous suspension, and fill up to a final volume of 100 ml. Store solution on ice.
3. 15% (w/v) sucrose: Add 30 g sucrose to ~180 ml water, mix into a homogeneous suspension, and fill up to a final volume of 200 ml. Store on ice.
4. 15% (v/v) Percoll: Dissolve 15 g sucrose in ~50 ml water, mix with 15 ml Percoll, adjust to a final volume of 100 ml with water and store on ice.
5. 40% (v/v) Percoll: Dissolve 15 g sucrose in ~40 ml water, mix with 40 ml Percoll, adjust to a final volume of 100 ml with water and store on ice.
6. Two or more 15 ml Corning glass centrifuge tubes.
7. Two 10 ml glass test tubes.
8. Glass rod.
9. Nylon mesh.
10. Glass micropipette.
11. Swinging bucket refrigerated centrifuge capable of sustaining 4000 × *g* and maintaining 4 °C.

12. 1 mg/ml fluorescein diacetate.
13. Fluorescence microscope equipped with differential interference or phase contrast microscopy.
14. Fluorescein or similar (FITC) fluorescence filter cube at excitation wavelength of ~494–495 nm and emission wavelength of 512–519 nm.
15. Ambion MEGAscript T7 Kit.
16. ND 1000 Nanodrop Spectrophotometer (Thermo Fisher Scientific).
17. Covaris S220 system (http://covarisinc.com).
18. GeneChip Two-Cycle cDNA Synthesis Kit (Affymetrix).
19. 57K Rice Genome GeneChip Microarray (Affymetrix).
20. GeneChip Fluidics Station.
21. GeneChip Scanner 3000.

3 Methods

3.1 Collection of Anthers

1. Collect fresh panicles of rice containing mature pollen in the morning before anthesis. Stems with unopened flowers may be cut and stored wrapped in moist toweling, which helps to delay anthesis, for collection during the day.
2. For collecting anthers use two fine forceps, one in each hand, to hold the floret and to dissect it. Florets may be dissected in situ or removed from the inflorescence prior to dissection, but not allowed to dry. To open the floret, grasp the distal palea in one forceps and the opposing lemma in the other and separate them until the floret is splayed open. Six distinctive yellow anthers are each suspended on slender, versatile filaments and can be separated from the floret taking care not to damage the integrity of the anther (*see* **Notes 1** and **2**). These are then dropped into a 1.5 ml microcentrifuge tube. Stems can be collected, but panicles should be draped in damp paper toweling to slow anthesis/prevent drying.
3. Collect anthers in a 1.5 ml microcentrifuge tube until a volume of approximately 1 ml is reached with typically three or four microcentrifuge tubes used for an isolation; this volume will allow sperm cells to be visible at critical times in preparation (*see* **Note 3**). The collected tubes may be stored in cool conditions for up to 4 h before processing. The fluorochromatic reaction (FCR) can be used as a sensitive assay of cellular quality (*see* Subheading 3.3) [8, 9]. We never retained the anthers for more than a few hours, as yields and FCR assays reflect loss of quality and intactness.

3.2 Separation of Sperm Cells

1. Clean at least two 15 ml Corning glass centrifuge tubes, two 10 ml glass test tubes, and a glass rod to be used in different stages of the isolation. Particular care should be taken if the sample will be used for RNA or other labile products. Poor isolations can arise from not using scrupulously cleaned glass, fresh water, poor staging, or trace contaminants.
2. Cut squares of 10, 30, and 100 μm nylon mesh to appropriate dimensions (typically 2.5 cm, 2.5 cm, and 5 cm, respectively), depending on the size of the collections.
3. Add sufficient 45% sucrose solution to unopened anthers to half fill a 10 ml round-bottomed glass test tube.
4. Introduce a glass rod into the test tube and crush anthers just enough to release whole, intact pollen grains with minimal damage to the pollen.
5. Pass this solution through a 100 μm nylon filter and collect the filtrate.
6. Transfer the pollen filtrate into another chilled glass centrifuge tube (4 °C) and centrifuge at 300 × *g* for 3 min to isolate the pollen fraction, which contains mostly intact pollen grains.
7. Resuspend the pollen grains in 15% chilled sucrose using ten times the prior volume of the filtrate and allow ~20 min for the pollen grains to release their sperm cells.
8. Rinse the mixed solution through a 30 μm nylon filter to remove intact pollen grains and pollen walls from the enriched sperm cells, which are in solution in the tube.
9. Add about 2–4 ml of chilled 40% Percoll to the base of a 15 ml centrifuge tube, followed by careful layering of 2–4 ml of chilled 15% Percoll on top of the 40% Percoll.
10. Carefully layer the obtained filtrate from the prior **step 8** on to the top of the 15% Percoll solution.
11. Without disturbing layers, carefully place the tube in a pre-cooled swinging bucket and centrifuge the tube at 4 °C and 4000 × *g* for 40 min.
12. Carefully remove the centrifuge tube taking care not to disturb the layered solutions. The sperm cell-rich fraction will form a band at the 15/40% interface (Fig. 1a).
13. Collect ~200 μl from this layer in each tube using a glass micropipette. If there are sufficient sperm cells, this layer will form a visible band; needless to say all of the material should be collected for further processing.
14. Add the sperm cell enriched fraction to a 4× volume of 15% sucrose solution and centrifuge at 900 × *g* at 4 °C for 10 min. Then collect 2 ml of the solution from the bottom of the

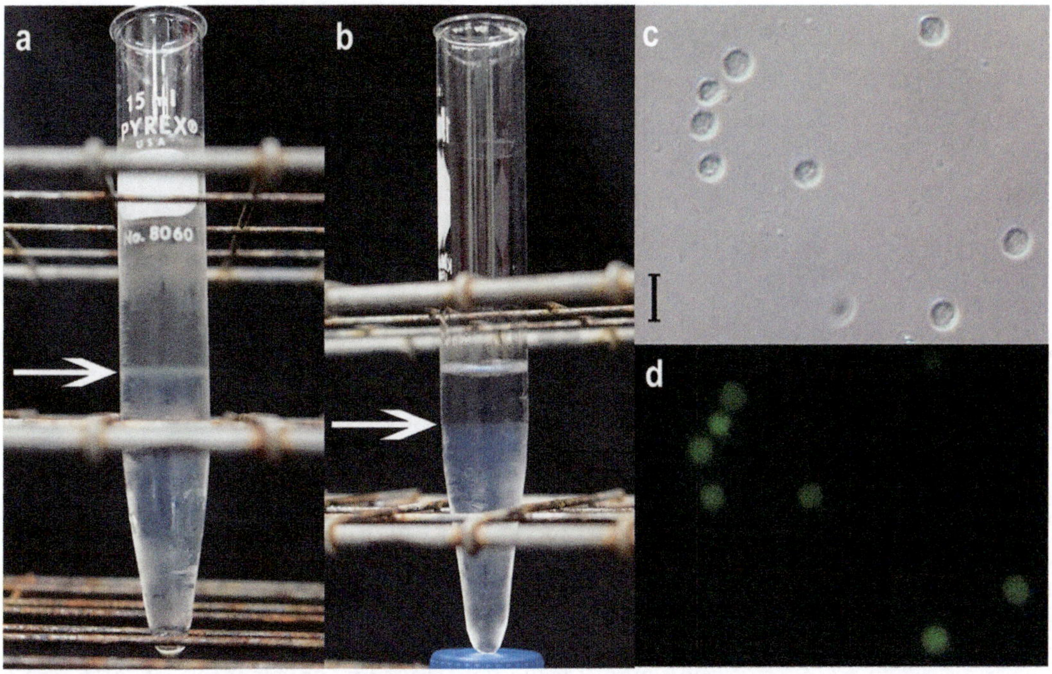

Fig. 1 (**a**) Sucrose pad after extended centrifugation. *Arrow* indicates the appearance of the band of isolated sperm cells at the 15%/40% Percoll layer. (**b**) Layering of sample (indicated by *arrow*) on top of fresh sucrose pad. (**c**) Intact sperm cells evident in differential interference contrast microscopy. (**d**) Epifluorescence microscopy of fluorescein diacetate-treated isolated sperm cells (bar = 5 μm). (**c** and **d** [9] are reproduced with permission of J.C. Wiley Press)

centrifuge tube. (The bottom fraction may include some aggregated pollen organelles.)

15. Resuspend the pellet in 15% sucrose solution and filter though a 10 μm nylon mesh.
16. Collect the filtrate containing the enriched sperm cells. The following steps (**steps 17–19**) will remove remaining contaminants, principally starch grains and cytoplasmic debris, but some sperm cells may be lost. Go directly to **step 20** if cell quantity is more important than cell purity.
17. Add 2 ml of chilled 40% Percoll in a 15 ml glass centrifuge tube. Layer filtrate from **step 16** on to the 40% Percoll layer without disturbing it.
18. Carefully replace the tube into the swinging bucket and centrifuge at 4000 × *g* at 4 °C for 10 min.
19. Collect the ~200 μl band between sample solution and basal 40% Percoll sucrose layer. The mixture contains most of the purified sperm cells at the boundary (Fig. 1b).

20. Place the collected band in an empty 15 ml glass centrifuge tube and add 4× volume of 15% sucrose solution to dilute the Percoll.
21. Centrifuge at 900 × *g* for another 10 min at 4 °C.
22. Collect about ~50–100 μl from the bottom-most of the centrifugation tube.
23. Use 10 μm nylon filter to remove the possibly aggregated cell debris if necessary.
24. Freeze the isolated sperm cell fraction and store at −70 °C until use.

3.3 Controlling Purity and Quality of Sperm Cell Isolations

Collected solutions will contain intact sperm cells (Fig. 1c), which may be tested using accumulation of fluorescein via the fluorochromatic reaction [8].

1. To assay sperm cells, add 1 μl of 1 mg/ml fluorescein diacetate in 1:20 dilution.
2. Green fluorescence is evident following acetate cleavage, through the action of endogenous esterases. Using a fluorescein cube (excitation/emission wavelengths of 494 nm/ 512 nm), the green fluorescence characteristic of fluorescein is readily detected. The FITC cube, with excitation/emission wavelengths of 495 nm/519 nm, can also be used.
3. Particulate contaminants can be screened visually using phase or differential interference microscopy.

3.4 RNA Isolation and Preparation for Transcriptional Analysis

1. Isolate total RNA by standard techniques, e.g., using Qiagen's RNeasy plant mini kit according to the manufacturer's instructions (Qiagen; http://www.qiagen.com/) (*see* **Notes 4** and **5**) [9–11]; the latter methods are sufficiently standardized that their preparation is identical to that in other cell types, with the exception of initial concentration.
2. For microarray analysis, RNA is amplified using a two-step process. This is recommended as total starting RNA will typically be low (in the range of 10–100 ng per sperm cell sample). Using the Ambion MEGAscript T7 Kit or the Affymetrix GeneChip Two-Cycle cDNA Synthesis Kit, follow the manufacturer's instructions for amplification and labeling of cRNAs. The Affymetrix 57K Rice Genome GeneChip oligonucleotide microarray can be used for hybridization (45 °C for 16 h) with a mixture containing 15 μg of fragmented cRNA. Subsequent washing and staining steps should be performed using a GeneChip Fluidics Station 450 and the chips can be scanned on a GeneChip Scanner 3000.
3. The preparation of control tissues (e.g., different sporophytic or gametophytic controls) is recommended (*see* **Note 6**).

4. Take special care in order to minimize experimental variability (*see* **Note 7**).
5. For RNA-Seq, cDNA can be synthesized using the NuGEN RNA-seq system V2 (NuGEN, http://www.nugeninc.com) using 1.5–10.0 ng of non-amplified RNA. The quantity of cDNA can be determined using a ND 1000 Nanodrop Spectrophotometer (Thermo Fisher Scientific), sheared into fragments using the Covaris S220 system according to their recommended settings to generate 200-bp fragments.
6. For the generation of reference DNA sequences (*see* **Note 8**).

4 Notes

1. For collecting specific stages of anther development, the floral order of the panicle has to be mapped and understood. Complexity of anthesis in a panicle is not always easy to predict. Rather than displaying a simple acropetal order as in racemes, because of their complex branching pattern of the panicle florets may appear to have mixed basipetal and acropetal sequences on different inflorescence branches. Typically, the floret at the tip of a branch flowers first. In secondary and tertiary branches, the terminal floret may open 1–2 days later, also blooming from the tip floret first. Higher order branches appear to flower later and the entire inflorescence may take up to a week to bloom, although apical florets are usually depleted within 3 days. Thus, a developmentally meaningful collection requires careful collecting.
2. Ability to assess anthesis florets improves with experience and floret stage may be readily identified with practice. Anthers appear to nearly fully occupy the floret prior to opening. When the floret is opened with the forceps, the anthers will hang free and are best grasped by the filament, which reduces damage to the anther sacs and their contents during dissection. If anthers from a specific day prior to anthesis are needed, particular attention is needed to assure a developmentally meaningful collection.
3. Collections required for sperm isolation used approximately 3–4 ml of anthers kept in 1.5 ml microcentrifuge tubes until enough isolated unopened anthers are gathered for processing. (We do not know the minimum amount required, but the quantity will presumably decrease as techniques become more sensitive.) Anthers may remain unopened up to a half day, but need to be processed promptly or viability may be harmed. Rice pollen that is present outside of the anther sac has an exceedingly short life span, so care has to be taken that anthers are not

damaged or stimulated to dehisce [6]. Humidity should be high but anthers should not get wet.

4. Concentration and quality of sperm cell RNA can be assessed using the Agilent RNA 6000 Pico Kit (http://www.genomics.agilent.com/). In a recent study, amplified libraries from low RNA inputs produced high diversity, high quality transcripts, which are indicative of intactness and integrity, without significant evidence of degradation [9, 11]. Relative concentration of transcripts was similar in two studies based on this material [9, 11], revealing the male gametes and pollen vegetative cell to be distinct and unique tissues in rice, as well as in other species of monocots and dicots [12, 13]. Concentration and quality of RNA of pollen and other tissues can be assayed using routine spectrophotometric methods [10].
5. For microarray use, sperm RNA extractions were pooled in a recent study and concentrations were not determined because of limited material [11].
6. In our previous study [11], control tissues were prepared using similar quantities of 10–100 ng per sample and processed similarly to the sperm cells. Standard precautions were taken for RNA processing, and intactness of RNA was verified using a Bioanalyzer. Controls included seedlings, pollen cytoplasm, female gametophytic cell sources, and various sporophytic controls. All experimental procedures strictly followed instructions specified in the Affymetrix GeneChip Expression Analysis Technical Manual. Instrument control and data collection were carried out with a GeneChip Operating Software (GCOS, ver. 1.1.1).
7. In order to minimize experimental variability, standardized microarray processing was performed by an experienced investigator throughout our study [11]. The quality and quantity of the original RNA samples and of the cRNA probes generated were determined with the Bioanalyzer 2100 (Agilent Technologies, Palo Alto, CA, USA, http://www.agilent.com/), and by spectrophotometric measurements at 260 and 280 nm on a Nanodrop ND-1000 spectrophotometer (Nanodrop Technologies, Wilmington, DE, USA, http://www.nanodrop.com/). The microarray data generated from all chips met quality control criteria set by Affymetrix.
8. In a previous study, reference DNA sequences were generated using sheared DNA in a 170-ng portion purified with Qiagen's MinElute Reaction Cleanup Kit. Libraries at this stage were prepared using NuGEN Ovation Ultralow DR Multiplex Systems 1–8. Multiplexed samples were run with six samples/lane on the Illumina HiSeq 2000 DNA sequencer [9].

Acknowledgements

We acknowledge research funding by the National Science Foundation (IOS 1128145), University of Oklahoma, and University of Melbourne, and for plant growth support from the University of Oklahoma and Dale Bumpers National Rice Research Center.

References

1. Dumas C, Knox RB, McConchie CA, Russell SD (1984) Emerging physiological concepts in fertilization. What's New Plant Physiol 15:17–20
2. Russell SD (1991) Isolation and characterization of sperm cells in flowering plants. Annu Rev Plant Physiol Plant Mol Biol 42:189–204
3. Williams JH, Taylor ML, BC O'M (2014) Repeated evolution of tricellular (and bicellular) pollen. Am J Bot 101(4):559–571. doi:10.3732/ajb.1300423
4. Friedman WE (1999) Expression of the cell cycle in sperm of Arabidopsis: implications for understanding patterns of gametogenesis and fertilization in plants and other eukaryotes. Development 126(5):1065–1075
5. Wada T, Ito T, Ito M, Takeoka Y (1992) Light microscopic observations on pollen and anther development in rice (Oryza sativa L.). II. Stages from early microspore to mature pollen. Jpn J Crop Sci 61:136–144
6. Song ZP, Lu BR, Chen JK (2001) A study of pollen viability and longevity in *Oryza rufipogon*, *O. sativa*, and their hybrids. Int Rice Res Notes 26(2):31–32
7. Engel ML, Holmes-Davis R, McCormick S (2005) Green sperm. Identification of male gamete promoters in Arabidopsis. Plant Physiol 138(4):2124–2133
8. Heslop-Harrison J, Heslop-Harrison Y, Shivanna K (1984) The evaluation of pollen quality; and a further appraisal of the fluorochromatic (FCR) test procedure. Theor Appl Genet 67:367–375
9. Anderson SN, Johnson CS, Jones DS, Conrad LJ, Gou X, Russell SD, Sundaresan V (2013) Transcriptomes of isolated rice gametes characterized by deep sequencing: evidence for distinct sex-dependent chromatin and epigenetic states before fertilization. Plant J 76:729–741
10. Russell SD, Bhalla PL, Singh MB (2008) Transcriptome-based examination of putative pollen allergens of rice (*Oryza sativa* ssp. *japonica*). Mol Plant 1:751–759
11. Russell SD, Gou X, Wong CE, Wang X, Yuan T, Wei X, Bhalla PL, Singh MB (2012) Genomic profiling of rice sperm cell transcripts reveals conserved and distinct elements in the flowering plant male germ lineage. New Phytol 195:560–573
12. Singh MB, Bhalla PL, Russell SD (2008) Molecular repertoire of flowering plant male germ cells. Sex Plant Reprod 21:27–36
13. Russell SD, Jones DS (2015) The male germline of angiosperms: repertoire of an inconspicuous but important cell lineage. Front Plant Sci 6:173

Chapter 18

Manual Isolation of Living Cells from the *Arabidopsis thaliana* Female Gametophyte by Micromanipulation

Maria Englhart, Lucija Šoljić, and Stefanie Sprunck

Abstract

The few-celled female gametophyte, or embryo sac, of flowering plants is not easily accessible as it is buried within the sporophytic tissues of the ovule. Nevertheless, it has become an attractive model system to study the molecular mechanisms underlying patterning and cell type specification, as well as fertilization of the two female gametes, the egg and the central cell. While female gametes, zygotes, and early embryos can be manually isolated from the embryo sacs in maize, wheat, tobacco, and rice by micromanipulation, this approach had been considered impossible for the much smaller embryo sac of the model plant *Arabidopsis thaliana*. Here, we describe a method to isolate living cells from the *Arabidopsis* female gametophyte by micromanipulation. The manual isolation of egg cells, central cells, and synergid cells is a technique that enables a number of important studies such as cell-type-specific transcriptional profiling or the analysis of DNA methylation profiles. It also offers the possibility to use isolated female gametes for in vitro fertilization studies.

Key words Egg cell, Central cell, Synergid cell, Embryo sac, Enzymatic treatment, Dissection, Micromanipulator

1 Introduction

The female gametophyte (embryo sac) of flowering plants is the site of female gamete formation. It is the haploid generation of a plant that develops in the ovule, the female reproductive organ of the flower which gives rise to the seed after fertilization. The female gametophyte contains four distinct cell types that develop from a single haploid spore through three rounds of mitosis in a syncytium, followed by nuclei migration and positioning, cellularization, and cell specification [1]. The mature female gametophyte of the model plant *Arabidopsis thaliana* is composed of one egg cell and two synergid cells, located near the entrance point of the pollen tube (the micropyle; Fig. 1a), a larger homo-diploid central cell that occupies the central region of the embryo sac and three antipodal cells, located at the opposite chalazal end. During double

Anja Schmidt (ed.), *Plant Germline Development: Methods and Protocols*, Methods in Molecular Biology, vol. 1669, DOI 10.1007/978-1-4939-7286-9_18, © Springer Science+Business Media LLC 2017

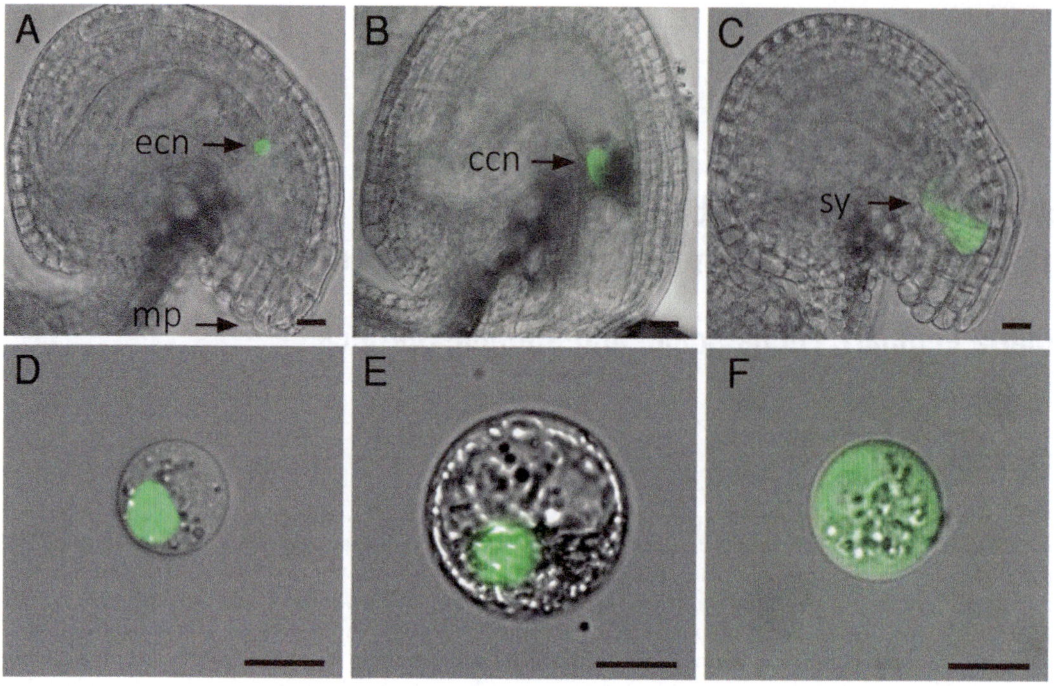

Fig. 1 Ovules of cell-type-specific marker lines and single embryo sac cells, manually isolated by micromanipulation. (**a** and **d**) Egg cell marker line EC1p:NLS-3xGFP-17A. (**b** and **e**) Central cell marker line At5g07850p:NLS-3xGFP. (**c** and **f**) Synergid cell marker line MYB98p:GFP. GFP fluorescence is visible in the nucleus of the egg cell (**a**, **d**) and the central cell (**b**, **e**) and in the cytoplasm of the synergid cell (**c**, **f**). Abbreviations: *ccn* central cell nucleus, *ecn* egg cell nucleus, *sy* synergid cell, *mp* micropyle. *Scale bars* 10 μm

fertilization, the egg cell and the central cell each fuse with one of the two sperm cells delivered by the pollen tube to form the embryo and the endosperm, respectively [2].

In order to investigate underlying molecular determinants of cell specification or zygotic genome activation, it is advantageous to have access to the individual cells of the female gametophyte. Although technically challenging, the manual isolation of female gametes, zygotes, and early embryos by micromanipulation was achieved in plant species such as maize, barley, wheat, tobacco, and rice, either with or without the help of cell wall-degrading enzymes [3–12]. Here, we describe a method to isolate intact, living egg cells, central cells, and synergid cells of the much smaller embryo sac of *Arabidopsis*, using fluorescent marker lines for the cell types of interest. The isolated cells are suitable for PCR-based gene expression studies. Furthermore, in combination with the rapidly evolving techniques for single-cell transcriptome and methylome studies, the manual isolation of *Arabidopsis* embryo sac cells by micromanipulation will provide important new insights into the epigenetic state of these cells and the molecular mechanisms underlying female gamete specification and zygotic genome activation.

2 Materials

2.1 Plant Material and Solutions for Cell Isolation

1. Transgenic marker lines for cell types of the female gametophyte: Our *Arabidopsis thaliana* (ecotype Col-0) plant lines EC1.1p:NLS-3xGFP-17A and At5g07850p:NLS-3xGFP, and the plant line MYB98p:GFP [13] can be used to manually isolate egg cells, central cells, and synergid cells by micromanipulation (Fig. 1) (*see* **Note 1**).
2. Wash solution: 0.45 M mannitol, 730 mOsm/L (*see* **Note 2**), pH 7.0. Add about 30 mL RNase free H_2O to a 50 mL falcon tube. Weigh 4.099 g mannit and transfer to the falcon tube. Mix and adjust the pH with 1 M NaOH. Make up to 50 mL with RNase free H_2O. Check the osmotic concentration with a cryoscopic osmometer. Filter-sterilize the wash solution and store at room temperature.
3. Enzyme solution: 1% cellulase (chromatographically purified, ≥45 units per mg dry weight), 0.3% macerozyme R-10 (for protoplast isolation), 0.05% pectolyase Y23 (highly purified), 0.45 M mannitol, 730 mOsm/L, pH 7.0. Add 6.16 mL wash solution and 3.84 mL RNase free H_2O to a 10 mL falcon tube. Weigh 100 mg cellulase, 30 mg macerozyme R-10 and 5 mg pectolyase Y23 and transfer to the falcon tube. Mix well and filter sterilize (syringe filter, pore size 0.2 μm). Freeze in 100 μL aliquots liquid N_2 and store at −20 °C or at −80 °C.
4. Paraffin oil, viscous.
5. ≥99.8% Ethanol, p.A., absolute.

2.2 Equipment

1. Stereomicroscope with 5:1 zoom minimum.
2. Inverse fluorescence microscope equipped with a multiple slide holder microscopy stage (Fig. 2a, b) and high-resolution 20× and 40× microscope objectives suitable for fluorescence applications.
3. Dewar with liquid nitrogen.
4. Petri dishes: diameter 35 mm.
5. Glass microscope slides.
6. Double-sided tape.
7. High precision forceps (extremely pointed, tip width 0.1 mm, length 110 mm).
8. Disposable hypodermic needles (diameter 0.40 mm, length 13 mm).
9. Disposable capillary pipettes with a ring mark, volume 50 μL.
10. Optional: Borosilicate glassrods, diameter 2 mm, to prepare glass needles.

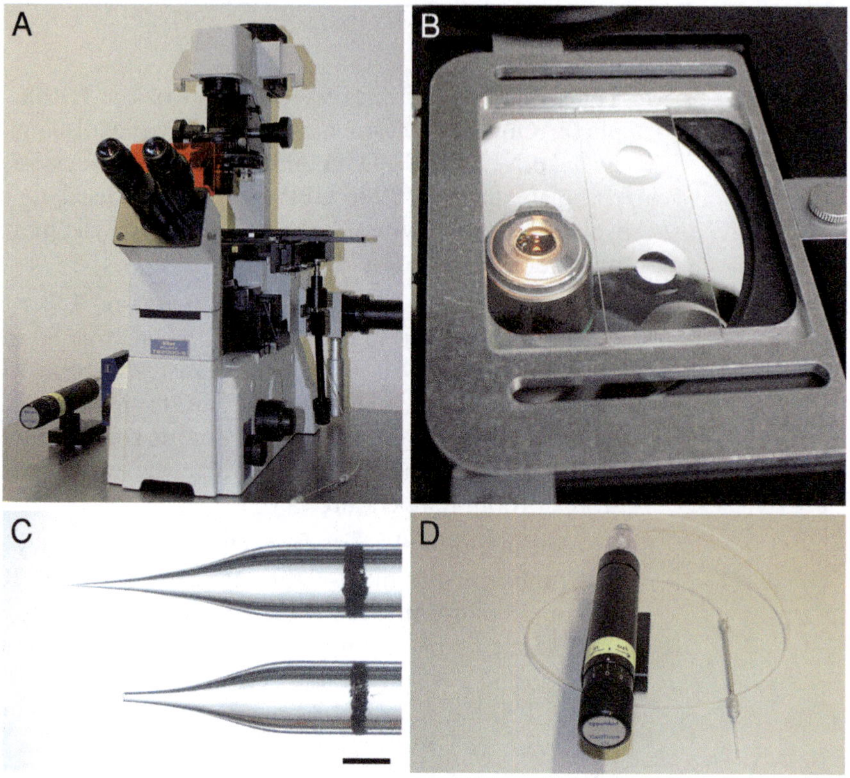

Fig. 2 Equipment for the manual isolation of embryo sac cells by micromanipulation. (**a**) Inverse fluorescence microscope, e.g., Eclipse TE200-S (Nikon). (**b**) Customized multiple slide holder microscopy stage with two slides on xy stage. (**c**) Pulled glass capillary for egg cell isolation before (*top*) and after (*bottom*) breaking off the tip. (**d**) Manual micromanipulator with glass capillary mounted to the metal holder. *Scale bar* 1 mm

2.3 Special Equipment for Micromanipulation

1. Cryoscopic osmometer, e.g., Osmomat 030® (Gonotec GmbH).
2. Programmable micropipette puller, e.g., P-97 Flaming/Brown type micropipette puller (Sutter Instruments®).
3. Manual microinjector to aspirate and dispense cells, e.g., CellTram® Oil (Eppendorf) (Fig. 2d).
4. Glass capillaries: pull disposable capillary pipettes with a programmable micropipette puller. Use forceps to break the tip of the capillary under the stereomicroscope. The capillary should have a diameter of 150–200 μm and a precisely defined breaking edge (Fig. 2c).
5. Optional: Very fine glass needles (*see* Subheading 3.3, **step 1**).
6. Soda-Lime glass microscope slides (76 × 26 × 1.5 mm) with single cavity and with two (or three) cavities (cavity diameter 15 mm, depth 0.7 mm) (Fig. 3a).
7. Humid chamber (*see* Subheading 3.2, **step 1**).

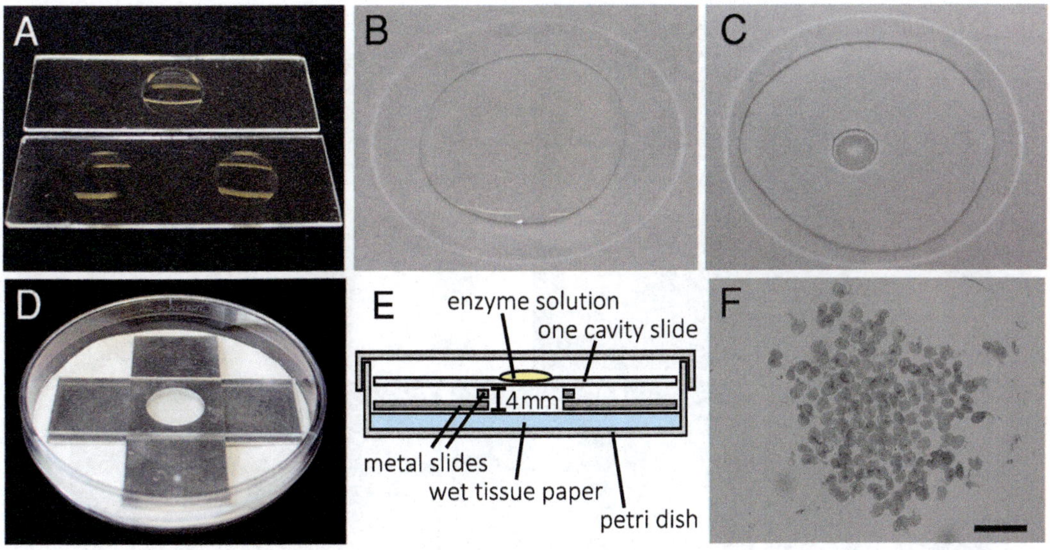

Fig. 3 Special slides for cell isolation and set up for the enzymatic digestion of ovules. (**a**) Glass microscope slides with one (*top*) and two cavities (*bottom*). (**b**) Enzyme solution (50 μL) in the cavity of the glass microscope slide. (**c**) Wash solution droplet (5 μL) in a 75 μL-drop of paraffin oil. (**d**) Humid chamber used for enzymatic digestion of ovules. (**e**) Scheme showing the assembly of the humid chamber. (**f**) Dissected ovules in the enzyme solution. *Scale bar* 500 μm

2.4 Material for RT-PCR and Blot

1. Dynabeads® mRNA DIRECT™ Micro Purification Kit (Thermo Fisher Scientific).
2. Hybond N+ nylon membranes.
3. High-quality reverse transcriptase that lacks RNase H.
4. RNase Inhibitor.
5. Primers (*see* Subheading 3.4, **steps 3** and **4**).

3 Methods

Carry out all procedures at room temperature unless otherwise specified.

3.1 Plant Growth and Emasculation of Flower Buds

1. Sow seeds of these transgenic marker lines (*see* Subheading 2.1) in a soil/sand mix (9:1), stratify for 2 days at 4 °C and germinate and grow at short-day conditions (8 h light, 22 °C, 70% humidity). After 4 weeks transfer the plants to long-day conditions (16 h light, 22 °C, 70% humidity).
2. To prepare for emasculation of flower buds (**steps 3–5**), put a piece of double-sided tape on an objective glass slide under the stereomicroscope.

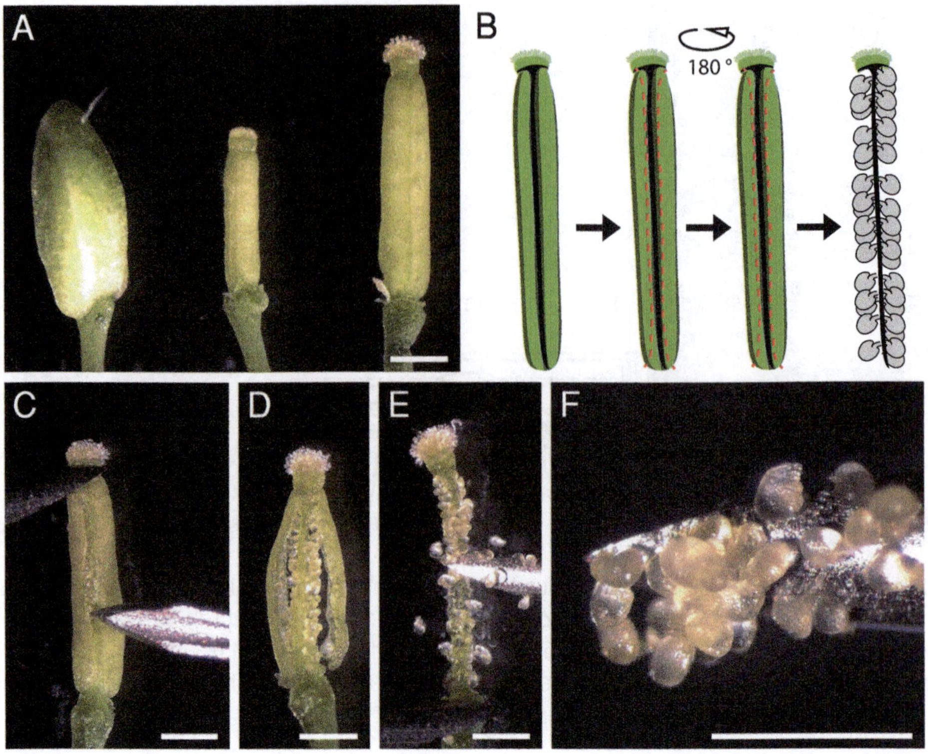

Fig. 4 Pistil preparation. (**a**) Flower bud and emasculated pistil. The mature pistil 2 days after emasculation is shown on the *right.* (**b**) Schematic overview of pistil preparation. The pistil is cut longitudinally along both sides of the septum. Afterward, the pistil is turned by 180° and cut again longitudinally along both sides of the septum. Carpels are removed. (**c** and **d**) Preparing the pistil with two hypodermic needles. (**e**) Ovules are carefully scraped off the placenta and remain attached to the tip of the hypodermic needle (**f**). *Scale bars* 0.5 mm

3. Tip over the plant pot and place one inflorescence, still connected to the plant, in horizontal position on the double-sided tape.
4. Carefully remove all siliques, open flowers, flower buds that show white petals and too young flower buds from the inflorescence using the fine forceps or hypodermic needles.
5. Carefully open the remaining 2–3 flower buds that show no white petals yet (flower stage: 2 days before anthesis) and remove their anthers using the fine forceps or hypodermic needles. If necessary, cut off all sepals and petals with the hypodermic needles (Fig. 4a). It is important to avoid injuries to the pistils (*see* **Note 3**).
6. Put plants back into the growth chamber for 2 days (*see* **Note 4**).

3.2 Ovule Preparation

1. To prepare a humid chamber, cover the bottom of a petri dish with tissue paper or pure cellulose filter paper. Do not place the soda-lime glass slide with one cavity directly on the filter paper

but elevate it. For example, stack two metal slides with a hole in the middle below the glass slide (Fig. 3d, e). Wet the tissue and cover the petri dish with the lid to create a humid atmosphere.

2. Thaw an aliquot of enzyme solution and put a droplet of 50 μL into the cavity of a one-cavity glass microscope slide in the humid chamber (Fig. 3b). Keep the humid chamber closed to avoid evaporation.
3. Take one emasculated mature pistil from the plant and put it onto a glass slide.
4. Hold the pistil in place with one hypodermic needle (with the opening showing downward).
5. Use another hypodermic needle (with the opening showing upward) to cut the pistil longitudinally along both sides of the septum. Turn the pistil by 180° and cut again longitudinally along both sides of the septum as shown in Fig. 4.
6. Remove the carpels and gently scrape off the ovules with the hypodermic needle, the opening showing downward (Fig. 4e, f).
7. Transfer the ovules into the drop of enzyme solution in the humid chamber (*see* **Note 5**).
8. Repeat the **steps 3–7** until you have approximately 100–200 ovules in the enzyme solution (Fig. 3f).
9. Add another 25 μL enzyme solution.
10. Close the humid chamber and place it on a horizontal shaker at 70 rpm for 30 min (*see* **Note 6**).

3.3 Isolation of Living Embryo Sac Cells

1. Pull glass needles by hand from Borosilicate glassrods, diameter 2 mm, by using a heating device. Introduce a curve near the tip of the needle.
2. Connect a glass capillary to the manual microinjector with the help of Parafilm® to keep it airtight (*see* **Note** 7).
3. Put a drop of 100 μL wash solution in the middle of a 35 mm petri dish and close the lid.
4. Fill another 35 mm petri dish with absolute ethanol and close the lid.
5. Put a drop of 75 μL paraffin oil in each cavity of a two-cavity (or three-cavity) glass slide.
6. Pipet a droplet of 5 μL wash solution into each drop of paraffin oil, onto the bottom of the glass slide (Fig. 3c).
7. Wash the glass capillary on the in and outside with absolute ethanol by repeated aspirating and dispensing.

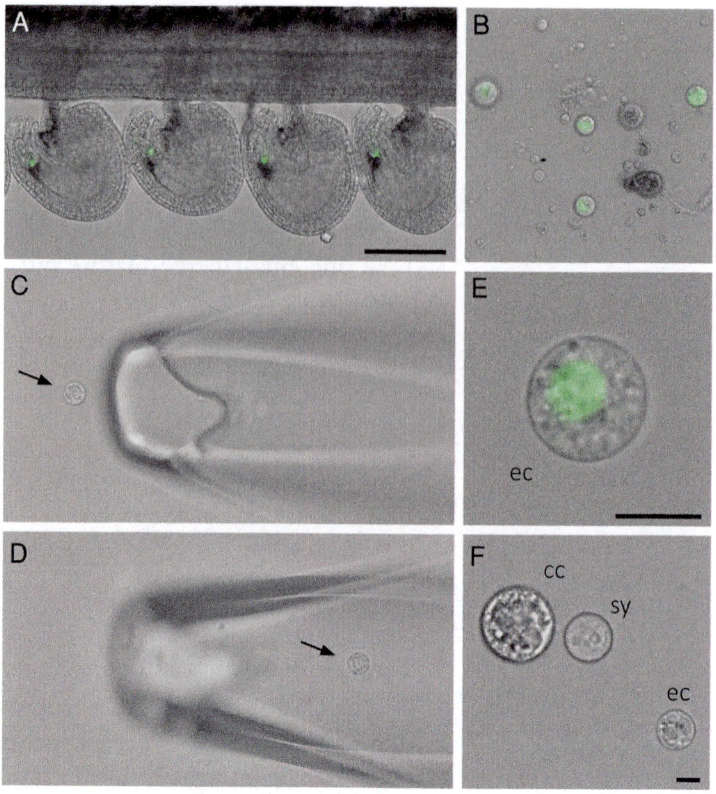

Fig. 5 Egg cell isolation. (**a**) Ovules of the homozygous marker line EC1.1p:NLS-3xGFP-17A. (**b**) Protoplasts released from the ovules of the marker line after 30 min incubation in the enzyme solution. Released egg cells are recognized by their green fluorescence. (**c** and **d**) Egg cell (*arrow*) aspirated by the glass capillary. (**e**) Single egg cell in the wash solution, showing the GFP signal in the nucleus. The diameter of isolated egg cells is approximately 13–15 μm. (**f**) Isolated central cell, egg cell, and synergid cell. The cell types show morphological differences. Abbreviations: *cc* central cell, *ec* egg cell, *sy* synergid cell. *Scale bars* **a** 100 μm, **e** and **f** 10 μm

8. Wash the glass capillary twice by aspirating fresh wash solution from the 100 μL drop and dispensing into the petri dish (*see* **Note 8**).
9. Aspirate about 5 μL of fresh wash solution (*see* **Note 9**).
10. Put the slide with the ovules under the microscope, on the multiple slide holder microscopy stage.
11. Put the two-cavity (or three-cavity) slide with the wash solution droplets in oil next to it (Fig. 2b).
12. Search the bottom of the cavity with enzyme solution for single cells (protoplasts) of the female gametophyte, which are released from the ovules. Check their identity by their fluorescence (*see* **Note 10**) (Fig. 5b).

13. Take the glass capillary and dip into the liquid, while observing the ovules under the microscope. Gently push ovules and cellular debris away from the cells of interest by dispensing minute quantities of wash solution (*see* **Note 11**). Alternatively take a glass needle to push ovules away and to detach and separate the cells of the female gametophyte.
14. Aspirate the single cell (protoplast) of interest with the glass capillary, while observing under the microscope.
15. Move the xy stage of the microscope to the first of the wash solution drops, push the tip of the capillary through the paraffin oil and dispense the single cells slowly into the droplet of wash solution, at the very bottom (*see* **Note 12**).
16. Repeat the **steps 8–15** until about 10–20 cells of interest are collected in the first droplet of washing solution (*see* **Note 13**).
17. Wash the capillary with ethanol and wash solution as described and aspirate approximately 5 μL of wash solution. Gently aspirate the cells that have been collected in the droplet of wash solution and transfer them to the second drop of wash solution in paraffin oil (Fig. 5c, d).
18. Check the identity of the cells by fluorescence.
19. Wash the capillary with ethanol and wash solution and aspirate approximately 2 μL of wash solution. Aspirate as many cells as possible at once from the second droplet of wash solution to keep the volume of liquid at a minimum.
20. Gently dispense the isolated cells into a suitable reaction tube (*see* **Note 14**) and freeze immediately in liquid nitrogen.

3.4 Validation of Collected Cell Types by RT-PCR

1. Pool 5–15 cells of the cell type of interest in one reaction tube for mRNA isolation.
2. Isolate poly(A) mRNA from pooled cells using the Dynabeads® mRNA DIRECT™ Micro Purification Kit.
3. Perform first-strand cDNA synthesis without eluting the mRNA from the beads. Use additional Oligo$(dT)_{23}$ primers and a high-quality reverse transcriptase that lacks RNase H activity to synthesize the cDNA in a 20 μL reaction volume, following the manufacturer's instructions and with the addition of a RNase Inhibitor.
4. Perform a PCR using 2 μL cDNA template and primers for the constitutively expressed *CB5-E* (At5g53560) as a control (*see* **Note 15**). Primers for cell-type-specific expressed genes such as the egg cell-specific *EC1.2/DD45* (At2g21740) [14, 15], the central cell-specific *DD65* (At3g10890), and the synergid-specific *DD31* (At1g47470) [15] are suitable to verify the identity of collected cells. Primer sequences for RT-PCR are as follows:

CB5-E_fw	AGGCGATGAAGTCTTGTTGTCC
CB5-E_rev	CCTTTGGCTTCTTCTAGTCTTTCT
EC1.2_fw	AACTCTCCCGGAGACGGAAG
EC1.2_rev	AAGAACGTTGGTTTCATCAGAG
DD65_fw	AAAACCCTTACGGCTTGGATGAAT
DD65_rev	TTGCGTTGAACACTGCGTCTCT
DD31_fw	CAGGAGGTGCCAAAAGATGT
DD31_rev	AAAAGAGACCGGAGCACCAAT

5. Prepare 2% agarose gels and analyze the PCR products by gel electrophoresis (*see* **Note 16**).

4 Notes

1. The use of homozygous transgenic plant lines expressing a fluorescent marker protein in the nucleus or the cytoplasm of the target cell is beneficial to identify it among all the other protoplasts released from the ovules after enzymatic treatment and to recognize damaged cells. Southern blot analysis confirmed a single-copy T-DNA integration in the homozygous plant line 17A, expressing the reporter construct *EC1.1p:NLS-3xGFP*.
2. It may be necessary to slightly adjust the osmotic concentration and the composition of the enzyme solution, depending on the plant ecotype and growth conditions.
3. The ovules of damaged pistils will dry out and cannot be used for cell isolation.
4. The method described is also suitable for the isolation of *Arabidopsis* zygotes. In this case, emasculated pistils need to be hand-pollinated 2 days after emasculation with a suitable paternal marker line to clearly distinguish the protoplasts of unfertilized egg cells and zygotes during the cell isolation procedure.
5. Due to surface tension and attached air bubbles, the ovules tend to stay on the surface of the enzyme solution. Avoid this by pushing the needle with the ovules relatively fast and steep into the drop. As the ovules can stick quite tight to the needle, peel them off by turning the needle while touching the bottom of the cavity. Ovules that float on the surface can be transferred into the enzyme solution using the needle to push them very fast and repeatedly. All the remaining floating ovules should be

removed from the surface. They cannot be used and will later interfere with the micromanipulation process.

6. Gentle horizontal shaking supports the embryo sac cells to become released from the ovule. It also keeps the released protoplasts in the center of the cavity. This helps tremendously in finding them under the microscope.
7. To avoid uncontrolled aspiration of the liquid, it is important to keep the connection airtight when mounting the glass capillary to the micromanipulator.
8. Keep the initial drop of wash solution clean. Dispense the used wash solution with a noticeable gap to the fresh drop.
9. Take care that the glass capillary is filled with wash solution to the very tip. This will help to keep the turbulences as low as possible when dipping the capillary into the drop of enzyme solution. It also avoids the cells to get in contact with air while being aspirated. This is very important, as they get damaged otherwise. 5 μL of wash solution equate approximately 10 mm liquid level in the glass capillary. Keep in mind that the wash solution in the glass capillary starts to dry out quite fast. The increasing concentration of mannitol in the solution will cause crystallization and clogging of the capillary (Fig. 6b). Changing osmotic concentrations will also damage the embryo sac cells. It is therefore important to wash the capillary after each transfer of cells with ethanol and wash solution.
10. Keep the time of UV light exposure as short as possible to avoid phototoxic effects and damaging of the cells.
11. The ovules float relatively easy. By dispensing wash solution not at the very bottom but at the height of the ovules, only these will float and not the single cells.
12. The single cells should be monitored while moving out of the capillary into the droplet of washing solution. When the dispense speed is too fast, the cells will be swirled up or attach to the phase interface of wash solution and oil.
13. Take care to stop the cell isolation after 60 min of enzyme treatment. The embryo sac cells tend to start budding after too long enzyme treatment (Fig. 6c, d) or they shrink due to increasing osmotic concentration (Fig. 6e). Furthermore, embryo sac cells tend to fuse spontaneously after longer incubation in the enzyme solution (Fig. 6e–l). The fused cells still express the fluorescent reporter (here: *EC1.1p:NLS-3xGFP*) and look very similar to the target cell but are somewhat bigger in size. These hybrid cells contain nuclei and cytoplasm of two (Fig. 6g) or even three different cell types (Fig. 6k, l). The fused cell shown in Fig. 6k, l still shows marker fluorescence in the egg cell nucleus, but in addition the nuclei of the other cell

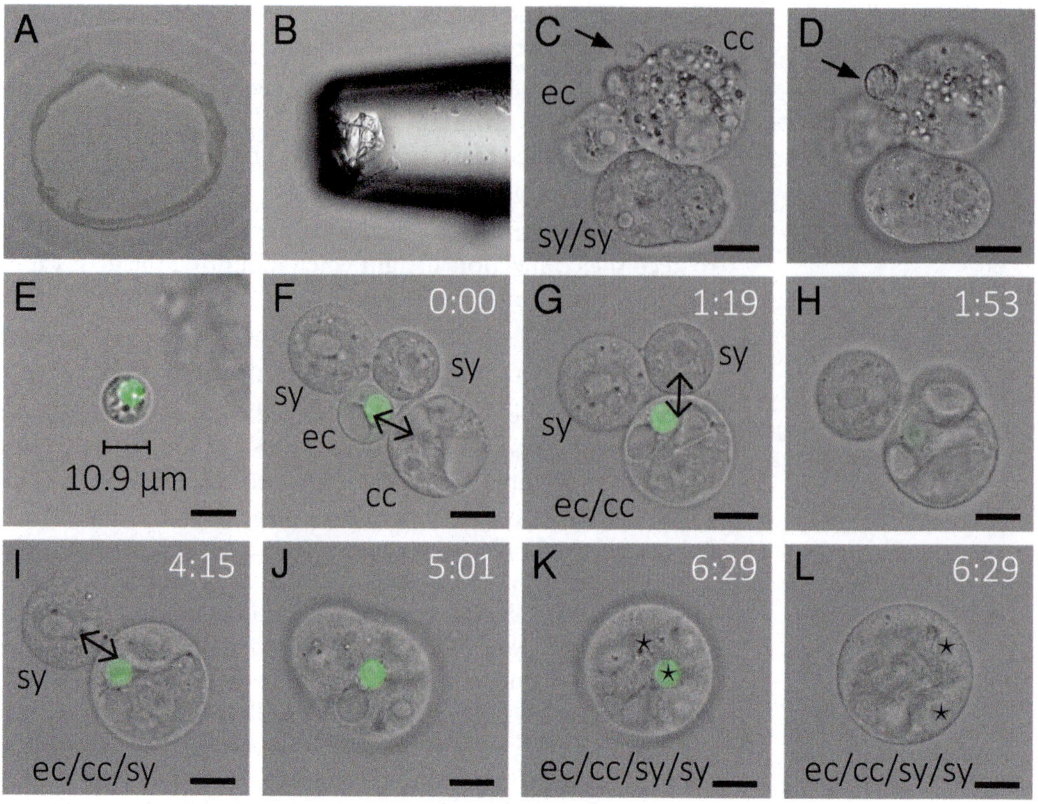

Fig. 6 Problems and pitfalls. (**a**) The drop of enzyme solution starts to form crystals. (**b**) Clogged glass capillary with crystals formed by the dehydrating mannitol solution. (**c** and **d**) Prolonged treatment in enzyme solution results in "budding" of the central cell (*arrows*) and spontaneous fusion of the synergid cells. (**e**) Egg cell, treated in a solution with too high osmotic concentration. Note the smaller size (<11 μm) and the lack of a clearly visible vacuole in this egg cell. (**f** to **l**) Time series of fusing embryo sac cells, observed after 90 min of enzyme treatment. Timestamps show min:s. *Double arrows* label the cells which will fuse shortly afterward. (**f**) Four cells of the embryo sac, still attached to each other. (**g**) The egg cell is fused with the central cell. Note that the egg cell nucleus (*green fluorescence*) and the vacuole of the egg cell are still visible within the fusion product. (**h**) One synergid cell fuses with the egg cell/central cell syncytium. (**i**) The fusion product is a syncytium of the egg cell, central cell, and synergid cell. (**j**) The second synergid cell fuses with the syncytium. (**k** and **l**) Two different optical sections showing the final product of three cell fusions, observed within 6:29 min. Four visible nuclei are indicated (*asterisks*). Abbreviations: *cc* central cell, *ec* egg cell, *sy* synergid cell. *Scale bars* 10 μm

types are visible, too. Keep in mind that you may not observe the fusion event itself, as it takes only seconds.

14. For downstream applications such as mRNA isolation or DNA extraction, we use 500 μL Eppendorf® LoBind tubes to collect the cells. For one-step RT-PCR, cells are collected in 200 or 500 μL thin-walled PCR tubes.
15. To detect genomic contaminations in the cDNA samples, use intron-flanking primers whenever possible.

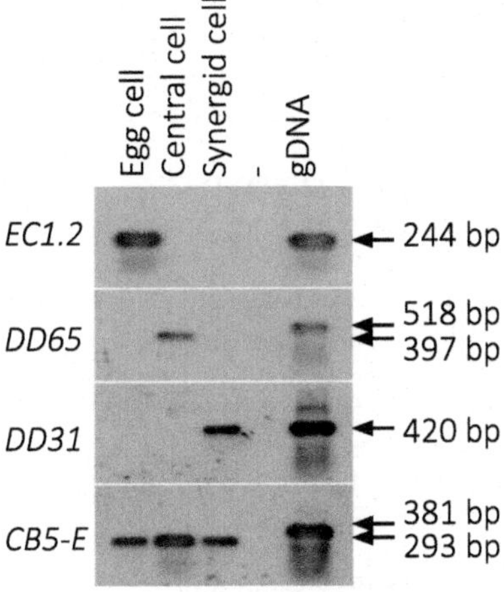

Fig. 7 Southern blot analysis of RT-PCRs to detect cell-type-specific expressed genes. The specificity of isolated cells and the quality of the cDNAs was verified by PCR analysis using cDNA from a 15-cell batch of each cell type, as indicated. Primers for the cell-specific-expressed genes *EC1.2/DD45* (egg cell), *DD65* (central cell), and *DD31* (synergid cell) were used for RT-PCR. After electrophoresis, the agarose gels were blotted and hybridized with respective radiolabeled probes to detect minor amounts of transcripts. Transcript sizes using cDNA and gDNA as a template are indicated. The control gene *CB5-E* encodes *Cytochrome b5 isoform E*. It is constitutively expressed in the cells of the female gametophyte

16. To increase the sensitivity of detection, blot the gels onto Hybond N+ nylon membranes. Hybridize the blots with radioactive or Digoxigenin (DIG)-labeled probes directed against *EC1.2/DD45*, *DD65*, and *DD31* as well as the constitutively expressed *CB5-E* (Fig. 7).

Acknowledgment

We gratefully acknowledge financial support by the German Research Council (DFG) grants SP 686/1-2 and SFB960 to S.S. We thank T. Dresselhaus for supporting this project and M. Gebert for her help in method development.

References

1. Sprunck S, Groß-Hardt R (2011) Nuclear behavior, cell polarity and cell specification in the female gametophyte. Sex Plant Reprod 24:123–136
2. Dresselhaus T, Sprunck S, Wessel GM (2016) Fertilization mechanisms in flowering plants. Curr Biol 26:R125–R139
3. Kranz E, Bautor J, Lörz H (1991) *In vitro* fertilization of single, isolated gametes of maize mediated by electrofusion. Sex Plant Reprod 4:12–16
4. Kranz E, von Wiegen P, Quader H, Lörz H (1998) Endosperm development after fusion of isolated, single maize sperm and central cells *in vitro*. Plant Cell 10:511–524

5. Holm PB, Knudsen S, Mouritzen P, Negri D, Olsen FL, Roue C (1994) Regeneration of fertile barley plants from mechanically isolated protoplasts of the fertilized egg cell. Plant Cell 6:531–543
6. Kumlehn J, Lörz H, Kranz E (1998) Differentiation of isolated wheat zygotes into embryos and normal plants. Planta 205:327–333
7. Sprunck S, Baumann U, Edwards K, Langridge P, Dresselhaus T (2005) The transcript composition of egg cells changes significantly following fertilization in wheat (*Triticum aestivum* L.) Plant J 41:660–672
8. Leljak-Levanić D, Juranić M, Sprunck S (2013) De novo zygotic transcription in wheat (*Triticum aestivum* L.) includes genes encoding small putative secreted peptides and a protein involved in proteasomal degradation. Plant Reprod 26:267–285
9. Sun MX, Moscatelli A, Yang HY, Cresti M (2000) *In vitro* double fertilization in *Nicotiana tabacum* (L.): fusion behavior and gamete interaction traced by video-enhanced microscopy. Sex Plant Reprod 12:267–275
10. Zhao J, Xin H, Qu L, Ning J, Peng X, Yan T, Ma L, Li S, Sun MX (2011) Dynamic changes of transcript profiles after fertilization are associated with de novo transcription and maternal elimination in tobacco zygote, and mark the onset of the maternal-to-zygotic transition. Plant J 65:131–145
11. Zhao J, Zhou C, Yang HY (2000) Isolation and *in vitro* culture of zygotes and central cells of *Oryza sativa* L. Plant Cell Rep 19:321–326
12. Uchiumi T, Komatsu S, Koshiba T, Okamoto T (2006) Isolation of gametes and central cells from *Oryza sativa* L. Sex Plant Reprod 19:37–45
13. Kasahara RD, Portereiko MF, Sandaklie-Nikolova L, Rabiger DS, Drews GN (2005) *MYB98* is required for pollen tube guidance and synergid cell differentiation in *Arabidopsis*. Plant Cell 17:2981–2992
14. Sprunck S, Rademacher S, Vogler F, Gheyselinck J, Grossniklaus U, Dresselhaus T (2012) Egg cell–secreted EC1 triggers sperm cell activation during double fertilization. Science 338:1093–1097
15. Steffen JG, Kang I-H, Macfarlane J, Drews GN (2007) Identification of genes expressed in the Arabidopsis female gametophyte. Plant J 51:281–292

Part V

Omics Technologies and Bioinformatic Analyses

Chapter 19

Isolation and Detection of Small RNAs from Pollen

German Martinez

Abstract

In the last decade, the accumulation and roles of small RNAs have been slowly uncovered. Recently, the RNA silencing pathways active in the male gametophyte of plants have started to be analyzed in depth. Although several studies have shed light on the small RNA populations present in the pollen, we still lack a clear picture of their regulatory activity. This chapter outlines an extraction method of mature pollen grains and its different downstream applications for the purification and detection of small RNAs.

Key words Small RNAs, RNA silencing, Pollen grain, Northern blot, Transposable elements, Sperm cells, *Arabidopsis thaliana*

1 Introduction

Angiosperms life cycle alternates from a diploid sporophytic form to a highly reduced haploid gametophytic form. The male gametophyte of angiosperms (the pollen grain) is a fascinating and very specialized structure [1]. It is responsible for the delivery of the paternal genomic information (sometimes upon physical and temporary obstacles) to the embryo sac in a process termed "double fertilization" [2, 3]. In this process, two symmetric sperm cells, contained within the pollen grain, fertilize the central cell and the egg cell in order to produce the endosperm and embryo respectively [3]. The obvious relevance of pollen grain in plant reproduction makes it an interesting model system not only for the study of the genetics and genomics of sexual reproduction but also for the study of cell cycle and development [2]. Together with this, it is becoming increasingly evident that the epigenetic information contained in the pollen grain/sperm cells is relevant for successful reproduction. This highlights the need for a detailed analysis of the RNA silencing pathways active in this cell in order to uncover the genetic and epigenetic processes taking place during gamete development.

Anja Schmidt (ed.), *Plant Germline Development: Methods and Protocols*, Methods in Molecular Biology, vol. 1669, DOI 10.1007/978-1-4939-7286-9_19, © Springer Science+Business Media LLC 2017

In the last 10 years, our knowledge of the RNA populations in the gametophytes has grown in parallel with the development of genomic and high-throughput sequencing techniques [2, 4–8]. This has allowed for the study of the transcriptomics of the pollen grain and sperm cells through microarrays [6, 9, 10] and recently through high-throughput RNA sequencing [11]. Together with this, the population of micro RNAs (miRNAs) [8, 12, 13] and small interfering RNAs (siRNAs) [14, 15] has been uncovered in the pollen grain of different plant species. These analyses have revealed that the miRNA and siRNA populations in the male gametophyte are completely different from sporophytic tissues, highlighting its specialized nature and an interesting epigenetic reprogramming of transposable element sequences taking place during gamete maturation [14–16].

Here, we describe a method for the extraction of high amounts of pure mature pollen grains that is suitable for the extraction of RNA or DNA for high-throughput sequencing purposes. This protocol has been successfully used for the high-throughput sequencing of small RNAs from the *Arabidopsis thaliana* Columbia-0 ecotype wild type and *dcl2/4* double mutant pollen grains [15]. It has also been applied to other species with slight modifications in the pollen extraction part [17].

2 Materials

Prepare all solutions using Milli-Q water. All glass material should be autoclaved before use. Radioactive material handling must follow all waste disposal regulations.

2.1 Plant Material, Pollen Isolation, and Total RNA Extraction

1. A flat of *Arabidopsis thaliana* ecotype Columbia-0 grown under long day conditions (*see* **Note 1**).
2. Pollen extraction buffer: Weight 36 g of sucrose in 200 ml Milli-Q water.
3. Autoclaved 50 ml falcon tubes and 1.5 ml microcentrifuge tubes.
4. Centrifuge with swinging bucket rotor and temperature control.
5. Centrifuge for microcentrifuge tubes with temperature control.
6. Trizol (Invitrogen).
7. Isopropanol.
8. Chloroform.
9. 75% Ethanol (v/v) solution.

2.2 Pollen Viability Analysis

1. Staining buffer for DAPI: 0.1% Nonidet P40 (v/v), 10% DMSO (dimethylsulfoxid) (v/v), 50 mM PIPES (piperazine-*N,N′*-bis(2-ethanesulfonic acid), pH 6.9) and 5 mM EGTA (Ethylene glycol-bis(2-aminoethylether)-*N,N,N′,N′*-tetraacetic acid, pH 7.5).
2. DAPI stock solution: Dissolve 1 mg of DAPI (4′,6-Diamidino-2-phenylindole) in 1 ml of DMSO 1 M.
3. DAPI staining solution: Mix 50 μl of the staining buffer for DAPI with 50 μl of the DAPI stock solution.

2.3 Denaturing Polyacrylamide Gel

1. TBE 5× running buffer: Weigh 54 g of Tris base, 27.5 g of boric acid and 20 ml of 0.5 M EDTA (Ethylene-diamine-tetraacetic acid, pH 8.0) and dissolve in 1 l of Milli Q-autoclaved water (*see* **Note 2**).
2. MOPS 10× running buffer: Add 41.8 g of MOPS (3-morpholinopropane-1-sulfonic acid) salt to 900 ml of water. Equilibrate the solution to pH 7 with NaOH and bring to a final volume of 1 l. Filter-sterilize and store in a dark place or wrapped with aluminum foil.
3. 15% Acrylamide solution for sRNA extraction: Mix 18.75 ml of 40% (19:1) Acrylamide/bis solution, 21 g of UREA and 5 ml of 5× TBE solution. Bring to a final volume of 50 ml with water. Dissolve in a water bath at 65 °C mixing gently by hand every 5 min until the urea is completely dissolved.
4. 15% Acrylamide solution for sRNA Northern blot: Same solution as in **item 3** but substituting 5× TBE for 10× MOPS.
5. Ammonium persulfate: 10% (w/v) solution in water.
6. TEMED (*N,N,N′,N′*-Tetramethylethan-1,2-diamin).
7. RNA Gel Loading Dye (10 ml of stock solution): 95% formamide, 18 mM EDTA, 0.025% SDS (w/v), 10 mg bromophenol blue, 10 mg of xylene cyanol.
8. Glass plates for vertical electrophoresis (*see* **Note 3**).
9. Power source.

2.4 Gel Transfer and Northern Blot

1. Neutral nitrocellulose membranes (*see* **Note 4**).
2. 3MM Whatman paper.
3. 0.5× TBE solution: Dissolve 50 ml of 5× TBE in 500 ml of water.
4. 500 ml of autoclaved Milli-Q water or autoclaved diethyl pyrocarbonate (DEPC)-treated water. For DEPC treatment, add 0.5 ml of DEPC to 500 ml of water, shake vigorously and incubate for 12 h at 37 °C.
5. Semi-dry blotting transfer apparatus.

6. EDC-chemical crosslinking solution: Add 0.753 g of EDC (1-ethyl-3-(3-dimethylaminopropyl) carbodiimide hydrochloride) to 20 ml of a 0.13 M 1-methylimidazole. Mix and make up to a total volume of 24 ml.
7. DNA primer complementary to the miRNA or siRNA sequence of interest or cDNA with the T7 or T3 promoter sequences flanking the siRNA-producing region of interest.
8. Gamma P32/P33 ATP or alpha P32/P33 UTP (*see* **Note 5**).
9. 10 units/μl T4 Polynucleotide Kinase (PNK) or T7/T3 (15 units/μl and 30 units/μl respectively) in vitro transcription kit.
10. Unlabeled nucleotides: 10 mM CTP, 10 mM GTP and 10 mM ATP (10 mM) for in vitro T7/T3 transcription of cDNA with P32/P33 labeled UTP.
11. Sephadex G25 or G50 (RNase-free) spin columns.
12. Sodium carbonate buffer: Dissolve 0.672 g of sodium bicarbonate in 80 ml of water. Next, dissolve 1.2 g of sodium carbonite. Bring to 100 ml final volume.
13. 3 M sodium acetate solution (pH 5.5): Dissolve 204.12 g sodium acetate-3H_2O (MW = 136.08) in 400 ml H_2O. Adjust to pH 5 with glacial acetic acid and bring to a final volume of 500 ml with water.
14. 2 units/μl DNase I.
15. Hybridization solution suitable for small RNAs (*see* **Note 6**).
16. Thermoblock.
17. Hybridization oven.
18. Phosporimager screen.
19. Phospor screen scanner.
20. Northern blot washing solution: 2× SSC and 0.2% (w/v) SDS.
21. Northern blot stripping solution: 0.1× SSC and 0.2% (w/v) SDS.
22. 20× SSC stock solution: Dissolve 175.3 g of sodium chloride in 800 ml of water. Next, dissolve 88.2 g of sodium citrate. Adjust pH to 7.2 and bring to 1 l final volume.

2.5 sRNA Precipitation and Elution

1. Glycoblue (Ambion).
2. 3 M sodium acetate solution (pH 5.5): Dissolve 204.12 g sodium acetate-3H_2O (MW = 136.08) in 400 ml H_2O. Adjust to pH 5 with glacial acetic acid and bring to a final volume of 500 ml with water. This solution must be autoclaved before use.

3. 25% (w/v) PEG solution: 25% (w/v) PEG (polyethylene glycol, MW = 8000) in autoclaved Milli-Q water. This solution must be autoclaved before use.
4. 5 M sodium chloride solution: Mix 146 g of NaCl into 400 ml of Milli-Q water. Mix well and adjust the volume to 500 ml. This solution must be autoclaved before use.
5. Phenol:Chloroform:Isoamyl Alcohol (24:24:1) solution.
6. 75% (v/v) ethanol solution.
7. 10% (w/v) SDS solution.
8. sRNA elution buffer: 0.3 M sodium chloride solution. This solution must be autoclaved before use.

3 Methods

3.1 Mature Pollen Extraction

The method described here is suitable for *Arabidopsis thaliana* and could be suitable for plants with dehiscent pollen grains. A graphical overview of the protocol is supplied in Fig. 1.

1. Prepare an ice bucket and place as many 50 ml falcon tubes as samples to be harvested. Avoid the ice touch the upper part of the tube.
2. Using flowering *Arabidopsis* plants, harvest the entire apical part of the shoots containing open and unopened flowers (*see* **Note** 7). Place these apical parts in the 50 ml falcon tubes. This step can take from 30 min to 1 h depending on the level of flowering of the plants.
3. Once enough inflorescences are collected (till the tube is full with approximately 100–150 inflorescences), add 20 ml pollen extraction buffer and mix gently by hand for 20 s.
4. Place the tubes in a vortex adaptor and mix for 5 min at medium-high revolutions.
5. Remove tubes from the vortex and place the tubes back in the ice bucket. Add additional 20 ml of pollen extraction buffer and balance the tubes for centrifugation.
6. Spin tubes at 3000 × *g* for 5 min in a swinging bucket rotor pre-cooled at 4 °C. After this step, the pollen grains will pellet to the bottom and the rests of somatic tissue will rise to the top. If the amount of starting material is not enough to get a clearly visible pollen pellet at this step, continue with the protocol and check the concentrated pollen pellet in the last centrifugation step with the 1.5 ml tubes (**steps 8** and **9**).
7. After centrifugation, pour off the extraction buffer together with the tissues from the tube gently but firmly. Pollen pellet is

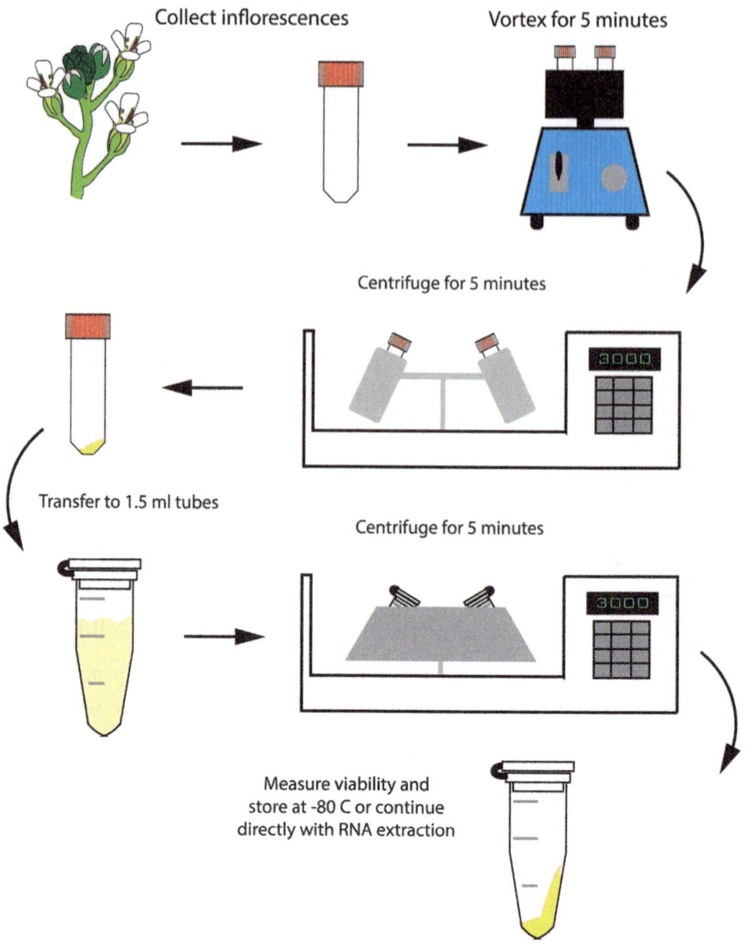

Fig. 1 Cartoon representation of the mature pollen extraction protocol

slightly disturbed by this process. Try to pour off the buffer through the opposite tube wall where the pollen pellet sits.

8. Add 1 ml of pollen extraction buffer to the pollen pellet and dissolve by pipetting gently. Transfer the solution in two steps of 0.5 ml in order to avoid precipitation of the pollen grains to a 1.5 ml microcentrifuge tube.
9. Centrifuge at 3000 × *g* for 5 min at 4 °C. The pollen should pellet in a clearly visible yellow mass.
10. Remove the buffer by pipetting, take a 5 μl sample to assess pollen quality and quickly freeze in liquid nitrogen or store at −80 °C

3.2 Pollen Viability

1. Place a 5 μl drop of extracted pollen in a microscope slide and add 5 μl of DAPI staining solution. Cover gently with a glass cover and (preferably) seal the slide (*see* **Note 8**).

2. Incubate for 30 min protected from light.
3. Visualize under an epifluorescence microscope (340–380 nm).
4. Determine the percentage of intact pollen grains. Usually around 85–90% of viable pollen grains are obtained with this method. If viability is lower than this in wild-type plants re-check the extraction protocol steps and make sure that temperature is kept low in all the centrifugation steps.

3.3 Small RNA Extraction

3.3.1 Total RNA Extraction

1. Prepare an autoclaved mortar and pestle. Clean with a 75% (v/v) ethanol solution and cool down with liquid nitrogen. Always use protection when working with liquid nitrogen.
2. Dissolve pollen pellet in 1 ml of Trizol (*see* **Note 9**). The pollen pellet should be frozen and can take up to 1 min to dissolve in the Trizol solution.
3. Pipette up and down the Trizol solution in order to avoid the sedimentation of the pollen grains, and transfer to the liquid nitrogen-cold mortar. Avoid to the minimum the pipetting of the Trizol as the pollen grains get attached to the pipette tip walls easily.
4. Trizol solution containing the pollen grains will freeze into a pink solid mass. Grind the frozen Trizol solid mass into a fine powder (*see* **Note 10**).
5. Once grinded to the maximum, melt the Trizol powder. Grind the liquid Trizol solution.
6. Transfer solution and follow the manufacturer's instructions for total RNA extraction.

3.3.2 Small RNA Enrichment

Once total RNA is obtained, in order to follow downstream procedures like sRNA library preparation or Northern blot, it is recommended to enrich the percentage of sRNAs present in the sample. This will help to estimate the real concentration of sRNAs and improve the library quality or detection. Two different enrichment methods, differential precipitation of sRNAs (**steps 1–9**) and small RNA purification from acrylamide gel (**steps 10–23**), are presented here (*see* **Note 11**).

1. Resuspend total RNA in a volume of 350 μl of autoclaved Milli-Q or DEPC treated water. Heat at 65 °C for 5 min and place on ice.
2. Add 100 μl of the 25% PEG solution and 50 μl of the 5 M NaCl solution. Mix well by vortexing.
3. Place on ice for 30 min.
4. Centrifuge at 10,000 × *g* for 10 min at 4 °C. After centrifugation, the supernatant contains the sRNA fraction and long RNAs are pelleted.

5. Transfer the supernatant to a new tube and add an equal volume of Phenol:Chloroform:Isoamyl alcohol (25:24:1, v/v/v) solution. Vortex for 5 min and centrifuge the emulsion for 10 min at 4 °C. This step removes PEG from the sample.
6. Transfer the supernatant to a new tube and add 3 volumes of ethanol. Mix by inverting the tube several times and place at −20 °C for at least 2 h.
7. Centrifuge at 10,000 × *g* for 15 min. The sRNA fraction should form a pellet in the bottom of the tube.
8. Wash pellet with a 75% (v/v) ethanol solution.
9. Resuspend in 25–50 μl of autoclaved Milli Q or DEPC-treated water.
10. Wash the electrophoresis apparatus, glass plates, combs, and spacers with hot water and SDS 10% (w/v). After washing clean with a 75% (v/v) ethanol solution and let evaporate for 5 min before assembling the plates.
11. Prepare 50 ml of 15% (w/v) acrylamide solution and place on ice for 15 min once the urea is dissolved.
12. Mix the 15% acrylamide/7 M urea solution with 250 μl of the 10% APS solution and mix by carefully inverting the tube. Avoid as much as possible the formation of bubbles.
13. Add 20 μl of TEMED and carefully mix by inverting the tube. Again, avoid the formation of bubbles.
14. Pour the mix in the assembled glass plates and place the comb. Let polymerize for 30 min (*see* **Note 12**).
15. After polymerization, remove the comb and place in the electrophoresis apparatus. Add 0.5× TBE running buffer to the assembled electrophoresis unit. If air bubbles appear at the bottom of the gel try to remove them with a syringe filled with 0.5× TBE buffer and an insulin needle.
16. Pre-run the gel at 200–300 V for 30 min.
17. While the gel is pre-running, prepare the RNA samples and the RNA size marker (*see* **Note 13**) for the electrophoresis. Prepare samples in a volume from 20 to 40 μl of water (*see* **Note 14**).
18. Add an equal volume of RNA gel loading dye solution to the RNA samples and place at 65 °C for 5 min. Immediately place on ice after the heat treatment.
19. Once the pre-run of the gel is completed, wash the wells thoroughly with a syringe loaded with 0.5× TBE buffer in order to remove the excess of urea.
20. Carefully load the samples and the sRNA size marker.
21. Run at 200–300 V for 4–5 h (*see* **Note 15**).

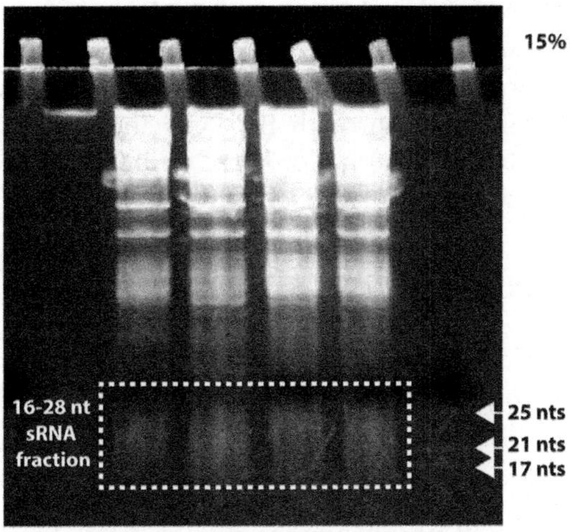

Fig. 2 Ethidium bromide-stained 15% acrylamide/7 M urea gel after total RNA running. The small RNA containing region is contained between the ladder marker and is highlighted in the image by *dashed lines*. For high-quality small RNA purification from pollen, this region must be isolated and eluted

22. Dismantle apparatus, mark gel in order to keep track of sample orientation and place on a clean plastic tray containing 0.5× TBE running buffer mixed with the appropriate amount of ethidium bromide or SYBR gold.
23. Stain the gel for 15 min with very soft movement.
24. Place on an UV transilluminator and identify the region of the gel containing the sRNA region present between the two sRNA size markers (Fig. 2).
25. Cut with a new razor blade that portion of the gel. Use a different razor blade for each sample in the gel.
26. Break into small pieces with the same razor blade or use tubes suitable for breaking acrylamide gels into pieces.
27. Cover the gel pieces with sRNA elution buffer.
28. Incubate overnight at 4 °C with rotation.
29. Proceed to sRNA precipitation by adding 1 μl of glycoblue and 10% sodium acetate 3 M (pH 5.5). Mix by vortexing or pipetting and add 1 volume of isopropanol.
30. Incubate at −20 °C for 2 h and centrifuge at 12,000 × *g* for 15 min.
31. Clean pellet with a 75% ethanol (v/v) solution.
32. Resuspend in 10–20 μl of autoclaved Milli-Q or DEPC-treated water or suitable volume for sRNA library preparation (*see* **Note 16**).

3.4 Northern Blot Detection of Small RNAs

Northern blot is an sRNA detection method with medium to low sensitivity, but allows the detection of multiple sRNAs with one transferred gel. There are other methods for the detection of miRNAs through RT-PCR and qRT-PCR with higher sensitivity, but Northern blot keeps on being one of the best methods for the detection of siRNAs with high to medium levels of accumulation.

1. Prepare a 15% acrylamide/7 M urea gel, load and run samples as in Subheading 3.3.2, **steps 10–23**, but use 1× MOPS instead of 0.5× TBE to prepare the gel acrylamide solution and the running buffer (*see* **Note 17**).
2. Dismantle apparatus, mark gel in order to keep track of sample orientation and place on a clean plastic tray containing autoclaved Milli-Q water.
3. Cut one piece of nitrocellulose neutral membrane and six pieces of 3MM Whatman paper with the same size as the gel to be transferred.
4. Wet the membrane and the six paper pieces in autoclaved Milli-Q water.
5. Place in the semi-dry electroblotting apparatus carefully and individually three sheets of paper, the membrane, the gel and three more sheets of paper as indicated in Fig. 3. Remove air bubbles carefully on each layer.
6. Transfer at the constant voltage for 60 min at 4 °C.
7. During the transfer prepare the EDC-chemical crosslinking solution and a plastic tray containing a sheet of Whatman paper approximately 1 cm larger than the membrane on each dimension. Pre-heat an oven at 60 °C.
8. After the transfer, soak the Whatman paper with the EDC-crosslinking solution and place the membrane on top of the

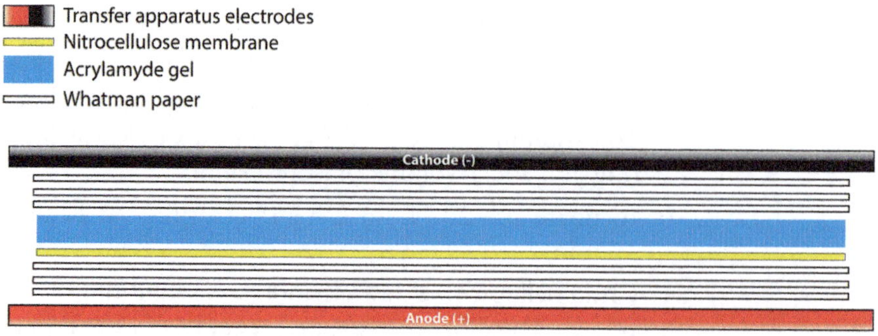

Fig. 3 Gel-membrane assembly for the electric transfer of sRNAs to the nitrocellulose membrane. The acrylamide gel is sandwiched between with the nitrocellulose membrane, 3MM Whatman paper, and the two electrodes of the blot transfer apparatus. The whole transfer assembly must be placed in a cold room at 4 °C to avoid excessive warming

saturated paper. The membrane must have the side where the RNA has been transferred facing up.

9. Wrap in Saran wrap to avoid excessive evaporation and place in the oven for 2 h. If continuing with Northern blot detection, pre-heat hybridization oven at 40 °C.
10. Once the crosslinking is done, wash the membrane with Milli-Q water and proceed to Northern blot detection or alternatively dry and store overnight.
11. Place the membrane on a hybridization tube with 10–20 ml of hybridization buffer at 40 °C for at least 1 h.

    Synthesize the probe for detection of sRNAs under study during the pre-hybridization time (Subheading 3.5).
12. Add denaturalized (or hydrolyzed) probe directly to the hybridization solution within the hybridization tube.
13. Hybridize overnight at 40 °C.
14. Wash twice with 2× SSC/0.2% SDS solution at 50 °C for 15 min each wash.
15. Cut a piece of Whatman paper slightly larger than the membrane, wet with 2× SSC/0.2% SDS and place membrane on top of it.
16. Cover with Saran wrap and expose in phosporimager screen from 1 h or up to 1 week.
17. Blots can be stripped and re-hybridize up to five times. For stripping, wash membrane twice in 100 ml of a 0.2% SDS, 0.1× SSC solution at 65 °C for 15 min. If the membrane retains radioactivity, increase temperature to 85 °C and increase stripping time up to 1 h.

3.5 Probe Synthesis for Northern Blot

Depending on the sequence to be detected, two different probes should be used. For miRNA (or specific siRNA) detection, it is recommended to end label a single DNA oligo with gamma P32 (or P33) ATP (Subheading 3.5.1). If the detection of siRNAs derived from longer regions is needed (e.g., a viral genome or a transposable element), it is recommended to synthesize a labeled RNA probe and degrade this to smaller fragments in order to enhance the detection of sRNAs. In the later case, start the synthesis of the probe with enough time to allow the right degradation of the probe in smaller fragments (Subheading 3.5.2).

3.5.1 miRNA Probe Synthesis

1. Mix 1 μl of a 10 μM oligo dilution with 16.5 μl of water, 2.5 μl of the T4 PNK buffer, 2 μl of T4 PNK and 3 μl of gamma P32 ATP (total of 30 μCi).
2. Incubate at 37 °C for 1 h.

3. After incubation clean the probe with a G25 column in order to get rid of unincorporated radioactivity.
4. It is recommended to measure the percentage of incorporation of the probe by measuring radioactive emissions with a Geiger counter before and after probe purification.
5. After probe synthesis, add 200 μl of hybridization buffer to the tube containing the probe, denaturalize at 95 °C for 5 min and place on ice for 20 s.

3.5.2 siRNA Probe Synthesis

1. Starting with a PCR fragment containing the T7/T3 promoters flanking the sequence of interest, mix 75 ng of that PCR product with water in order to have a final volume of 8 μl. Mix with 1 μl of ATP, 1 μl of CTP, 1 μl of GTP, 2 μl of transcription buffer, 2 μl of T7 or T3 polymerase and 5 μl of alpha P32 UTP.
2. Incubate at 37 °C for 1 h.
3. Add 1 μl of DNase I and further incubate for 15 min at 37 °C.
4. After incubation clean the probe with a G50 column in order to get rid of unincorporated radioactivity.
5. It is recommended to measure the percentage of incorporation of the probe by measuring radioactive emissions with a Geiger counter before and after probe purification.
6. Add 300 μl of sodium carbonate buffer and place at 60 °C to hydrolyze the probe. The time required for probe hydrolysis will depend on the initial length of the probe as well as its final length according to this formula:

$$\text{Time} = \frac{(\text{length of initial probe} - \text{length of final probe})}{0.11 \times (\text{length of initial probe} \times \text{length of final probe})}$$

7. After hydrolysis add 20 μl of 3 M NaOAc (pH 5).

4 Notes

1. Other species can be used for these experiments.
2. A concentrated 10× TBE solution can be prepared. In this case, it is recommended to pass the stock solution through a 0.22 μm filter to prevent the formation of precipitates.
3. Alternatively, mini PROTEAN gels can be used although it is recommended to increase the concentration of acrylamide to 20% if used.
4. Alternatively, positively charged membranes can be used.
5. There are several options available in the market for the synthesis of radioactive and non-radioactive small RNA probes. Radioactive probes have higher sensitivity than non-radioactive

probes, but instead cannot be reused. The use of radioactivity also requires special permits and certain skills/knowledge.

6. Several hybridization solutions suitable for sRNA detection are available including commercial ones. Based on authors' experience, Ambion's Ultrahyb-Oligo hybridization solution offers low levels of background signal, but homemade hybridization buffers can work equally well. Specifically, for DNA oligos it is recommended the use of the Church buffer [18].
7. If an inflorescence control is required for the experiment sample inflorescences before removing open flowers from bolts.
8. Sealing is optional but can help to avoid evaporation of the DAPI staining solution. For sealing apply a layer of sealing compound (nail polish, bee wax, or other commercial sealing reagents) around the edge of a coverslip and let it dry for a couple of minutes.
9. Trizol (Invitrogen) is recommended based on author's experience. Nevertheless, similar commercial products or RNA extractions buffers followed by phenol:chloroform extraction may work in a similar way.
10. Thoroughly grinding is recommended.
11. Both methods will cause losing up to 25% of the sRNAs present in the sample. If total RNA concentration is extremely low, it is recommended to follow downstream applications without following these enrichment procedures. Nevertheless, for pollen sRNA library preparation, it is highly recommended to follow the sRNA gel extraction enrichment strategy.
12. Save an aliquot of the mix in the falcon tube in order to monitor the polymerization of the solution.
13. It is recommendable to use an sRNA size marker in order to perform an accurate size selection of the sRNAs. There are several commercially available sRNA marker ladders (e.g., MicroRNA marker from NEB). Alternatively, RNA oligonucleotides of sizes between 28 and 18 nt can be used.
14. Volume to be loaded will be determined by the width of the gel and size of the wells. If unsure, measure the size of the well and width of the gel and calculate the cubic millimeter volume of the well ($1\ mm^3 = 1\ \mu l$).
15. A good indicative of the RNA size separation in the gel comes from the loading dyes that mark 10 nts (bromophenol blue) and 40 nts (xylene cyanol) in a 15% acrylamide gel.
16. If sRNAs are enriched for the preparation of sRNA libraries, it is highly recommended to use exclusively the gel extraction method and not the PEG enrichment or commercially available columns. As can be shown in Fig. 2, sRNA gel extraction can

help to remove unwanted sRNAs competing for RNA adapters during library preparation.

17. 0.5× TBE acrylamide/urea gels can be used for Northern blot detection, but 1× MOPS acrylamide/urea gels are recommended for the chemical crosslinking method explained here.

Acknowledgement

The author apologizes for the works that were not cited owing to length restraints. The author thanks Keith Slotkin, Claudia Köhler, Christopher DeFraia, and Saivageethi Nuthikattu for their mentorship and pollen expertise exchange. The author specially thanks Clément Lafon-Placette for the critical reading of this manuscript.

References

1. Wilson ZA, Zhang DB (2009) From Arabidopsis to rice: pathways in pollen development. J Exp Bot 60:1479–1492
2. Borg M, Brownfield L, Twell D (2009) Male gametophyte development: a molecular perspective. J Exp Bot 60:1465–1478
3. Berger F, Twell D (2011) Germline specification and function in plants. Annu Rev Plant Biol 62:461–484
4. Lee JY, Lee DH (2003) Use of serial analysis of gene expression technology to reveal changes in gene expression in Arabidopsis pollen undergoing cold stress. Plant Physiol 132:517–529
5. Becker JD, Boavida LC, Carneiro J, Haury M, Feijo JA (2003) Transcriptional profiling of Arabidopsis tissues reveals the unique characteristics of the pollen transcriptome. Plant Physiol 133:713–725
6. Honys D, Twell D (2003) Comparative analysis of the Arabidopsis pollen transcriptome. Plant Physiol 132:640–652
7. Honys D, Twell D (2004) Transcriptome analysis of haploid male gametophyte development in Arabidopsis. Genome Biol 5:R85
8. Borges F, Pereira PA, Slotkin RK, Martienssen RA, Becker JD (2011) MicroRNA activity in the Arabidopsis male germline. J Exp Bot 62:1611–1620
9. Pina C, Pinto F, Feijo JA, Becker JD (2005) Gene family analysis of the Arabidopsis pollen transcriptome reveals biological implications for cell growth, division control, and gene expression regulation. Plant Physiol 138:744–756
10. Borges F, Gomes G, Gardner R, Moreno N, McCormick S, Feijo JA, Becker JD (2008) Comparative transcriptomics of Arabidopsis sperm cells. Plant Physiol 148:1168–1181
11. Loraine AE, McCormick S, Estrada A, Patel K, Qin P (2013) RNA-seq of Arabidopsis pollen uncovers novel transcription and alternative splicing. Plant Physiol 162:1092–1109
12. Grant-Downton R, Le Trionnaire G, Schmid R, Rodriguez-Enriquez J, Hafidh S, Mehdi S, Twell D, Dickinson H (2009) MicroRNA and tasiRNA diversity in mature pollen of Arabidopsis thaliana. BMC Genomics 10:643
13. Grant-Downton R, Kourmpetli S, Hafidh S, Khatab H, Le Trionnaire G, Dickinson H, Twell D (2013) Artificial microRNAs reveal cell-specific differences in small RNA activity in pollen. Curr Biol 23:R599–R601
14. Slotkin RK, Vaughn M, Borges F, Tanurdzic M, Becker JD, Feijo JA, Martienssen RA (2009) Epigenetic reprogramming and small RNA silencing of transposable elements in pollen. Cell 136:461–472
15. Martinez G, Panda K, Kohler C, Slotkin RK (2016) Silencing in sperm cells is directed by RNA movement from the surrounding nurse cell. Nat Plants 2:16030
16. Borges F, Calarco JP, Martienssen RA (2012) Reprogramming the epigenome in Arabidopsis pollen. Cold Spring Harb Symp Quant Biol 77:1–5
17. Castellano M, Martinez G, Marques MC, Moreno-Romero J, Kohler C, Pallas V, Gomez G (2016) Changes in the DNA methylation pattern of the host male gametophyte of viroid-infected cucumber plants. J Exp Bot 67:5857–5868
18. Church GM, Gilbert W (1984) Genomic sequencing. Proc Natl Acad Sci U S A 81:1991–1995

Chapter 20

Analysis of Proteins Enriched in Rice Gamete

Takashi Okamoto

Abstract

In angiosperms, female gamete differentiation, fertilization, and subsequent zygotic development occur in embryo sacs deeply embedded in the ovaries. Despite their importance in plant reproduction and development, how the egg cell is specialized, fuses with the sperm cell, and converts into an active zygote for early embryogenesis remains unclear. Proteins enriched in each cell type are thought to reflect the biological function of the cells, since the composition of cellular proteins generally differs depending on cell type. Therefore, proteins enriched in gametes, egg and sperm cells, should have a role in reproductive and/or developmental processes, such as gamete differentiation, gamete fusion, egg activation, and early zygotic development. In this chapter, a method for identification of proteins enriched in rice gametes is first explained, and then functional analyses for these proteins using rice and *Arabidopsis* lines with mutations in genes encoding the putative gamete-enriched proteins are also described.

Key words Egg cell, Fertility, LC-MS/MS, Proteome, Rice, Sperm cell

1 Introduction

In the ovary of an angiosperm, the female gametophyte develops from a functional megaspore via three or more rounds of mitosis without cytokinesis. Subsequently, plasma membranes/cell walls are formed between the nuclei, resulting in a cellularized female gametophyte, generally known as an embryo sac. The embryo sac most commonly consists of one egg cell, one central cell, two synergid cells, and three antipodal cells [1, 2]. In the anthers, male microspores undergo an initial asymmetric mitotic division, and the smaller generative cell, which establishes the male germline, migrates into a larger vegetative cell. The generative cell divides into two sperm cells within the vegetative cell before or after germination of the pollen tube, which delivers sperm cells into the embryo sac [3]. Upon double fertilization, one sperm cell from the pollen tube fuses with the egg cell, and the resultant zygote develops into an embryo transmitting genetic material from the parents to the next generation. The second sperm cell fuses with the central cell to form a triploid primary endosperm cell,

Anja Schmidt (ed.), *Plant Germline Development: Methods and Protocols*, Methods in Molecular Biology, vol. 1669, DOI 10.1007/978-1-4939-7286-9_20,

which develops into the endosperm nourishing the developing embryo and later the seedling [3–5]. Within the embryo sac, the haploid egg cell, which is specially differentiated for fertilization and subsequent embryogenesis, fuses with the sperm cell and consequently converts into an active zygote [6, 7]. However, most mechanisms in such reproductive events remain unclear, despite their importance in plant reproduction and development.

Proteins enriched in each cell type are considered to reflect the biological function of the cells, since the composition of cellular proteins generally differs depending on cell type. Therefore, identification of the proteins enriched in gametes, egg and sperm cells, will provide basic knowledge of their character. Recently, single-cell proteomic approaches have been widely employed to dissect the functions of specific cells, because cellular-level information is diluted when organs or tissues, which comprise various differentiated cells, are used as starting materials [8]. Moreover, state-of-the-art proteomics technologies enable high-resolution and high-throughput analyses using limited numbers of cells, such as gametes isolated from flowers in angiosperms. Proteins expressed in egg and sperm cells isolated from rice flowers [9, 10] were globally detected by highly sensitive liquid chromatography coupled with tandem mass spectroscopy (LC-MS/MS) technology, and proteins that were preferentially expressed in female or male gametes were identified by comparison of protein expression profiles between gametes and somatic cells/pollen grains [11]. In addition, several rice or *Arabidopsis* lines with mutations in genes encoding the putative gamete-enriched proteins showed clear phenotypic defects in seed set or seed development. This suggests that the identification of gamete-enriched proteins and subsequent analyses using mutant plant lines are an effective experimental approach for understanding the mechanisms of reproduction and early development in angiosperms.

2 Materials

2.1 Isolation and Transfer of Gametes

1. Rice plants grown in an environmental chamber (*see* **Note 1**).
2. Inverted microscope.
3. Non-treated plastic dishes with a diameter of 3.5 cm.
4. Coverslips (24 × 40 mm), siliconized at the edges with 5% dichloromethylsilane in 1,1,1-trichloroethane (*see* **Note 2**).
5. Mineral oil (embryo culture tested grade).
6. Mannitol solution adjusted to 370 mOsmol/kg H_2O and autoclaved.
7. Sliding stage for the insertion of a coverslip and a plastic dish.

8. Glass capillaries made from 50 μL aspirator tubes (Drummond Scientific Co., Broomall, PA, USA), tip openings 150–250 μm (drawn by hand).
9. Glass needles with fine tips.
10. Manual handling injector for transfer of isolated egg cells (*see* **Note 3**).
11. 15% sucrose solution.
12. Percoll solutions containing sucrose: 60%, 20%, and 5% Percoll, containing 15% sucrose.
13. Refrigerated centrifuge for swing rotor.
14. 20-μm and 10-μm nylon bolting cloth.
15. Centrifugation tube (*see* **Note 4**).

2.2 Protein Extraction, SDS-PAGE/Silver-Staining, and Identification of Proteins

1. SDS-sample buffer: 2% (w/v) SDS, 25 mM Tris-Cl (pH 6.8), 30% (v/v) glycerol, 5% (v/v) 2-mercaptoethanol.
2. Compact slab gel electrophoresis system.
3. Small mold (50 × 60 × 1 mm).
4. Fixative: 50% (v/v) methanol, 7% (v/v) acetic acid.
5. Enhancement solution: 0.02% (w/v) sodium thiosulfate.
6. Silver staining solution: 1 mg/mL silver nitrate, 0.009% (w/v) NaOH, 0.047% (v/v) ammonia (*see* **Note 5**).
7. Color development solution: 2% (w/v) sodium carbonate, 0.04% (v/v) formalin.
8. 7% (v/v) acetic acid.
9. Calli derived from scutellum of rice seeds (*see* **Note 6**).
10. Rice seedlings with shoot lengths of 2.5–3 cm (*see* **Note** 7).
11. Acetoniterile.
12. 50% (v/v) acetoniterile, 5% (v/v) formic acid.
13. 100 mM Tris-Cl, pH 8.8.
14. Trypsin solution: 25 ng/μL trypsin dissolved with 100 mM Tris-Cl, pH 8.8.
15. A direct nanoflow LC-MS system equipped with an Orbi Trap XL (e.g., Thermo Scientific, Waltham, MA, USA).
16. Dataset of protein sequences in rice (Rice Annotation Project Database, Tsukuba, Japan; http://rapdb.dna.affrc.go.jp/download/irgsp1.html).
17. Mascot software (ver. 2.2.1 or higher, Matrix Science, Boston, MA, USA).

2.3 Analyses with Mutant Lines of Rice and Arabidopsis

1. Rice plants with insertion of Tos17 retrotransposon (*see* **Note 8**).
2. *Arabidopsis thaliana* plants with insertion of T-DNA (*see* **Note 9**).
3. Border primers for Tos17 and T-DNA insertions (*see* **Note 10**).
4. Primers specific for genes encoding gamete-enriched proteins (*see* **Note 11**).
5. Stereo microscope.

3 Methods

3.1 Isolation of Egg Cells and Preparation of Crude Extracts from Egg Cells

1. Collect panicles in which most flowers have matured and remain unflowered. Pick flowers from the panicles and dissect them. Isolate the ovaries and transfer into 3.5 cm plastic dishes filled with 3 mL of mannitol solution (370 mOsmol/kg H_2O).
2. Transfer the ovaries into new 3.5 cm plastic dishes filled with 3 mL of the mannitol solution, and remove the stigmas from ovaries (*see* **Note 12**). Sink the ovaries to the bottom of the dishes and cut them transversely with a razor blade at the middle (*see* **Note 13**). Transfer egg cells released from lower parts of the cut ovaries into a mannitol droplet on coverslips using the cell transfer system under an inverted microscope (*see* Fig. 1a, b).
3. Wash egg cells three times by transferring them into fresh droplets of mannitol solution on coverslips, and then transfer the cells into a 1 μL droplet of SDS-sample buffer. Transfer the egg cell lysate into a 0.2 mL tube, treat at 98 °C for 3 min, and store at −80 °C until use.

3.2 Isolation of Sperm Cells and Preparation of Crude Extracts from Sperm Cells

1. Collect approximately 120 anthers from mature and unflowered rice flowers in 3.5 cm plastic dishes filled with 3 mL of 370 mOsmol/kg H_2O mannitol. Then, transfer the anthers to four plastic dishes filled with 3 mL of 15% sucrose (approximately 30 anthers per dish), and break the tissues with forceps to free the pollen grains. After gentle shaking for 30 min, filter the sucrose solution, in which pollen grains released their sperm cells, through 20-μm then 10-μm nylon bolting cloth.
2. Add an equal volume of 60% Percoll containing 15% sucrose to the filtrate, mix gently, and transfer 4 mL of the mixture into a centrifugation tube (*see* **Note 4**). Over the mixture, gently add 20% and 5% Percoll containing 15% sucrose and 15% sucrose to form a discontinuous Percoll gradient in the tube.
3. After centrifugation at 3000 × *g* for 30 min at 4 °C, collect the interface between the 5% and 20% Percoll layers in a new 1.5 mL sample tube, and add an equal volume of 15% sucrose to the

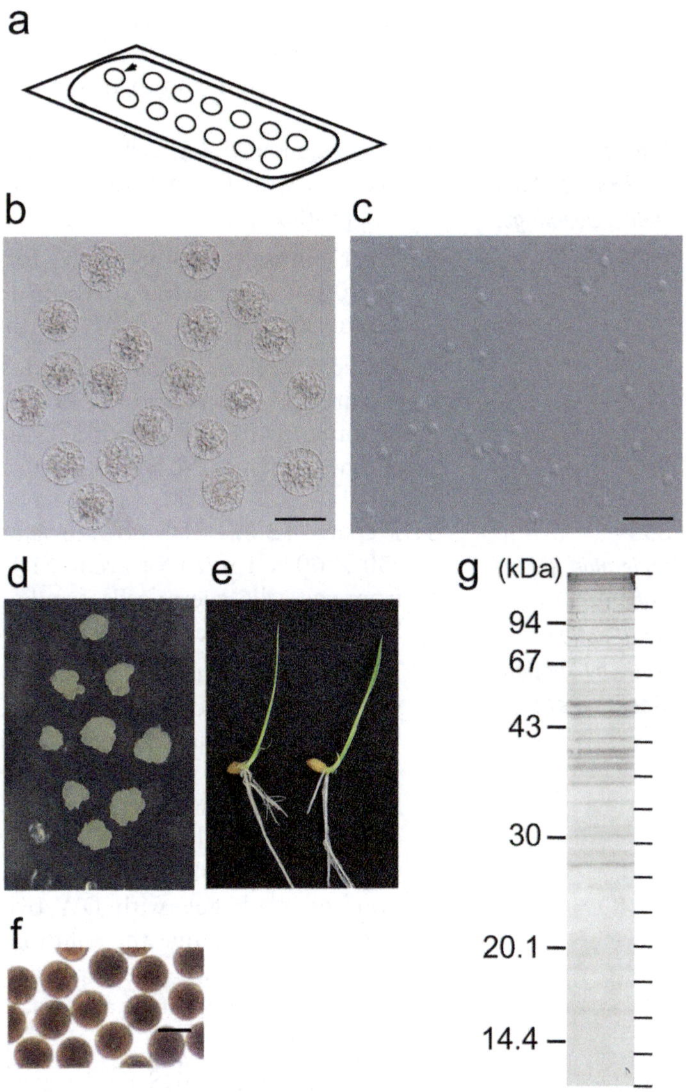

Fig. 1 Plant materials subjected to SDS-PAGE. (**a**) An illustration of the mannitol droplets on a coverslip covered with mineral oil. The *arrow* indicates a mannitol droplet. Approximately ten isolated egg cells were transferred to a drop. Empty droplets are used for washing cells. (**b**) Isolated egg cells. (**c**) Isolated sperm cells fraction. (**d**) Callus derived from scutellum of rice seeds. (**e**) Rice seedlings. (**f**) Pollen grains. (**g**) SDS-PAGE of lysates from 500 egg cells. After silver-staining, the gel was cut into 15 pieces and used for protein identification by LC-MS/MS. Panels b and c were cited from Abiko et al. [11] and are published under the Creative Commons Attribution (CC BY) license (http:/creativecommons.org/licenses/by/4.0/). Bars = 50 μm

collected fraction. After gentle mixing, centrifuge at 5000 × *g* for 4 min at 4 °C. Remove most of the supernatant, but leave the bottom 15 μL in the tube, in which isolated sperm cells are concentrated (*see* Fig. 1c and **Note 14**). Add an equal volume of SDS-sample buffer to the sperm cell fraction, treat the mixture at 98 °C for 3 min, and then store at −80 °C until use.

3.3 Preparation of Crude Extracts from Rice Callus, Seedlings, and Pollen Grains

1. Homogenize 0.3 g callus (*see* Fig. 1d) or five seedlings (*see* Fig. 1e) with 0.4 mL of SDS-sample buffer using a mortar and pestle.
2. Treat homogenates at 98 °C for 5 min, then centrifuge at 10,000 × *g* for 5 min at room temperature. Use the supernatants as lysates for SDS-PAGE and store at −80 °C until use.
3. Harvest pollen grains released from 10–20 anthers from mature and unflowered rice flowers (*see* Fig. 1f), and homogenize them with 0.5 mL of SDS-sample buffer using a mortar and pestle. Treat the homogenate as in Subheading 3.3, **step 2**.

3.4 SDS-Polyacrylamide Gel Electrophoresis, and Identification of Proteins by Tandem Mass Spectrometry

1. Prepare 12.5% SDS-polyacrylamide gels in a small mold (50 × 60 × 1 mm) for use in SDS-PAGE. Separate cell lysates from egg cells, sperm cells, seedlings, calli, and pollen grains in SDS-sample buffer using gel electrophoresis (*see* **Note 15**).
2. Incubate the separated SDS-PAGE gels in fixative for 10 min with shaking. Discard fixative, add 50% methanol, and incubate 10 min with shaking. Discard 50% methanol and then incubate gels with distilled water (DW) for 10 min with shaking. Discard DW and add enhancement solution. Incubate for 3 min with shaking and wash gels with DW twice. Add silver staining solution and incubate for 30 min. Discard silver staining solution and wash gels with DW twice. Add color development solution, and change the solution 5–10 times when the solution turns to brown color. Stop color development by exchanging the solution with 7% acetic acid. Gels can be stored at 4 °C in 1% acetic acid.
3. After cutting the SDS-PAGE gels into 15 pieces (*see* Fig. 1g), transfer each gel piece into a 0.6 mL sample tube, and crush it by centrifugation at 15,000 × *g* for 3 min at room temperature. Add 0.4 mL of water and incubate for 10 min. After centrifugation, remove water from the tube and repeat the wash of gels with water once. Add 0.2 mL of 100% acetoniterile to the gel precipitate and leave the sample for 10 min at room temperature. After centrifugation, remove acetoniterile and dry precipitates completely by vacuum. Add 10 μL of trypsin solution, and leave the sample for 15 min on ice. Add 20 μL of 100 mM Tris-Cl (pH 8.8) and incubate at 37 °C overnight.
4. Add 50 μL of 50% acetoniterile/5% formic acid, and leave the sample for 10 min at room temperature. After centrifugation at 5000 × *g* for 1 min, transfer the supernatant into new sample tube (Sup 1). Repeat this step once. Add 200 μL of 100%

acetoniterile to the gel precipitate, and incubate for 10 min at room temperature. After centrifugation at 5000 × *g* for 1 min, transfer the supernatant into the Sup 1 tube, and mix the content in the tube. After the centrifugation at 5000 × *g* for 1 min, transfer the supernatant to a new sample tube, and concentrate into the volume of approximately 50 μL by vacuum. Store the sample at 4 °C.

5. Apply samples of Subheading 3.4, **step 4** to a direct nanoflow LC-MS system equipped with an Orbi Trap XL mass spectrometer [12]. Collect all MS/MS spectra by search engine, Mascot program, against the dataset of protein sequences obtained from the Rice Annotation Project Database. In case of search, set parameters as follows: the fixed modification was propionamide (Cys) and variable modification parameters are pyro-Glu, acetylation (protein N-terminus), and oxidation (Met). Set the maximum missed cleavage at 3 with a peptide mass tolerance of +/− 15 ppm, and allow peptide charges from +2 to +4 states and MS/MS tolerances of +/− 0.8 Da. Determine the criteria for peptide identification based on the vendor's definitions (expectation value <0.05, Matrix Science), and assign the protein "identified" if at least two peptides were identified from the protein (*see* **Note 16**).

3.5 Identification of Proteins Enriched in Rice Gametes

1. Select proteins with >2 identified spectra in egg cells and none in other cell types (sperm cell, callus, seedlings, pollen grains) to screen proteins enriched in egg cells. In addition, chose proteins with >3 identified spectra in egg cells and one in another cell type as candidates for egg cell-enriched proteins. In Table 1, proteins putatively enriched in rice egg cells with five identified spectra are listed [11].
2. To screen for proteins enriched in sperm cells, pick-up proteins with >2 identified spectra in sperm cells and none in other cell types (egg cell, callus, seedlings, pollen grains) and with >3 identified spectra in sperm cells and one in another cell type. Proteins putatively enriched in rice sperm cells with five identified spectra are listed in Table 2 [11].

3.6 Fertility of Rice Plants with Mutations in Genes Encoding Gamete-Enriched Proteins

1. Search rice Tos17 lines in which Tos17 transposon is inserted in genes encoding gamete-enriched proteins with Rice GBrowse (http://rapdb.dna.affrc.go.jp/viewer/gbrowse/irgsp1/?name=chr01%3A1..50000) [13]. Obtain seeds for the Tos17 lines (*see* **Note 8**), and grow into seedlings (*see* **Note 7**).
2. Conduct genomic PCR using small leaf pieces from rice Tos17 seedlings with gene-specific and border primers according to the instructions of the Tos17 database (http://tos.nias.affrc.go.jp/index.html.en). Select homozygous seedlings and grow them in an experimental field.

Table 1
Proteins enriched in egg cells with >5 identified spectra. The number of identified spectra in all cell types are presented. *EG* egg cell, *SP* sperm cell, *C* callus, *S* seedling, *P* pollen grain. The table is cited from Abiko et al. [11] and are published under the Creative Commons Attribution (CC BY) license (http:/creativecommons.org/licenses/by/4.0/)

		Number of identified spectra					
cDNA accession	Gene locus	EG	SP	C	S	P	Annotations[a]
AK106474	Os06g0602400	17	0	0	0	1	Similar to DEAD-box protein 3 (DEAD-box RNA helicase DEAD3)
AK065887	Os03g0283100	15	0	1	0	0	Similar to In2–1 protein
Os06t0706700-01	Os06g0706700	14	0	0	0	0	Similar to PsAD1
AK101183	Os05g0168800	11	0	1	0	0	KIP1-like domain containing protein
AK063589	Os05g0115600	10	0	0	0	0	Protein of unknown function DUF674 family protein
AK106371	Os03g0276800	9	0	0	0	0	Heat shock protein Hsp70 family protein
AK107844	Os05g0143600	9	0	0	0	0	Similar to Jasmonate-induced protein
AK067215	Os01g0698000	8	0	0	0	0	Conserved hypothetical protein
AK063560	Os12g0600100	8	0	0	0	0	Tetratricopeptide-like helical domain containing protein
AK121612	Os02g0717400	8	0	0	0	1	Tetratricopeptide-like helical domain containing protein
AK073477	Os01g0369200	7	0	0	0	0	Similar to Cullin-1
AK058611	Os01g0895100	7	0	0	0	0	Similar to Membrane-associated 30 kDa protein, chloroplast precursor
AK106478	Os01g0771100	7	0	1	0	0	Mitochondrial glycoprotein family protein
AK072587	Os05g0164900	6	0	0	0	0	Galactose oxidase/kelch, beta-propeller domain containing protein
AK067210	Os04g0504800	6	0	1	0	0	Similar to Poly(A)-binding protein
AK072334	Os03g0583900	5	0	0	0	0	DEAD-like helicase, N-terminal domain containing protein
AK119521	Os06g0175800	5	0	0	0	0	Similar to Cystathionine beta-lyase, chloroplast precursor
AK069281	Os09g0471100	5	0	0	0	0	Similar to Peroxidase 17 precursor
AK072719	Os10g0574800	5	0	0	0	0	Similar to ARF GAP-like zinc finger-containing protein ZIGA2
AK064995	Os12g0197500	5	0	0	0	0	Putative Zinc finger, XS and XH domain containing protein
Os01t0876900-00	Os01g0876900	5	0	1	0	0	Conserved hypothetical protein

[a]Protein annotations are referred from The Rice Annotation Project Database (RAPDB)

Table 2
Proteins enriched in sperm cells with >5 identified spectra. The number of identified spectra in all cell types are presented. *EG* egg cell, *SP* sperm cell, *C* callus, *S* seedling, *P* pollen grain. The table is cited from Abiko et al. [11] and are published under the Creative Commons Attribution (CC BY) license (http:/creativecommons.org/licenses/by/4.0/)

		Number of identified spectra					
cDNA accession	Gene locus	EG	SP	C	S	P	Annotations[a]
AK071495	Os11g0255300	0	10	0	0	0	Cysteine endopeptidase
Os01t0267600-01	Os01g0267600	1	10	0	0	0	Sad1/UNC-like, C-terminal domain containing protein
AK065231	Os01g0323100	0	7	0	0	0	Similar to Pto kinase interactor 1
AK099178	Os02g0726000	0	7	0	0	0	FAS1 domain containing protein
AK071561	Os05g0163700	0	7	0	0	0	Similar to Acyl-coenzyme A oxidase 4, peroxisomal
AK107034	Os02g0185200	0	6	0	0	0	Cytochrome P450 family protein
AK066587	Os03g0220100	0	6	0	0	0	Similar to Very-long-chain fatty acid condensing enzyme CUT1
Os04g0611200-00	Os04g0611200	0	6	0	1	0	Similar to OSIGBa0152L12.11 protein
AK065311	Os06g0174400	0	6	0	1	0	Similar to Vesicle-associated membrane protein 712
AK069025	Os04g0569000	1	6	0	0	0	Similar to Activator 1 40 kDa subunit (Replication factor C 40 kDa subunit)
AK105867	Os02g0608900	0	5	0	0	0	Similar to Epstein-Barr virus U2-IR2 domain encoding nuclear protein
AK071196	Os05g0399700	0	5	0	0	0	Chitinase
AB087745	Os05g0595100	0	5	1	0	0	Similar to UDP-glucose-4-epimerase
AK069984	Os02g0775200	1	5	0	0	0	Similar to Activator 1 36 kDa subunit (Replication factor C 36 kDa subunit)

[a]Protein annotations are referred from The Rice Annotation Project Database (RAPDB)

3. Check fertility for mutant rice plants by counting the numbers of developed and undeveloped seeds after harvesting the fully grown plants (*see* Fig. 2a).

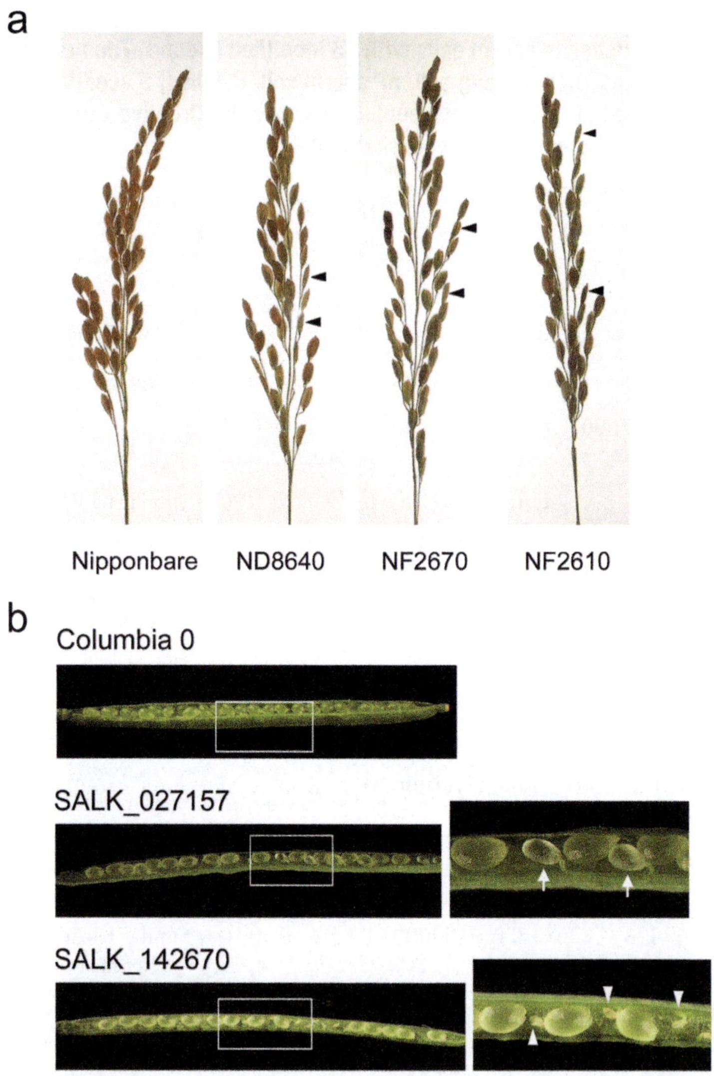

Fig. 2 Rice Tos17 mutants (**a**) and *Arabidopsis* T-DNA mutants (**b**) showing defects in seed set or seed development. (**a**) Panicles of Tos17 mutants showing fertility phenotypes. In panicles of Nipponbare, more than 95% of the seeds developed fully with a *light brown* color. In the three Tos17 mutants, undeveloped seeds were often observed. Two typical undeveloped seeds are indicated by *arrowheads* in each panel. (**b**) Dissected developing siliques of *Arabidopsis* mutants. Panels on right are magnifications of the boxed areas in the left panels. *White arrowheads* and *arrows* indicate failed ovules and seeds arrested at immature stages, respectively. Images were cited from Abiko et al. [11] and are published under the Creative Commons Attribution (CC BY) license (http:/creativecommons.org/licenses/by/4.0/)

3.7 Seed Development of Arabidopsis Plants with Mutations in Genes Orthologous to Rice Genes Encoding Gamete-Enriched Proteins

1. Identify the *Arabidopsis* genes orthologous to rice genes encoding gamete-enriched proteins using the Surveyed Conserved Motif Alignment Diagram and the Associating Dendrogram (SALAD) database version 1.0 (http://salad.dna.affrc.go.jp/salad/) (*see* **Note 17**) [14].
2. Search mutant lines in which T-DNA is inserted in the orthologous genesOrthogous gene using *Arabidopsis* GBrowse (https://gbrowse.arabidopsis.org/cgi-bin/gb2/gbrowse/arabidopsis/). Obtain seeds of *Arabidopsis* plants with a T-DNA insertion for the orthologous genes, and germinate and grow them into young plants (*see* **Note 9**).
3. Conduct genomic PCR using small leaf pieces from young *Arabidopsis* plants with gene-specific and border primers according to the instructions of the Salk Institute Genomic Analysis Laboratory (http://signal.salk.edu/), and confirm homozygous and heterozygous mutations in *Arabidopsis* mutants.
4. Grow them further until they set seeds. Check seed-set and/or seed development by dissecting developing siliques and observing the seeds in the siliques (*see* Fig. 2b).

4 Notes

1. Rice plants (*Oryza sativa* L. cv Nipponbare) should be grown in environmental chambers at 26 °C in a 13/11 h light/dark cycle with a photosynthetic photon flux density of 150–300 μmol photons m^{-2} s^{-1}. Under these growth conditions, flowers can be obtained throughout all seasons.
2. Coverslips should be non-coated, as using coated coverslips will result in attachment of the cells to the surface of the coverslip. Non-coated coverslips supplied from Fisher Scientific are recommended.
3. Injector with 30–50 μL delivery volume per a syringe rotation is recommended.
4. Tube size is 1.5 cm (diameter) × 9.6 cm (height). The tube should be clear to view the layer of sperm cells after Percoll density centrifugation.
5. Prepare as follows: 4.1 mL of 0.36% NaOH + 0.3 mL of 25% ammonia (Solution A), 160 mg silver nitrate dissolved with 0.8 mL water (Solution B). Add 0.7 mL of Solution B into Solution A with shaking. Solution B must be added drop by drop, since a brown precipitate occurs when Solution B is added all at once. Add H_2O to the mixture to a final volume of 160 mL, and use as silver-staining solution.

6. Rice seeds need to be cultured on N6D medium containing 2,4-D (2,4 dichlorphenoxyacetic acid) for 7 days at 30 °C under continuous light [15], and the callus derived from the scutellum of rice seeds needs to be harvested.
7. Rice seeds should be sown in water and grown at 26 °C in darkness for 4 days, and then the germinated seeds should further be grown at 26 °C with a 13/11 h light/dark cycle for 7 days.
8. Tos17 insertional rice plants can be obtained from the Rice Genome Resource Center, National Institute of Agrobiological Sciences (NIAS, Tsukuba, Ibaraki, Japan), and can be grown under field conditions.
9. *Arabidopsis thaliana* ecotype Columbia and T-DNA insertion lines obtained from the Arabidopsis Biological Resource Center (ABRC, Columbus, OH, USA) should be grown in an air-conditioned room at 23 °C with a 16/8 h light/dark cycle.
10. Border primer for Tos17 line (5′-ATTGTTAGGTTGCAAGTTAGTTAAGA-3′), for SALK T-DNA lines (5′-ATTTTGCCGATTTCGGAAC-3′), and for SAIL T-DNA lines (5′-TAGCATCTGAATTTCATAACCAATCTCGATACAC-3′).
11. Gene-specific primers of rice Tos17 insertion lines can be designed with "https://tos.nias.affrc.go.jp/cgi-bin/tos17/seqtools.cgi?action=Primer1." Gene-specific primers of *Arabidopsis* T-DNA insertion lines can be designed with "http://signal.salk.edu/tdnaprimers.2.html."
12. Without removing the stigmas, ovaries always float on the mannitol solution. To isolate egg cells, sinking ovaries into the mannitol solution is essential. Usually, approximately 30 ovaries are put into a dish.
13. Usually, 4–10 egg cells are automatically released from approximately 30 cut ovaries. Gentle pushing of the basal portion of the lower part of the cut ovary with a glass needle will produce additional egg cells.
14. The number of sperm cells in the fraction can be estimated using the sperm cell fraction (1 μL) and a hemocytometer.
15. In our study [11], lysates of 500 egg cells, approximately 3×10^4 sperm cells, 1 μg of callus proteins, 1.2 μg of seedling proteins, and 0.6 μg of pollen proteins were separated by SDS-PAGE for LC-MS/MS analyses.
16. Analyzing for these egg and sperm cell lysates, 2138 and 2179 proteins were detected, respectively, in our study [11]. Among these proteins, 1276 and 1076 proteins were assigned by at least two peptides. In calli, seedlings, and pollen grains, 2877, 2473, and 2246 proteins were detected, respectively.

17. When *Arabidopsis* orthologs were searched for 100 genes encoding gamete-enriched proteins with >3 identified spectra, 34 putative orthologous genes were identified [11].

References

1. Huang B-Q, Russell SD (1992) Female germ unit: organization isolation, and function. Int Rev Cytol 140:233–292
2. Drews GN, Yadegari R (2002) Development and function of the angiosperm female gametophyte. Annu Rev Genet 36:99–124
3. Russell SD (1992) Double fertilization. Int Rev Cytol 40:357–390
4. Nawaschin S (1898) Revision der Befruchtungsvorgange bei Lilium martagon und Fritillaria tenella. Bull Acad Imp Sci St Petersburg 9:377–382
5. Guignard ML (1899) Sur les antherozoides et la double copulation sexuelle chez les vegetaux angiosperm. Rev Gen Bot 11:129–135
6. Mori T, Kuroiwa H, Higashiyama T, Kuroiwa T (2006) GENERATIVE CELL SPECIFIC 1 is essential for angiosperm fertilization. Nat Cell Biol 8:64–71
7. Sprunck S, Rademacher S, Vogler F, Gheyselinck J, Grossniklaus U, Dresselhaus T (2012) Egg cell–secreted EC1 triggers sperm cell activation during double fertilization. Science 338:1093–1097
8. Dai S, Chen S (2012) Single-cell-type proteomics: toward a holistic understanding of plant function. Mol Cell Proteomics 11:1122–1130
9. Uchiumi T, Komatsu S, Koshiba T, Okamoto T (2006) Isolation of gametes and central cells from Oryza sativa L. Sex Plant Reprod 19:37–45
10. Uchiumi T, Uemura I, Okamoto T (2007) Establishment of an in vitro fertilization system in rice (Oryza sativa L.) Planta 226:581–589
11. Abiko M, Furuta K, Yamauchi Y, Fujita C, Taoka M, Isobe T, Okamoto T (2013) Identification of proteins enriched in rice egg or sperm cells by single-cell proteomics. PLoS One 8(7):e69578
12. Yoshikawa H, Komatsu W, Hayano T, Miura Y, Homma K et al (2011) Splicing factor 2-associated protein p32 participates in ribosome biogenesis by regulating the binding of Nop52 and fibrillarin to preribosome particles. Mol Cell Proteomics 10(8):M110.006148
13. Miyao A, Tanaka K, Murata K et al (2003) Target site specificity of the Tos17 retrotransposon shows a preference for insertion within genes and against insertion in retrotransposon-rich regions of the genome. Plant Cell 15:1771–1780
14. Mihara M, Itoh T, Izawa T (2010) SALAD database: a motif-based database of protein annotations for plant comparative genomics. Nucleic Acids Res 38:D835–D842
15. Toki S, Hara N, Ono K, Onodera H, Tagiri A et al (2006) Early infection of scutellum tissue with Agrobacterium allows high-speed transformation of rice. Plant J 47:969–976

Chapter 21

Phosphoprotein Enrichment from Tobacco Mature Pollen Crude Protein Extract

Jan Fíla and David Honys

Abstract

Protein phosphorylation was repeatedly shown to be the most dynamic post-translational modification mediated by a huge orchestra of protein kinases and phosphatases. Upon landing on a stigma, pollen grain dehydration and activation are accompanied by changes in protein phosphorylation together with the translation activation of stored mRNAs. To enable studies of the total phosphoproteome, it is usually necessary to apply various enrichment techniques. In this chapter, one of these protocols that worked previously well on tobacco mature pollen is presented in more detail. The method comprises of three basic steps: (1) picking flowers from the flowering tobacco plants (*Nicotiana tabacum* cv. Samsun), and collection of the shed pollen grains; (2) extraction of total proteins by TCA/acetone; (3) phosphoprotein enrichment by MOAC with aluminum hydroxide matrix. Taken together this protocol describes how to isolate phosphoproteins out of tobacco mature pollen.

Key words Phosphoprotein enrichment, Metal-oxide/hydroxide affinity chromatography, MOAC, Male gametophyte, Pollen grain, Aluminum hydroxide, TCA/acetone protein extraction

1 Introduction

Angiosperm mature pollen has an extremely desiccated cytoplasm, which is surrounded by a tough cell wall. After reaching the papillary cells of a stigma, pollen cytoplasm becomes re-hydrated, the metabolism is activated, and one particular pollen aperture is used for pollen tube growth (reviewed in [1]). Such change from a quiescent to an activated structure during progamic phase is accompanied by the translation activation of mRNAs stored in EDTA/puromycine-resistant particles (EPPs; [2, 3]), and also post-translational modifications of the existing proteins, phosphorylation in particular [4, 5]. Such re-hydration-related phosphorylation was described also in the xerophyte *Craterostigma plantagineum* [6], and in the leaf growth zone in maize (*Zea mays*) [7].

Protein phosphorylation represents one of the most dynamic post-translational modifications that plays a crucial role in plenty of

Anja Schmidt (ed.), *Plant Germline Development: Methods and Protocols*, Methods in Molecular Biology, vol. 1669, DOI 10.1007/978-1-4939-7286-9_21, © Springer Science+Business Media LLC 2017

cellular processes and structures (e.g., [8–11]). Protein phosphorylation is mediated by protein kinases, whereas the phosphate removal is performed by protein phosphatases. The studies of protein phosphorylation usually apply one of the available enrichment protocols [12]. The enrichment techniques are mostly essential since (1) only part of the phosphoproteins is phosphorylated at a time in a cell, (2) the phosphorylated protein can co-exist with the non-phosphorylated isoform of the same species, (3) the technical aspects of mass spectrometry make the identification of phosphorylated peptides in the mixture where non-phosphorylated species dominate very unlikely.

The enrichment step can be performed either at the level of intact phosphoproteins or at the level of peptides that were acquired by proteolytic digest of the whole proteins (reviewed in [12]). Both of these approaches showed their advantages as well as limitations. In this chapter, we focus on phosphoprotein enrichment by metal oxide/hydroxide affinity chromatography (MOAC) with aluminum hydroxide matrix [13]. The presented protocol starts with the collection of pollen grains from flowering tobacco (*Nicotiana tabacum* cv. Samsun) plants [14]. After the pollen grains are collected, the extraction of total proteins is carried out. In tobacco pollen, it was shown that the proper homogenization approach together with the selection of extraction protocol is vitally important [15]. In this chapter, the TCA/acetone protein extraction protocol [16] was adopted for mature pollen since it was the optimal one prior to the phosphoprotein enrichment. The obtained total protein extract was subjected to phosphoprotein enrichment by MOAC with aluminum hydroxide matrix [13]. Further separation steps, such as SDS-PAGE or 2D IEF-PAGE, were not subjected to a detailed description in this chapter. Taken together, this protocol comprises three parts: (1) flower collection, and pollen grain isolation, (2) extraction of total proteins, and (3) MOAC phosphoprotein enrichment (for a scheme, *see* Fig. 1).

2 Materials

2.1 Pollen Grain Collection

Prepare the following items to collect pollen grains:

1. Grown flowering tobacco plants (*Nicotiana tabacum* cv. Samsun).
2. Petri dish, 15–20 cm in diameter.
3. Roundly shaped filtration paper of the size matching inside the Petri dish.
4. Sieve from a stocking fixed onto a glass roll (*see* **Note 1**).
5. Laboratory scales.
6. 1.5 ml microcentrifuge tubes.

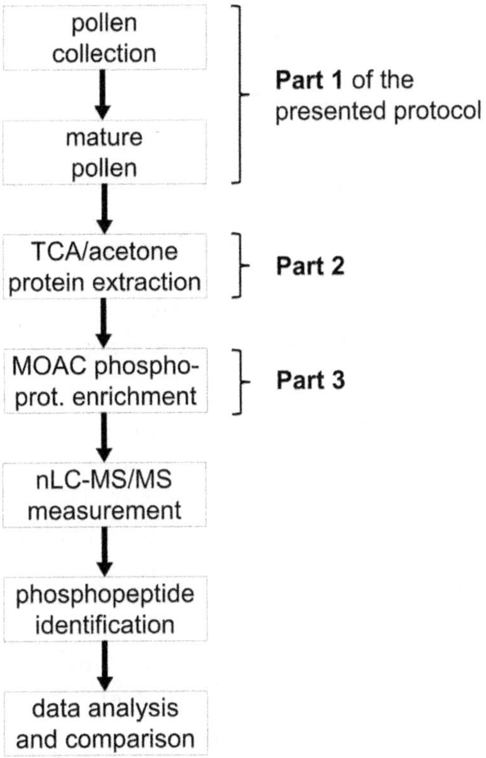

Fig. 1 Working scheme of a phosphoproteomics experiment. The parts presented in this manuscript are highlighted

2.2 Protein Extraction by TCA/Acetone

Prepare the following solutions freshly before each extraction together with the following laboratory equipment:

1. 500 mg tobacco mature pollen (from Subheading 3.1) (*see* **Note 2**).
2. Extraction buffer 10% TCA, and 1% DTT in acetone: weigh 1 g trichloroacetic acid (TCA), and 100 mg dithiothreitol (DTT) on laboratory scales; transfer the weighted chemicals into a 50 ml Falcon tube and top up to 10 ml by acetone (*see* **Note 3**).
3. Washing buffer 1% DTT in acetone: weigh 150 mg dithiothreitol (DTT), put it to a 50 ml Falcon tube and fill in with acetone to 15 ml (*see* **Note 4**).
4. Pestle and mortar.
5. 2 ml microcentrifuge tubes.
6. A 15 ml Falcon tube.
7. Cooled centrifuge (4 °C).
8. Freezer −20 °C.
9. Ultrasonic bath.

10. Liquid nitrogen.
11. Glass rod.
12. Vortex.
13. Vacuum pump.
14. Injection needle (*see* **Note 5**).

2.3 Phosphoprotein Enrichment by MOAC

Prepare the following solutions and laboratory equipment prior to the enrichment. The stock solutions can be stored for a longer period whilst the working solutions should be prepared freshly before each use and any rest of them should be discarded afterward.

1. Incubation buffer stock solution: 30 mM MES, 20 mM imidazole, 0.2 M L-aspartic acid potassium salt, 0.2 M L-glutamic acid sodium salt (*see* **Note 6**), 8 M urea, pH 6.1. Dissolve 120.2 g urea in 100 ml distilled water (*see* **Note 7**). After urea becomes completely solubilized, add 1.599 g MES, 8.56 g L-aspartic acid potassium salt, 8.455 g L-glutamic acid sodium salt, and 0.3404 g imidazole. Every time, wait with the addition of the subsequent substance until the previous one is fully dissolved. Finally, set the pH of the buffer to 6.1 with hydrochloric acid.
2. Incubation buffer solution ready to use: 2.5% CHAPS, phosphatase inhibitors in incubation buffer stock solution. Prepare 30 ml buffer per enriched sample: dissolve 0.75 g CHAPS (3-[(3-cholamidopropyl)dimethylammonio]-1-propanesulfonate) together with 0.3 ml phosphatase inhibitor cocktail to a final volume of 30 ml incubation buffer stock solution.
3. Elution buffer stock solution A: 100 mM KH_2PO_4. Dissolve 0.6805 g KH_2PO_4 in a final volume of 50 ml distilled water.
4. Elution buffer stock solution B: 100 mM K_2HPO_4. Dissolve 4.355 g in a final volume of 250 ml distilled water.
5. Elution buffer stock solution AB: 8 M urea in elution buffer stock solution B (100 mM K_2HPO_4), pH 9.0. Dissolve 48.08 g urea in 50 ml elution buffer stock solution B (100 mM K_2HPO_4) (*see* **Note 7**), top the volume up to 100 ml elution buffer stock solution B, and set the pH of the final solution to 9.0 with buffer stock solution A (100 mM KH_2PO_4).
6. Elution buffer ready to use: stock solution AB plus phosphatase inhibitors. Prepare freshly before each use 2 ml elution buffer per enriched sample: add 20 μl phosphatase inhibitor cocktail into 2 ml elution buffer stock solution AB and mix well.
7. 160 mg aluminum hydroxide in a 15 ml Falcon tube (per enriched sample).
8. 15 ml Falcon tubes (*see* **Note 8**).

9. 1.5 ml microtubes.
10. Thermoblock (set to 37 °C).
11. Vortex.
12. Ultrasonic bath.
13. Cooled centrifuge suitable for Falcon tubes.
14. Rotator in a cold place.

3 Methods

3.1 Pollen Grain Collection

1. Grow the tobacco plants until they start flowering. Cut the overgrown flowers and also whole old branches almost without any flowers to encourage the tobacco plants in setting new flower buds.
2. From the flowering tobacco plants, pick the flowers 1 day before anthesis (*see* **Note 9**) (Fig. 2).
3. Remove the anthers from the flowers and put them onto a filtration paper in a Petri dish. Do not close the Petri dish.
4. Let the anthers dehisce overnight at room temperature (*see* **Note 10**), they will open and pollen grains will be shed out of them.
5. Separate pollen grains from the debris, represented mainly by the green parts of the anthers, by a home-made sieve prepared from a stocking fixed onto a glass roll.
6. Transfer the filtered pollen grains (*see* **Note 11**) into 1.5 ml microcentrifuge tubes, weigh them and store them at −20 °C (*see* **Note 12**).

Fig. 2 Flowering tobacco plant (*Nicotiana tabacum* cv. Samsun). The anther 1 day before anthesis for the collection is labeled

3.2 Protein Extraction by TCA/Acetone

Before the extraction, cool down the centrifuge to 4 °C since all centrifugation steps will be carried out at 4 °C in this part of the protocol.

1. Fill in the mortar with liquid nitrogen and let it evaporate, and also cool down the pestle in liquid nitrogen.
2. Fill in the mortar again with liquid nitrogen and add 500 mg pollen grains (*see* **Note 13**) into it.
3. Grind the pollen grains under liquid nitrogen and transfer the homogenized pollen into a 15 ml Falcon tube.
4. Resuspend the transferred sample with tenfold volume (i.e., 5 ml per 500 mg pollen grains) extraction buffer (10% TCA + 1% DTT in acetone), and vortex thoroughly.
5. Transfer the resuspended sample (*see* **Note 14**) by a 1 ml pipetman with a cut tip to four 2 ml microtubes (1.25 ml sample per each microcentrifuge tube).
6. Snap freeze the samples in liquid nitrogen.
7. Incubate the samples at −20 °C for 45 min. Mix the suspension by turning the tubes upside down several times after 5, 10, and 15 min of incubation.
8. Centrifuge the samples (15 min, 25,000 × *g*, 4 °C), and immediately after the centrifugation, discard the supernatant by using an injection needle (*see* **Note 15**) connected to the vacuum pump.
9. Resuspend the pellet by 1.5 ml washing buffer (1% DTT in acetone), and incubate for 5 min in an ultrasonic bath.
10. In case the pellet is still compact, disturb it by a glass rod.
11. Snap freeze the samples in liquid nitrogen.
12. Centrifuge the samples (10 min, 25,000 × *g*, 4 °C), and immediately after the centrifugation, discard the supernatant by an injection needle (*see* **Note 15**) connected to the vacuum pump.
13. Resuspend the pellet in 1.5 ml washing buffer (1% DTT in acetone), and incubate for 5 min in an ultrasonic bath.
14. In case the pellet is still compact, disturb it by a glass rod.
15. Snap freeze the samples in liquid nitrogen.
16. Centrifuge the samples (10 min, 25,000 × *g*, 4 °C), and immediately after the centrifugation discard the supernatant by an injection needle (*see* **Note 15**) connected to the vacuum pump.
17. Dry the pellet in vacuum, and store the dried pellet at −20 °C for a short-term storage, and at −80 °C for a long-term storage.

3.3 Phosphoprotein Enrichment by MOAC

Prior to the enrichment, warm the thermoblock to 37 °C, and place the rotator to the cold room.

1. Add 50 μl incubation buffer per 1 mg TCA/acetone precipitate from **step 2** (*see* **Note 16**).
2. Vortex the sample briefly, and incubate for 5 min in an ultrasonic bath.
3. Put the sample onto the thermoblock (1 h, 37 °C).
4. Centrifuge the sample (25,000 × *g*, 15 min, 22 °C), and immediately transfer the supernatant into a fresh tube (*see* **Note 17**).
5. Estimate protein quantity (*see* **Note 18**).
6. Dilute the samples in the incubation buffer to have 3 ml sample of 0.66 mg ml^{-1} concentration for the subsequent steps. The rest of the sample can be stored at −20 °C.
7. Prepare 160 mg aluminum hydroxide into a 15 ml Falcon tube.
8. Add 2 ml incubation buffer, and vortex vigorously to moisten the aluminum hydroxide powder (*see* **Note 19**).
9. Spin the aluminum hydroxide particles down (9000 × *g*, 2 min, 4 °C), and discard supernatant.
10. Put 3 ml of the diluted sample to the aluminum hydroxide powder, and vortex briefly to get a homogenous suspension.
11. Incubate 30 min in the cold room on a rotator (*see* **Note 20**).
12. Centrifuge the samples (9000 × *g*, 2 min, 4 °C) to pellet the aluminum hydroxide particles (*see* **Note 21**), and transfer the supernatant to a fresh tube (*see* **Note 22**).
13. Wash the pellet with 3.2 ml incubation buffer, and vortex thoroughly for 1 min.
14. Centrifuge the tube (9000 × *g*, 2 min, 4 °C) to pellet the aluminum hydroxide particles and keep supernatant (*see* **Note 23**).
15. Repeat **steps 13** and **14** four times.
16. Wash the pellet with 1.6 ml incubation buffer, and vortex thoroughly for 1 min.
17. Centrifuge the tube (9000 × *g*, 2 min, 4 °C) to pellet the aluminum hydroxide particles and keep supernatant.
18. Add 1.6 ml elution buffer to the pellet, and thoroughly vortex for 1 min.
19. Incubate 20 min in the cold room on a rotator (*see* **Note 20**).
20. Spin the aluminum hydroxide particles down (9000 × *g*, 2 min, 4 °C).

21. Keep the supernatant containing phosphoprotein-enriched fraction, and discard the pellet with aluminum hydroxide particles.

4 Notes

1. This type of sieve is used in our laboratory. However, it can be replaced by any other sieve that will efficiently separate debris from mature pollen grains.
2. The amount of used pollen grains does not have to be exactly 500 mg but can be slightly higher or lower since TCA/acetone volume can be adjusted accordingly.
3. Be careful while manipulating with solid TCA; wear gloves since it is a corrosive compound.
4. The final volume of washing buffer was calculated for four microcentrifuge tubes containing aliquots of protein extract. In case of planning more aliquots due to higher amount of starting material used for protein extraction, increase the volume of prepared washing buffer by 3 ml per additional microcentrifuge tube.
5. It is important to have a thin enough needle that will enable sucking the supernatant on one hand but will not remove any parts of the pellet on the other hand.
6. The concentrations of L-aspartic acid potassium salt and L-glutamic acid sodium salt can be optimized since too low concentration causes a higher non-specificity of the enrichment (i.e., lower selectivity) whereas too high concentration reduces the sensitivity of the method [6].
7. Warm up the urea solution until urea crystals become completely dissolved.
8. Falcon tubes can be replaced by any other tubes suitable for centrifugation. In case Falcon tubes are used, check their suitability for centrifugation, otherwise Falcon tube cracking might occur.
9. It is necessary to pick the flowers exactly 1 day before anthesis, and not the younger ones since their anthers will be less developed and will not open until the following day. On the other hand, the older flowers do not contain much pollen grains since they were already shed.
10. The temperature in the room has to be above 20 °C. A warmer temperature up to 30 °C causes no harm to the drying anthers. However, a temperature lower than 20 °C is not warm enough to enable anther dehiscence and shedding of the pollen grains.

11. It is critical that all remnants of the anthers will be removed from the pollen grains. Repeat the filtration step in case several anther pieces remained in the pollen sample.
12. Do not store the pollen grains under other temperatures. The temperature of −80 °C would diminish pollen grain germination rate notably. Moreover, avoid the stored samples from multiple freeze–thaw cycles and aliquot the pollen grains per several 1.5 ml microcentrifuge tubes, of which every tube will be dedicated for a single use only.
13. Avoid re-freezing of the pollen grains. Usually the whole amount of pollen grains inside one microtube is used.
14. The resuspended sample has to be really homogenous; ensure sample homogeneity by mixing the sample thoroughly before pipetting every aliquot.
15. It is important to use a thin injection needle to prevent any losses of the homogenized pollen grains.
16. The TCA/acetone pellet should be weighted before resuspension, usually the amount of approximately 40–50 mg should be sufficient for one round of phosphoprotein enrichment.
17. The centrifuge should not be cooled during this step in order to avoid any precipitation of salts in the incubation buffer.
18. We typically apply the 2D Quant kit (GE Healthcare) using the manufacturer's instructions for the estimation of protein concentration, but in principle any method that tolerates 8 M urea in the incubation buffer should be feasible.
19. It is of vital importance to make the whole volume of the aluminum hydroxide particles wet.
20. Other ways of keeping the aluminum hydroxide and the proteins in a homogenous mixture can be applied. In any case it is important not to have the samples in warmer temperatures than 8 °C.
21. The phosphoproteins should be bound to the aluminum hydroxide particles whereas non-phosphorylated species should be in the supernatant.
22. The supernatant should contain non-phosphorylated proteins. Nevertheless, it should be kept to enable its analysis by 1D SDS-PAGE in order to compare protein spectra from the supernatant and the phosphoprotein-enriched fraction.
23. This supernatant together with all following supernatants containing wash-out proteins should also be kept to enable their analysis by 1D SDS-PAGE.

Acknowledgment

The authors gratefully acknowledge the financial support from the Czech Science Foundation (grants no. 13-06943S and 15-16050S).

References

1. Hafidh S, Fíla J, Honys D (2016) Male gametophyte development and function in angiosperms: a general concept. Plant Reprod 29:31–51
2. Honys D, Combe JP, Twell D, Čapková V (2000) The translationally repressed pollen-specific ntp303 mRNA is stored in non-polysomal mRNPs during pollen maturation. Sex Plant Reprod 13:135–144
3. Honys D, Reňák D, Feciková J et al (2009) Cytoskeleton-associated large RNP complexes in tobacco male gametophyte (EPPs) are associated with ribosomes and are involved in protein synthesis, processing, and localization. J Proteome Res 8:2015–2031
4. Fíla J, Matros A, Radau S et al (2012) Revealing phosphoproteins playing role in tobacco pollen activated *in vitro*. Proteomics 12:3229–3250
5. Fíla J, Radau S, Matros A et al (2016) Phosphoproteomics profiling of tobacco mature pollen and pollen activated *in vitro*. Mol Cell Proteomics 15:1338–1350
6. Röhrig H, Colby T, Schmidt J, Harzen A, Facchinelli F, Bartels D (2008) Analysis of desiccation-induced candidate phosphoproteins from *Craterostigma plantagineum* isolated with a modified metal oxide affinity chromatography procedure. Proteomics 8:3548–3560
7. Bonhomme L, Valot B, Tardieu F, Zivy M (2012) Phosphoproteome dynamics upon changes in plant water status reveal early events associated with rapid growth adjustment in maize leaves. Mol Cell Proteomics 11:957–972
8. Ballesta JPG, Rodriguez-Gabriel MA, Bou G, Briones E, Zambrano R, Remacha M (1999) Phosphorylation of the yeast ribosomal stalk. Functional effects and enzymes involved in the process. FEMS Microbiol Rev 23:537–550
9. Swain PK, Hicks D, Mears AJ et al (2001) Multiple phosphorylated isoforms of NRL are expressed in rod photoreceptors. J Biol Chem 276:36824–36830
10. Augustine RC, Vidali L, Kleinman KP, Bezanilla M (2008) Actin depolymerizing factor is essential for viability in plants, and its phosphoregulation is important for tip growth. Plant J 54:863–875
11. Heberle-Bors E, Voronin V, Touraev A, Testillano PS, Risueno MC, Wilson C (2001) MAP kinase signaling during pollen development. Sex Plant Reprod 14:15–19
12. Fíla J, Honys D (2012) Enrichment techniques employed in phosphoproteomics. Amino Acids 43:1025–1047
13. Wolschin F, Wienkoop S, Weckwerth W (2005) Enrichment of phosphorylated proteins and peptides from complex mixtures using metal oxide/hydroxide affinity chromatography (MOAC). Proteomics 5:4389–4397
14. Petrů E, Hrabětová E, Tupý J (1964) The technique of obtaining germinating pollen without microbial contamination. Biol Plant 6:68–69
15. Fíla J, Čapková V, Feciková J, Honys D (2011) Impact of homogenization and protein extraction conditions on the obtained tobacco pollen proteomic patterns. Biol Plant 55:499–506
16. Méchin V, Damerval C, Zivy M (2006) In: Thiellement H, Zivy M, Damerval C, Méchin V (eds) Total protein extraction with TCA-acetone, Methods in molecular biology. Springer, New York, pp 1–8

Chapter 22

Identification of *Cis*-Regulatory Modules that Function in the Male Germline of Flowering Plants

Benjamin Peters, Jack Aidley, Murray Cadzow, David Twell, and Lynette Brownfield

Abstract

The male germline of flowering plants develops within the vegetative cell of the male gametophyte and displays a distinct transcriptional profile. Key to understanding the development of this unique cell lineage is determining how gene expression is regulated within germline cells. This knowledge impacts upon our understanding of cell specification, differentiation, and plant fertility. Here, we describe methods to identify *cis*-regulatory modules (CRMs) that act as key regulatory regions in the promoters of germline-expressed genes. We detail the complimentary techniques of phylogenetic footprinting and the use of fluorescent reporters in pollen for the identification and verification of CRMs.

Key words *Cis*-regulatory module, Fluorescent reporter, Male germline, Phylogenetic footprinting, Pollen, Sperm cells, Microscopy, Transcription factor binding site

1 Introduction

The male germline of flowering plants is initiated when haploid meiotic products, the microspores, undergo a highly asymmetric division. The segregation of the smaller generative or germ cell establishes the short male germ cell lineage leading to the formation of the twin sperm cells needed for double fertilization [1]. Meanwhile, the larger vegetative cell plays a supportive role and forms the pollen tube through which the sperms cells are transported. During pollen (male gametophyte) development, the generative cell is engulfed within the vegetative cell cytoplasm where it divides to produce the two sperm cells. In species such as Arabidopsis (*Arabidopsis thaliana*), which shed tricellular pollen, this division occurs before pollen is released, while in species which shed bicellular pollen, the division occurs during pollen tube growth. Transcriptome studies of isolated sperm cells from Arabidopsis and rice (*Oryza sativa*) have shown that sperm cells possess hyper-differentiated profiles with many transcripts being enriched

Anja Schmidt (ed.), *Plant Germline Development: Methods and Protocols*, Methods in Molecular Biology, vol. 1669, DOI 10.1007/978-1-4939-7286-9_22, © Springer Science+Business Media LLC 2017

in, or specific to, the male germline cells [2, 3]. Thus, elucidating the mechanisms through which transcription is regulated in the germline is key to understanding the specification and differentiation of the male germline, and the vital role of the male gametes in plant fertility.

In plants, transcription is largely controlled by the binding of transcription factors (TFs) to transcription factor binding sites (TFBS) or *cis*-elements that are short DNA sequences (~6–10 bp) in gene promoter regions [4]. Precise spatial and temporal transcriptional control relies on the activity and interactions of multiple TFs, often in multi-protein complexes. Thus, gene expression is commonly regulated by *cis*-regulatory modules (CRMs), which represent organized arrays of TFBSs that may extend over fifty to several hundred base pairs.

Identification of CRMs which function in the plant male germline is challenging due to the embedded location of the germ cells within the pollen vegetative cell. For many molecular techniques, such as qRT-PCR, this makes identifying transcriptional events in the germline difficult due to the large contribution of the vegetative cell to samples of isolated pollen or developing gametophytes. In addition, sample preparation can be time-consuming and give low yield, due to the tough pollen wall being lysis-resistant, meaning large amounts of starting material may be required.

Here, we describe a two-step approach for the identification and verification of CRMs that regulate an identified male germline-expressed gene (the gene-of-interest) using Arabidopsis as a model. This approach has led to the successful characterization of a CRM that regulates the expression of the male germline-specific transcription factor DUO1 [5]. We first detail our approach to use online resources to identify candidate CRMs through phylogenetic footprinting. This is a general approach that is not limited to searching for CRMs in male germline genes. We go on to describe how to use fluorescent reporters for the identification and verification of CRMs, which function to control transcription during male germline development.

2 Materials

2.1 Phylogenetic Footprinting

1. Computer with Internet access.
2. Notepad and Microsoft Excel (optional).

2.2 Fluorescent Reporter Expression to Analyze CRM Activity

1. Vectors and cloning reagents to create promoter:NLS-fluorescent reporter constructs and plant transformation reagents to stably transform Arabidopsis.
2. Harvest Buffer: 0.1 M sodium phosphate, pH 7, 1 mM Na_2EDTA (Ethylenediaminetetraacetic acid disodium salt),

0.1% Triton X-100, 0.4 mg ml-1 DAPI (4′,6-Diamidino-2-phenylindole dihydrochloride), high grade) [6].

3. 53 μm nylon or stainless steel filter.
4. Bench-top centrifuge with a rotor suitable for 15 ml centrifuge tubes.
5. Microscope slides and coverslips (number 1 thickness).
6. Nail varnish.
7. Epifluorescence microscope, with camera and NIS Elements BR software (Nikon) or other image analysis software.
8. The statistical analysis program R [7].
9. Primers, high fidelity Taq polymerase, PCR machine.
10. Material for agarose gel electrophoresis of DNA.
11. Selected restriction enzymes (each about 20,000 units/ml) and 10× buffer.
12. PCR clean-up kit.
13. 4000,000 U/ml T4 ligase and 10× buffer.
14. Heat blocks.
15. QuikChange site-directed mutagenesis kit (Agilent Technology).

3 Methods

3.1 Phylogenetic Footprinting to Identify Candidate CRMs

Phylogenetic footprinting works on the premise that selective pressure will lead to regulatory regions of DNA evolving at a lower rate when compared to non-functional regions. As a result, conserved sequences in the promoters of orthologous genes are likely candidates for CRMs [8, 9]. As alterations in the transcriptional regulation of male germline-expressed genes are likely to impact upon plant fertility, selective pressure is expected to conserve regulatory sequences active in the germline. Thus, phylogenetic footprinting is a suitable approach to search for CRMs in the promoters of male germline genes. Supporting this approach is the increasing access to plant genome sequences through resources such as Phytozome [10] and the Arabidopsis 1001 Genome Consortium [11]. These resources enable "evolutionary windows" of different sizes (representing different groups of descendent taxa) to be explored for sequence similarity and thus to identify candidate CRMs. Using sliding "evolutionary windows" means that candidate CRMs that may be evolving at faster or slower rates can be identified by comparing more closely (e.g., Arabidopsis accessions, a single family) or distantly related plants (e.g., eudicots). Furthermore, repeating the analysis at selected phylogenetic levels can help to refine candidate CRMs and identify candidate TFBSs within a CRM.

Here, we describe how to identify a candidate CRM in the promoter of an identified male germline-expressed gene-of-interest by searching for conserved sequences within different flowering plants (angiosperms; *see* Subheading 3.1.1) and between Arabidopsis accessions (*see* Subheadings 3.1.2 and 3.1.3).

3.1.1 Finding Sequence Conservation in Flowering Plants

This analysis looks for over-represented sequences in the promoters of orthologous genes as these are likely to be under selection, and are thus candidate CRMs.

1. In Phytozome (https://phytozome.jgi.doe.gov/pz/portal.html) [10] select BLAST from the Tools dropdown menu.
2. Use the protein sequence encoded by the gene-of-interest as input into a TBLASTN search against selected angiosperm genomes to identify orthologous proteins (*see* **Note 1**).
3. Open the gene view in Phytozome for each ortholog. This can be achieved by clicking on "feature" on the BLAST results page and then clicking on "transcript" on the following page.
4. Under the sequences tab, expand the genomic sequence window.
5. In the "show flanking sequence" "upstream" box enter 1000 and press the submit button.
6. Highlight the text including the identifier in the top row up to the translation initiation site (TIS) and copy this text.
7. Paste text to generate a single FASTA file. This can be done in Notepad by pasting all sequences into a single document, with each sequence entered on a new line, and saving the document with the suffix ".fasta", or by using sequence analysis programs such as CLC or Geneious.
8. Open the MEME tool (http://meme-suite.org/tools/meme) from the MEME suit of programs (http://meme-suite.org/) [12].
9. Select the motif discovery mode: Normal mode.
10. Input the primary sequence by uploading the FASTA file from **step 7** or by cutting and pasting the sequences.
11. Set the site distribution to zero or one occurrence per sequence.
12. Set the number of motifs to 10.
13. Under the advanced options, set "How wide can the motifs be" to 10–100 bp.
14. Press "Start Search".
15. The MEME output will provide the number of motifs requested. To assess the level of conservation within the motif view the Position Specific Scoring Matrix and the

E-value. Additionally, the MEME output will map the motifs onto the input sequences. Look for patterns in the occurrence of the motif with regard to the order in which they appear and their proximity to the start codon. A candidate CRM may be detected as several motifs in a similar order (*see* **Note 2**).

16. Once candidate CRMs are identified, visually inspect this region of the input sequences to look for further conservation (*see* **Note 3**).

3.1.2 Finding Sequence Conservation in Arabidopsis Accessions Using SNPFreqPlot

This analysis looks for regions in the promoter of the gene-of-interest that have no, or very few, single nucleotide polymorphisms (SNPs) in Arabidopsis accessions, indicating selective pressure and candidate CRMs. This involves the downloading of VCF (Variant Call Format) files containing information on all SNPs in a specified genomic region of the Arabidopsis genome from the Arabidopsis 1001 Genome Consortium [11]. The VCF file can be processed in multiple ways with two options that require little bioinformatics experience either using SNPFreqPlot as described below, or by processing in Microsoft Excel (*see* Subheading 3.1.3).

1. For downloading VCF files, find the genomic coordinates for the promoter region of the gene-of-interest (typically 1000 bp upstream of the TIS) through TAIR or Araport, e.g., Chr3:22342429–22343491.
2. At the 1001 Genome Tools (http://tools.1001genomes.org/) select the VCF subset tab (http://tools.1001genomes.org/vcfsubset/#select_strains).
3. Under the "Select Strains" tab select all ecotypes by checking the top box (*see* **Note 4**).
4. Under the "Select Loci" tab enter the genomic coordinates for the region.
5. Under the "Select Files" tab choose Full genome VCF and VCF.
6. Download the VCF file.
7. To process the VCF files, we have developed the SNPFreqPlot application to enable easy visualization of the frequency at which SNPs occur across a genomic region of all Arabidopsis accessions by extracting key information from the VCF file (*see* **Note 5**).
8. Upload the VCF file into the SNPFreqPlot web application available at http://snpfreqplot.otago.ac.nz/SNPFreqPlot/.
9. Visually inspect the plot to identify areas with low SNP frequencies as these are candidate CRMs.
10. To optimize the output in SNPFreqPlot the strand direction can be changed based on the orientation of the gene-of-interest, the

window size can be altered to reflect the possible size of a CRM, a marker can be added for the start position based on the chromosomal coordinates and a table exported of conserved areas and a graph for publication.

11. Compare the SNPFreqPlot output to the results from Subheading 3.1.1 to see if there is overlap, which would provide further support for a candidate CRM in this region.

3.1.3 Finding Sequence Conservation in Arabidopsis Accessions Using Microsoft Excel

Microsoft Excel can be used as an alternative to SNPFreqPlot to visualize SNP frequency in Arabidopsis accessions. This is achieved by extracting information from the VCF files and creating a table of single numbers, which represent presence or absence of a SNP at each position for each accession and using this to calculate the SNP frequency at each position.

1. Download the VCF file as described in Subheading 3.1.2, **steps 1–6**.
2. Open the VCF file in Excel by selecting open file and navigating to the location of the VCF file. Ensure the "all files" is selected.
3. A text import wizard will pop up. In **Step 1**, ensure that the "Delimited" tab is checked and the "Start import at row" is set to one. Press Next.
4. In **Step 2** check "tab" is selected under "delimiters". Press Next.
5. In **Step 3** ensure "General" is selected under "Column data format".
6. Rename the worksheet from the long string of numbers and letters to something shorter, e.g., SNP.
7. In a new sheet type "=left(SNP!J17,1)" in cell J17 (*see* **Notes 6** and 7).
8. Copy this formula to the full extent of the table (*see* **Note 8**).
9. Make a new sheet and copy the newly generated table into it using Paste Special, Values to only copy the text. This may take a few minutes.
10. A small yellow box with an exclamation mark should appear at the top left corner of the copied region. Select the drop down option to convert the stored text to numbers. This may take a few hours.
11. Using Edit, Replace change any numbers that are greater than 1 (starting at 2) and replace them with the number 1 (*see* **Note 9**).
12. Sum across each row in the right most column for each position. This will provide the number of accessions with a SNP at this position (*see* **Note 10**).

13. Divide the numbers in the summed column by the number of accessions (1135 if all available accessions were used) and multiply by 100. This will calculate the percentage of accessions with a SNP at each location.
14. Assign base positions for each of these SNP frequencies. This will depend on the direction of the gene-of-interest. If the gene was in the forward direction, the top number is −1000 and if the gene was on the reverse strand then −1 is at the first position. −1 should reflect the base immediately upstream of the TIS. This can be achieved using the Excel Fill Feature.
15. Plot the position by SNP percentage. A scatterplot with straight lines is the easiest to read.
16. Visually inspect the plot to identify regions with low SNP frequencies, as these are candidates for CRMs. Compare this with the results from Subheading 3.1.1 to see if there is overlap, which would provide further support for the region as a candidate CRM.

3.2 Fluorescent Reporter Expression to Analyze CRM Activity

Along with phylogenetic footprinting, experimental work is a key component in detecting and verifying CRMs in the plant male germline. As the positioning of the germline cells within the vegetative cell makes it difficult to use molecular approaches, we use fluorescent reporters under the control of different promoter regions to analyze transcriptional activity in germline cells of stably transformed Arabidopsis plants. By observing and measuring fluorescence levels through microscopy, this approach can distinguish transcriptional activity in the germline from that in the vegetative nucleus and provides quantitative information on the transcriptional activity of the promoter region.

There are a large number of cloning methods, plant expression vectors, and fluorescent proteins that are suitable as reporters. We have typically used Multisite Gateway Cloning with plant expression vectors [13], but other cloning methods and vectors are equally applicable. An important element in the design with any cloning technique is the inclusion of a nuclear localization signal on the fluorescent reporter. Nuclear localization of the fluorescent reporter enables transcriptional activity in germ cells and the vegetative cell to be easily distinguished, and aids quantification (*see* Subheading 3.2.1). In particular, a nuclear signal provides a defined region to analyze rather than the more diffuse signals observed, for example, in cytoplasmic extensions of the sperm cells. We use Arabidopsis Histone H2B to provide nuclear localization of the reporter. As there are a range of cloning methods and vectors we do not include precise details here, but instead describe methods for the analysis of reporter lines (*see* Subheading 3.2.1) and how fluorescent reporters can be used to identify and characterize CRMs which are active in plant male germ cells (*see* Subheading 3.2.2).

3.2.1 Quantification of Reporter Fluorescence Levels in Plant Male Germ Cells

An important aspect of monitoring transcription through reporters is measuring fluorescence intensity as an indication of the level of transcription driven by different promoter regions. While the presence/absence of fluorescence can be used to identify regions that are essential for transcription, alterations in fluorescence level can be used to identify regions that repress or enhance expression. Our approach uses a nuclear-targeted fluorescent protein reporter and fluorescent microscopy to quantify expression in stably transformed Arabidopsis plants. However, using Agrobacterium to transfer T-DNA into plant genomes results in random insertion and the site of insertion influences the level of expression [14]. This effect introduces random variation into the expression levels of a given gene construct. To allow for this and to enable rapid quantification in the first generation of transformed (T1) plants, we detail a technique to quantify fluorescence in a large number of pollen grains from pooled flowers harvested from a population of T1 plants (*see* **Note 11** and Fig. 1). We now describe the collection and imaging of pollen with germline fluorescent reporters (**steps 1–11**) and the quantification of fluorescence intensity in sperm cells (**steps 12–21**).

1. Collect two open flowers from each of approximately 40 T1 Arabidopsis plants into a 15 ml centrifuge tube containing 5 ml of harvest buffer (*see* **Note 12**).
2. Filter through a 53 μm nylon (or stainless steel) filter.
3. Centrifuge at 2900 × *g* for 2 min to pellet the pollen.
4. Extract 3.6 μl by pushing the tip of a disposable P20 pipette tip into the pollen pellet and quickly releasing the plunger to draw pollen into the tip.
5. Pipette onto a clean microscope slide (*see* **Note 13**).
6. Allow pollen to settle for 30–45 s before placing a coverslip over the sample and sealing the edges with nail varnish to prevent drying of the sample during observation.
7. View pollen using a 40× objective lens under the fluorescent microscope with the appropriate excitation wavelength and filter settings for the fluorescent protein.
8. Using a control reporter (e.g., full-length promoter) construct that has the highest expected or observed fluorescence intensity determine the appropriate exposure time, ISO sensitivity and pixel size using the NIS Elements BR software or alternative imaging software (*see* **Note 14**).
9. Capture images with the determined settings until more than 100 pollen grains with a fluorescent signal in an in-focus sperm cell (*see* Fig. 2) have been captured. Save as TIFF files (*see* **Note 15**).

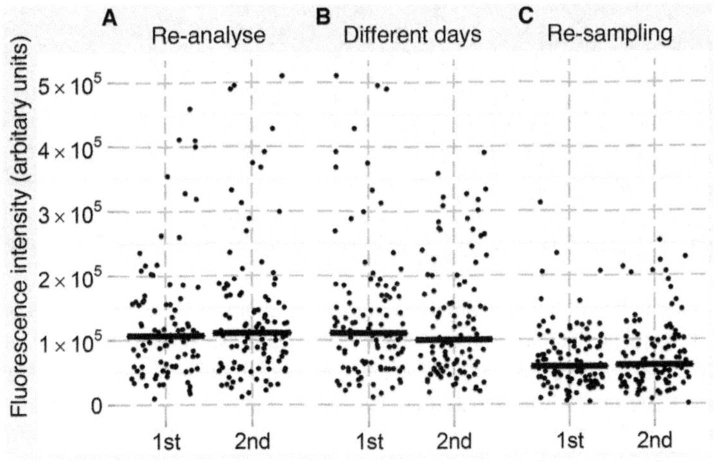

Fig. 1 Method validation for the measurement of fluorescence from reporters in plant germ cells. (**a–c**) Scatter plots for three sets of repeated measurements performed to validate the fluorescence quantification technique. Each point is a single measurement in arbitrary units. The horizontal line shows the median of each set of measurements. (**a**) Repeated measurement of the same set of images of pollen on different days. (**b**) Measurements from a representative construct for pollen collected and imaged on different days. (**c**) Measurements from two slides prepared from the same pollen collection. (**a**) and (**b**) were performed with the same construct, while (**c**) used a different, lower activity, construct. There were no statistically significant differences between the repeated measurements in **a**, **b**, or **c** using a Mann-Whitney U test

10. Prepare slides (**steps 1–6**) and capture images with the same settings for all the constructs to be tested for that experiment (e.g., a deletion series, *see* Subheading 3.2.2).
11. Optional: To add confidence in the analysis, repeat the analysis for each construct by preparing slides (**steps 1–6**) and capturing images with the same settings on separate days. This will generate measurements from 200 pollen grains for each construct (*see* **Note 16**).
12. To measure fluorescence intensity, open the saved TIFF image files in NIS Elements BR software or an appropriate image analysis program (*see* **Note 17**).
13. Remove background by selecting background regions of the vegetative cell cytoplasm for three expressing pollen grains in each image (if only one or two expressing grains are visible in an image, use them all) and selecting the automatic background removal tool (*see* Fig. 2a).
14. Draw a bounding region-of-interest (ROI) around one sperm cell in each expressing pollen grain (*see* Fig. 2a). Use only in focus sperm cells and only one sperm cell in any given pollen grain. Where two sperm cells are in focus, use the brighter cell (*see* Fig. 2b–d).

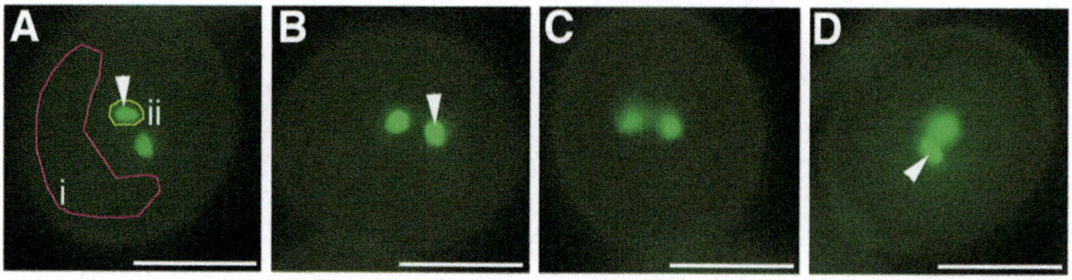

Fig. 2 Selection of sperm cells for fluorescence intensity measurement. Representative mature pollen grains with a germline fluorescent reporter are shown. (**a**) A background region of vegetative cell cytoplasm selected for automatic background removal in NIS Elements BR is highlighted in pink and labeled (i) while a ROI around a single sperm cell for measurement by "SumIntensity" is shown in yellow and labeled (ii). (**b–d**) Selection of sperm cells for measurements of fluorescence intensity, with selected sperm cells indicated by arrowheads. (**b**) Both sperm cells are in focus, with only one (the brighter of the two) selected for analysis. (**c**) Both sperm cells are out of focus and would not be included in intensity measurements. (**d**) The in-focus sperm cell is selected for measurement. To enhance visibility, contrast was automatically increased for these images. Scale bar = 10 μm

15. Measure the total fluorescence by using the automated measurement tool to collect the "SumIntensity" (*see* **Note 18**).
16. Repeat **steps 14** and **15** until 100 sperm cell nuclei have been measured for each construct (or 200 if two sets of images have been collected).
17. Additionally, measure 100 non-expressing pollen using the same-sized ROI as used in the measurement of the expressing pollen, pooled from across all constructs. Average these measurements and subtract it from the measured fluorescence values for each sperm cell. This serves as a further correction factor to account for background.
18. Working in R, calculate the median for the measurements for each construct (*see* **Note 19**).
19. Normalize the median value of each of the test constructs to the control construct in R (*see* **Note 20**).
20. Use Bootstrap re-sampling to estimate 95% confidence intervals on the normalized expression level in R (*see* **Note 21**).
21. Test the significance of changes in expression level between consecutive reporter constructs in a deletion series or between and a control and an altered construct with a pairwise Mann-Whitney U test in R (*see* **Notes 22** and **23** and Fig. 3).

3.2.2 Using Fluorescent Reporters in a Deletion Series to Identify Male Germline CRMs

Fluorescent reporters can be used in a deletion series to aid in the identification of candidate CRMs. A deletion series involves testing progressively shorter fragments of a promoter fused to the fluorescent reporter to identify regions of a promoter involved in transcriptional regulation. When fluorescence is lost or the intensity

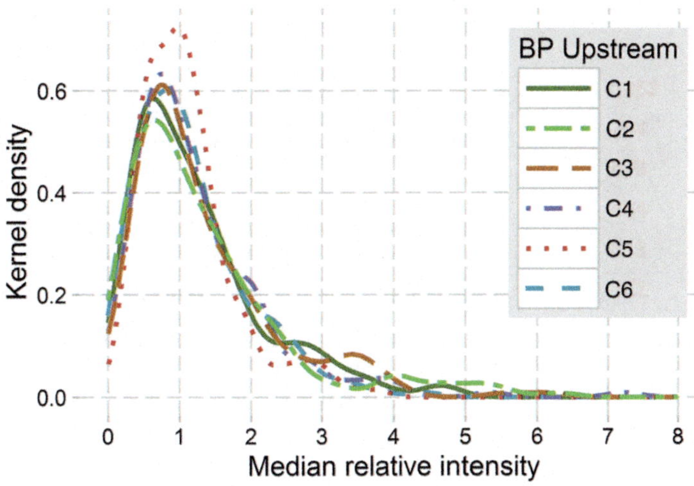

Fig. 3 Measurement distribution of construct intensity levels. Kernel density estimates for six reporter constructs (C1–C6) in sperm cells. The expression level has been normalized to the medians to allow the shape of the distribution to be visualized without respect to their absolute values. Five constructs (C1–C5) had high population penetrance, while one construct (C6) has low population penetrance, but the constructs show a similar distribution

altered upon removal of a promoter fragment, this indicates that the fragment contained sequences that influence transcription. A deletion series can be used in conjunction with the phylogenetic footprinting approaches described in Subheading 3.1 to identify candidate CRMs by providing evidence that a conserved region is important for function, or to narrow down regions for further footprinting analyses.

1. Create an expression vector using the cloning method of choice with a full-length promoter driving expression of a nuclear-localized fluorescent reporter. Stably transform Arabidopsis plants and confirm expression (*see* **Note 24**).
2. Create expression vectors with progressively smaller promoter fragments, removing approximately 150–200 bp each time and transform Arabidopsis plants (*see* **Note 25**).
3. Select and grow T1 generation plants. View mature pollen with a fluorescence microscope and quantify expression level relative to the full-length promoter as described in Subheading 3.2.1.

3.2.3 Construction of Tetramer Fluorescent Reporters to Verify Candidate CRMs

Once phylogenetic footprinting and/or a deletion series have identified a candidate CRM, further experimental verification is required to confirm a role in transcriptional regulation. The region of DNA predicted to function as a CRM is placed upstream of a minimal promoter that provides a transcription start site (TSS) and a 5'UTR, and the hybrid promoter is used to drive expression of the nuclear-localized fluorescent reporter. While this can involve a single copy of the region, a single copy is often not sufficient to provide reliably detectable fluorescence [5, 15]. To increase fluorescence levels of individual CRMs, we recommend and describe a method to make a tetramer (four copies) of the region using restriction enzyme digests and ligation.

1. Design suitable primers to amplify four copies of the candidate CRM by PCR each with different cloning and restriction enzyme sites (*see* **Note 26** and Fig. 4).
2. Amplify individual fragments from genomic DNA with a high-fidelity Taq polymerase in a 50 μl PCR and analyze 10 μl of the PCR product on a 2% agarose gel.
3. Pool the remaining PCR products (~40 μl each) and remove excess primers using a PCR clean up kit or precipitation. Elute/resuspend in 16.5 μl milliQ H_2O (*see* **Note 27**).
4. Digest the PCR products by adding 0.5 μl of each restriction enzyme and 2 μl restriction digest buffer, giving a total volume of 20 μl. Incubate at the temperature and time indicated by the enzyme manufacturer. Heat-inactivate by incubating at 65 °C for 10 min.
5. Clean up the restriction digest as in **step 3** and elute/suspend in 25 μl milliQ H_2O.
6. Add 3 μl T4 ligase buffer and 2 μl T4 ligase to the 25 μl digested PCR fragments and incubate at room temperature

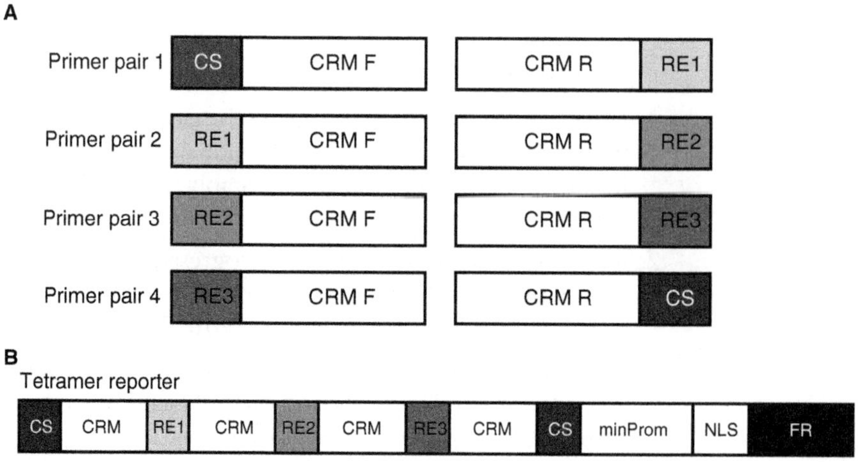

Fig. 4 Construction of a CRM tetramer fluorescent reporter. (**a**) The four primer pairs that are used to PCR amplify four copies of a candidate CRM flanked by different combinations of cloning sites (CS) and recognition sequences for three restriction enzymes (RE1–RE3). (**b**) The final tetramer reporter. The four PCR fragments are joined in a ligation following a restriction enzyme digest. The cloning sites are then used to clone the four copies of the CRM upstream of a minimal promoter (minProm) to drive expression of the fluorescent reporter (FP) with a nuclear localization signal (NLS)

for 1 h in a 500 ml beaker of room temperature water. Place the beaker at 4 °C overnight to slowly cool the reaction. Inactivate the enzyme by incubating at 65 °C for 10 min.

7. Clone the fragment into an appropriate cloning vector and verify by sequencing (*see* **Note 28**).
8. Transfer the verified tetramer region to an appropriate expression vector upstream of a minimal promoter to drive expression of the nuclear-localized fluorescent reporter (*see* **Note 29**).
9. Transform into Arabidopsis plants, select and analyze expression in multiple T1 lines as described in Subheading 3.2.1.

3.2.4 Identification of TFBSs within CRMs

Once a CRM has been identified and verified, candidate TFBSs within the CRM can be identified. These might be based on short sequences with very high sequence conservation across a wide evolutionary window in the phylogenetic footprinting (*see* Subheading 3.1), or by searches of databases containing known TFBSs using the CRM sequence. In principle the activity of an individual TFBS could be tested by placing it upstream of a minimal promoter and a nuclear-localized reporter. However, the practicality of this approach is limited. A single TFBS is unlikely to provide sufficient fluorescence to be reliably detected and multiplexing short sequences (often less than 10 bp) is technically challenging. Additionally, as TFBSs within a CRM may act in conjunction with other TFBSs, the absence of fluorescence does not necessarily mean that the TFBS is not used to control expression. It may instead have an enhancing role. Therefore, a suitable approach to verify if regions of a CRM are acting as TFBSs is site-directed mutagenesis of promoter fragments driving expression of a nuclear-localized fluorescent reporter.

1. Generate the control promoter:fluorescent reporter. This is a promoter fragment with the active CRM, the TSS, and 5′-UTR. The promoter fragment can be newly generated or may be already cloned as part of a deletion series.
2. Design the alterations required to test TFBS activity. This will involve altering 4–6 bases thought to be essential for activity of the TFBS to sequences with little resemblance to the initial sequence.
3. Create the altered versions of the promoter in a cloning vector. This can be done using the QuikChange site-directed mutagenesis kit or through chemical synthesis.
4. Create expression vectors with the promoter upstream of a nuclear-localized fluorescent reporter and transform into Arabidopsis.
5. Grow T1 plants and measure fluorescence relative to the unaltered promoter region as described in Subheading 3.2.1.

4 Notes

1. It is important to ensure that orthologous proteins that have retained the same function and expression pattern are used in the subsequent analysis. Identifying a functional ortholog in a distant species can be difficult when using a protein that is a member of a large family, such as many TFs. If the protein is known to have a distinguishing feature, then manually verify that all selected proteins have this feature. Another approach is to initially search closely related species (e.g., within Brassicaceae) to look for distinguishing sequences, and then repeat the searches, expanding the phylogenetic distance (e.g., eudicots). Additionally, transcriptomic data (where available) can help by confirming that selected genes have similar expression patterns and/or are expressed in pollen in different plants.
2. MEME finds motifs by searching for over-represented regions in the uploaded sequences. The output is very dependent on the input settings, so we suggest running the program multiple times with different parameters, such as the site distribution, number of motifs, and motif width. Additionally, run the analysis with different "phylogenetic windows." Also, if there are multiple orthologous sequences for a single species remove those that show the least conservation to get down to a single sequence for each species (unless you are confident that there are multiple functional orthologs). Re-running the analysis will allow for the detection of a range of motifs and the refinement of any motifs that are discovered.
3. By visually inspecting the motifs identified in MEME, manually adding gaps and making other manual modification, candidate CRMs can be refined and candidate TFBSs identified. This is because CRMs from orthologous genes may have the same TFBS but with altered spacing or orientation and MEME only searches for ungapped motifs.
4. At the time of writing, sequences for 1135 accessions were available and this number is likely to increase in future.
5. We provide access to the SNPFreqPlot application through http://snpfreqplot.otago.ac.nz/SNPFreqPlot/. However, if the application is likely to be used frequently, we request that you download it from Github (https://github.com/BrownfieldPlantLab/SNPFreqPlot) and use it through RStudio. While we use SNPFreqPlot for VCF files from The 1001 Genomes Consortium, it can be used for VCF files from other sources. Additionally, SNPFreqPlot will use default SNP calls so if more stringent limits are required on when a SNP is called, the VCF file will need to be modified.

6. In the formula "=left(SNP!J17,1)" the SNP refers to the name of the first sheet. If it has been renamed to a name other than "SNP," then use that name.
7. This command will take the information from cell J17 in the SNP sheet and return only the first character in the new sheet. J17 is the first cell in which the genotype SNP data appears in the VCF file. This first character of the VCF file indicates the presence or absence of a homozygous SNP. A 0 indicates that the nucleotide is the same as Col and a number >0 indicates a different nucleotide. As all the genomes in the 1001 genomes project are highly inbred, all the SNPs should be homozygous. To check use the find function and search for "0|1", as this indicates a heterozygous SNP.
8. The full extent of the table will be the number of bp selected (1000) on the vertical scale by the number of accessions in the VCF file (1135) on the horizontal.
9. The different numbers above 0 represent different SNPs, e.g., 1 might represent A to T, 2 might represent A to G, etc. Replacing numbers greater than 1 with 1 is reducing the complexity of the data to just report absence (0) or presence (1) of a SNP in that location.
10. If the computer is struggling, copy this column (Paste Special, Value) to a new workbook. Only take the characters, not the formula used to generate the numbers. This means the computer only has to deal with a 1000 character list instead of three 1000 × 1135 tables and formulas.
11. This method has been validated by three tests to ensure the methodology gave consistent results and the sampling method was representative. Firstly, the analysis was repeated on different days with the same images, to ensure that the method of analysis was robust for a single user (*see* Fig. 1a). Secondly, for each construct, the summed intensity of roughly 100 pollen grains was measured on two separate days (*see* Fig. 1b). Finally, the analysis was performed twice on separate slides taken from the same pool of pollen collected from 40 T1 plants (*see* Fig. 1c). As in each case there was no significant difference between the two analyses, it shows that the results obtained are consistent for a pool of plants, are fairly sampled from the population and are not overly influenced by differing subjective factors in measurement with a single operator.
12. It is best to choose the most recently opened flowers from inflorescence stems with 4–6 open flowers from plants that are not yet starting to senesce (i.e., from plants at stage 6.1–6.3 in the life cycle [16]).

13. Before pipetting the sample, hand-polish the slide with a Kimwipe after breathing moisture onto it, to prevent the sample spreading.
14. Exposure time is determined to give signals that are just below saturation level. We have captured images using NIS Elements BR version 4.00.03 (build 775) from Nikon, and determined saturation by turning on the “Pixel Saturation Indication” button to check for pollen nuclei with saturated signal. The exposure time was then adjusted so the signal intensity was just below the saturation level of the brightest pollen grains. The highlighting color can be selected from the nearby pull-down menu. Other image analysis software are likely to have a similar function. Once determined, these settings must not be changed between reporter constructs in a single experiment.
15. If no fluorescence is obvious, the inclusion of DAPI in the buffer allows the location of sperm cells to be determined, and these regions closely inspected in the matching fluorescent images to determine whether expression is completely absent or present at very low levels.
16. Our verification experiment sampling on different days gave similar results (*see* **Note 11** and Fig. 1b), thus repeating the measurement is not essential. However, we still suggest repeating the analysis to provide increased confidence in the conclusions and to detect any unexpected anomalies that may occur.
17. We describe measuring fluorescent intensity using NIS Elements BR software. However, the measurement can be conducted in other image analysis software, for example, in Image J. Here, the background noise can be removed using a rolling ball radius of 500 pixels under subtract background. The ROI is hand drawn around in focus sperm cells and the fluorescent intensity measured. To eliminate background variation, the ROI is dragged to a cytoplasmic region of the same pollen grain and intensity measured again. The background is then subtracted from the value for each pollen grain and the values processed as described.
18. “SumIntensity” is used as it is not normalized to the area measured and so the inclusion of any dark regions around the brightly fluorescing sperm cell nuclei in the ROI does not substantially alter the measured fluorescence level. Thus, subjective effects of where the ROI is precisely drawn are reduced.
19. We have found the distribution of fluorescence intensity is non-normally distributed with a heavily left-skew and a long tail (*see* Fig. 3). The non-normality of the dataset was confirmed by the D’Agostino K^2 test [17]. Due to the non-normality we use median values rather than means. This has the additional advantage of higher robustness in response to outlying data points and thus is insensitive to rare high brightness sperm cells

we observed with 5–7 times higher brightness than the median.

20. The normalization is conducted as the measured values are in arbitrary units. In addition, normalizing to the same control construct also enables separate experiments to be compared.

21. To generate an estimated confidence interval by bootstrap resampling, the population is repeatedly randomly sampled with replacement to generate a large number of virtual populations [18]. A 95% interval on the range of medians of these virtual populations provides an estimate as to the confidence interval of the true population. This can be achieved with the following R function:

```
BootstrapCIOnMedian = function(v)
{
v <- na.omit(v)
bootstrap.repeats <- 10^4
bootmed = apply(
matrix(
 sample(v,rep=TRUE,bootstrap.repeats*length(v)),
 nrow=bootstrap.repeats),1,median)
return (quantile(bootmed,c(.025,0.975)))
}
```

22. A pairwise Mann-Whitney U test is used because the data are non-normal and because the ordered nature of the deletion series means the differences are expected between sequential deletions.

23. We observed that in a weakly fluorescing construct, there was a decline in population penetrance. One possible explanation of this decline is that the level of fluorescence was falling below the detectable level of expression. In this case, a plot of the fluorescence intensity should show truncation to the left. To test this we normalized the expression data from this construct and other brighter constructs to the median and then constructed a kernel density estimate (*see* Fig. 3). This approximates the distribution of the points that were sampled for each construct enabling a comparison to be made and any truncations to be observed. For our data, no difference in distribution is observable (*see* Fig. 3). It can therefore be concluded that the fall in penetrance is real and not an artifact of minimum detectable fluorescence.

24. The full-length promoter refers to the longest fragment tested and providing reporter expression. The length of the promoter is somewhat arbitrarily decided upon, usually based on distance from the TIS or other genomic features such as the intergenic region. Generally, a region of around 1 kb is sufficient with most regulatory elements active in male germ cells being close

to the TIS. As there is flexibility in size, the precise fragment is often determined by the position of a site for a high-quality primer for amplifying the region. The 3′ end of the fragment is just before the ATG of the TIS (or the ATG of the reporter can be replaced) meaning that the construct includes the transcription start site (TSS) and 5′-UTR of the target promoter.

25. Removing fragments of around 150–200 bp per fragment enables a balance between the number of promoter:reporter constructs to analyze and the generation of useful data. The precise size is flexible and often determined by the location of sequences which provide good primers for the amplification of the region by PCR. If phylogenetic footprinting has highlighted conserved regions, avoid splitting this region across different fragments.

26. Four copies of the candidate CRM are amplified with different cloning sites on one end of two copies, and restriction enzyme sites on all copies, to enable then to be joined (*see* Fig. 4). The cloning sites depend on the vector and cloning system used. As we use Gateway cloning, the cloning sites are *attB* sites for a subsequent BP reaction. When considering which restriction enzymes to use try to find: (1) enzymes that are compatible in the same buffer (e.g., CutSmart buffer from NEB) and require the same incubation temperature, (2) enzymes that do not cut the CRM or cloning sites, (3) high fidelity enzymes that are heat inactivated if possible. For many enzymes, additional bases will also be required to ensure efficient cutting.

27. It is optimal to remove the primers as they will also be cut in the digest and interfere with the subsequent ligation. We use a PCR clean-up kit, but before a particular kit is used, check the size limits as some CRM PCR products may be too small to be retained on the column. If so, the PCR product can be cleaned using a phenol/chloroform isolation followed by an ethanol precipitation, although primers may also precipitate.

28. We conducted a BP reaction with the pDONRP4-P1R. The efficiency of this reaction was increased by adding an extra PCR step using the products of the ligation and primers that amplify only from the cloning sites at the ends of the tetramer fragment (e.g., *attB* sites).

29. We used the *CaMV 35S* minimal promoter (−46 to +18) as both the full-length and minimal *CaMV 35S* regions do not activate transcription in Arabidopsis pollen or in male germ cells [15, 19]. While there are a number of ways this construct can be made we have used Multisite Gateway Cloning with the tetramer in a pENTRP4-P1R vector, a pENTER221:min35S: H2B fusion and a fluorescent reporter in a pENTRP2R-P3 vector recombined into the DESTINATION vectors pB7m34GW or pB7m34GW [13].

Acknowledgments

This work was supported by a Fast Start grant from the Marsden Fund of the Royal Society of New Zealand (12-UOO-043) to L.B. and by research grants from the Biotechnology and Biological Sciences Research Council (BB/I011269/1; BB/N005090/1) to D.T. B.P. was supported by a University of Otago Doctoral Scholarship.

References

1. Berger F, Twell D (2011) Germline specification and function in plants. Annu Rev Plant Biol 62:461–484
2. Borges F, Gomes G, Gardner R, Moreno N, McCormick S, Feijó JA, Becker JD (2008) Comparative transcriptomics of Arabidopsis sperm cells. Plant Physiol 148:1168–1181
3. Russell SD, Gou X, Wong CE, Wang X, Yuan T, Wei X et al (2012) Genomic profiling of rice sperm cell transcripts reveals conserved and distinct elements in the flowering plant male germ lineage. New Phytol 195:560–573
4. Phillips T, Hoopes L (2008) Transcription factors and transcriptional control in eukaryotic cells. Nat Edu 1:119
5. Peters B, Casey J, Aidley J, Zohrab S, Borg M, Twell D, Brownfield L (2017) A conserved cis-regulatory module determines germline fate through activation of the transcription factor DUO1 promoter. Plant Physiol 173:280–293
6. Durbarry A, Vizir I, Twell D (2005) Male germ line development in Arabidopsis. *duo pollen* mutants reveal gametophytic regulators of generative cell cycle progression. Plant Physiol 137:297–307
7. R development core team (2008) R: a language and environment for statistical computing. R foundation for statistical computing, Vienna, Austria. isbn:3-900051-07-0. URL http://www.R-project.org
8. Tagle DA, Koop BF, Goodman M, Slightom JL, Hess DL, Jones RT (1988) Embryonic ε and γ globin genes of a prosimian primate (*Galago crassicaudatus*). Nucleotide and amino acid sequences, developmental regulation and phylogenetic footprints. J Mol Biol 203:439–455
9. Blanchette M, Tompa M (2002) Discovery of regulatory elements by a computational method for phylogenetic footprinting. Genome Res 12:739–748
10. Goodstein DM, Shu S, Howson R, Neupane R, Hayes RD, Fazo J et al (2012) Phytozome: a comparative platform for green plant genomics. Nucleic Acids Res 40:D1178–D1186
11. The Consortium 1001 Genomes (2016) 1,135 genomes reveal the global pattern of polymorphism in *Arabidopsis thaliana*. Cell 166:481–491
12. Bailey TL, Boden M, Buske FA, Frith M, Grant CE, Clementi L, Ren J, Li WW, Noble WS (2009) MEME suite: tools for motif discovery and searching. Nucleic Acids Res 37(Web Server):W202–W208. doi:10.1093/nar/gkp335
13. Karimi M, De Meyer B, Hilson P (2005) Modular cloning in plant cells. Trends Plant Sci 10:103–105
14. Peach C, Velten J (1991) Transgene expression variability (position effect) of CAT and GUS reporter genes driven by linked divergent T-DNA promoters. Plant Mol Biol 17:49–60
15. Zhang J, Yuan T, Duan X, Wei X, Shi T, Li J, Russell SD, Gou X (2016) *Cis*-regulatory elements determine germline specificity and expression level of an isopentenyltransferase gene in sperm cells of Arabidopsis. Plant Physiol 170:1524–1534
16. Boyes DC, Zayed AM, Ascenzi R, McCaskill AJ, Hoffman NE, Davis KR, Görlach J (2001) A model for high throughput functional genomics in plants. Plant Cell 13:1499–1510
17. D'Agostino R, Pearson E (1973) Tests for departure from normality. Empirical results for the distributions of b2 and $\sqrt{}$ b1. Biometrika 60:613–622. doi:10.2307/2335012
18. Efron B (1979) Bootstrap methods: another look at the jacknife. Ann Stat 7:1–26
19. Wilkinson J, Twell D, Lindsey K (1997) Activities of *CaMV 35S* and *nos* promoters in pollen: implications for field release of transgenic plants. J Exp Bot 48:265–275

Chapter 23

RNA-Seq Data Analysis: From Raw Data Quality Control to Differential Expression Analysis

Weihong Qi, Ralph Schlapbach, and Hubert Rehrauer

Abstract

As a revolutionary technology for life sciences, RNA-seq has many applications and the computation pipeline has also many variations. Here, we describe a protocol to perform RNA-seq data analysis where the aim is to identify differentially expressed genes in comparisons of two conditions. The protocol follows the recently published RNA-seq data analysis best practice and applies quality checkpoints throughout the analysis to ensure reliable data interpretation. It is written to help new RNA-seq users to understand the basic steps necessary to analyze an RNA-seq dataset properly. An extension of the protocol has been implemented as automated workflows in the R package ezRun, available also in the data analysis framework SUSHI, for reliable, repeatable, and easily interpretable analysis results.

Key words RNA-seq, Quality control, Read alignment, Gene expression quantification, Differential gene expression

1 Introduction

RNA-sequencing (RNA-seq) combines simultaneous transcript identification and quantification of a large number of genes in a single assay. It has many different applications in life sciences, ranging from identification of differentially expressed genes and transcripts, analysis of alternative splicing and polyadenylation, to detection of fusion genes and post-transcriptional events [1]. Consequently, the actual RNA-seq data analysis also has many variations, depending on the applications and studied organisms. In this chapter, we describe the data analysis steps in a typical RNA-seq experiment where the aim is to identify differentially expressed genes, including read alignment, quantification of gene expression levels, and differential gene expression analysis. Meanwhile, three quality control (QC) steps, read QC and pre-processing, alignment QC, and count QC, are also described. As pointed out in the recently published RNA-seq data analysis best practice [2], these checkpoints are essential for producing reliable analysis results.

Anja Schmidt (ed.), *Plant Germline Development: Methods and Protocols*, Methods in Molecular Biology, vol. 1669, DOI 10.1007/978-1-4939-7286-9_23, © Springer Science+Business Media LLC 2017

Read QC tools analyze sequence quality and GC content of the sequencing reads, and check the presence of adaptors and other contaminants. Based on the results, reads can be processed accordingly to trim adaptors, remove low quality bases, and discard short and/or low quality reads. Alignment QC monitors key measures of aligned data quality, such as mapping rate, mapping quality, mapped strand, transcript coverage, and so on. Count QC measures the quality of counts by checking count distributions according to GC content, gene length, and/or annotated gene features, as well as by analyzing reproducibility among biological replicates. These checkpoints are essential for researchers to make informed decisions about data interpretation and analysis strategies [3]. For example, quality-controlled reads can improve the alignment results by removing sequencing artifacts and errors that may contribute to incorrect interpretation of data [4]. The mapped strand inferred from alignment QC is an important parameter for accurate counting of the reads [5]. The length and/or GC bias of counts, or batch effects, if detected during the count QC, can be corrected by applying corresponding correcting methods [6]. This protocol is written with the aim to assist new RNA-seq users to understand the basic steps necessary to analyze an RNA-seq dataset properly and can be applied to the analysis of differential gene expression in comparisons of two conditions. An extension of the protocol has been implemented as automated workflows in the R package ezRun (https://github.com/uzh/ezRun), available also through the data analysis framework SUSHI [7], with the aim to provide better workflow documentation, as well as reliable, repeatable and easily interpretable analysis results.

2 Materials

1. RNA-seq Dataset: The *Arabidopsis thaliana* dataset generated from sorted cells within the medial domain of the gynoecium (GSE74458) [8] are used. For simplicity, three biological replicates without technical replicates are selected for each of the two conditions (Table 1).
2. Reference Genome: *Arabidopsis thaliana* genome sequence and updated annotation data (https://www.araport.org/downloads) are used as the reference.
3. Computing Software: Analysis of RNA-seq data consists of multiple steps and is typically done with the help of open source tools (https://omictools.com/rna-seq-category), with each tool performing a single step. The collection of software tools used in this protocol is listed in Table 2.
4. Computing Hardware: Almost all open source software packages for RNA-seq data analysis are developed for Linux

Table 1
RNA-seq read dataset selected from the RNA-seq analysis of *A. thaliana* medial domain of the gynoecium (GSE74458) [8]

Sample name	Run	BioSample	Condition	Biological replicate	Source name	Tissue cell type
GSM1921011	SRR2850572	SAMN04221482	YFP-NEG	BioRep1	Cells from inflorescence_YFP-NEG	Cells from inflorescence
GSM1921013	SRR2850574	SAMN04221484	YFP-NEG	BioRep2	Cells from inflorescence_YFP-NEG	Cells from inflorescence
GSM1921015	SRR2850576	SAMN04221486	YFP-NEG	BioRep3	Cells from inflorescence_YFP-NEG	Cells from inflorescence
GSM1921019	SRR2850580	SAMN04221490	YFP-POS	BioRep1	YFP-POS_Medial domain, Arabidopsis gynoecium	Medial domain, Arabidopsis gynoecium
GSM1921021	SRR2850582	SAMN04221492	YFP-POS	BioRep2	YFP-POS_Medial domain, Arabidopsis gynoecium	Medial domain, Arabidopsis gynoecium
GSM1921023	SRR2850584	SAMN04221494	YFP-POS	BioRep3	YFP-POS_Medial domain, Arabidopsis gynoecium	Medial domain, Arabidopsis gynoecium

Table 2
Open source software packages used in this protocol

Name	Hyperlink to the project home
UCSC utility scripts	http://hgdownload.soe.ucsc.edu/admin/exe/linux.x86_64/
NCBI SRA Toolkit	https://trace.ncbi.nlm.nih.gov/Traces/sra/sra.cgi?view=software
FastQC	http://www.bioinformatics.babraham.ac.uk/projects/fastqc/
Trimmomatic	http://www.usadellab.org/cms/?page=trimmomatic
STAR	https://github.com/alexdobin/STAR
SAMtools	http://samtools.sourceforge.net/
RSeQC	http://rseqc.sourceforge.net/
featureCounts	http://bioinf.wehi.edu.au/featureCounts/
R	https://www.r-project.org/
ezRun	https://github.com/uzh/ezRun
SUSHI	https://github.com/uzh/sushi

Follow the hyperlinks for details about the packages, such as installation and usage

or Unix-based operating systems. Among the multi-step RNA-seq data analysis workflow, the most computing extensive task is alignment of reads, which determine the suitable computing environment. To speed up the analysis, multi-threading is usually supported by these software tools. Consequently, a multi-core Linux/Unix computer server/cluster is needed. Within this protocol, multi-threading is set with 8, the total disk space needed (the raw data, intermediate files, and final results) is about 100 GB. The memory usage varies and its peak is at around 5 GB.

3 Methods

The RNA-seq data analysis protocol described here consists of three major steps: mapping of reads to the reference genome, counting mapped reads for annotated genes, and finding differentially expressed genes. At each step, there is also a quality control (QC) checkpoint: read QC and preprocessing, alignment QC, and count QC, respectively. Before describing each step in details, we start with setting up the computing environment:

3.1 Preparation of the Workspace and Data Files

1. Set up the working space (*see* **Note 1**). Create an empty directory, which will be the working directory. Then create two subdirectories within the working directory, as containers for the reference genome and raw reads, respectively:

```
mkdir tutorial
cd tutorial
mkdir reference
mkdir reads
```

2. Download and unpack the reference genome and annotation data:

```
cd reference
wget https://www.araport.org/downloads/TAIR10_genome_release/
assembly/TAIR10_Chr.all.fasta.gz
wget https://www.araport.org/download_file/Araport11_latest/
annotation/Araport11_GFF3_genes_transposons.201606.gff.gz
wget https://www.araport.org/download_file/Araport11_latest/
annotation/Araport11_GFF3_genes_transposons.201606.gtf.gz
gunzip -d TAIR10_Chr.all.fasta.gz
gunzip -d Araport11_GFF3_genes_transposons.201606.gff.gz
gunzip -d Araport11_GFF3_genes_transposons.201606.gtf.gz
```

3. Transform the annotation data from gff format to bed format, which will be needed during the alignment QC. We use utility scripts from UCSC (http://hgdownload.soe.ucsc.edu/admin/exe/linux.x86_64/) (*see* **Note 2**) for the file format converting:

```
gff3ToGenePred Araport11_GFF3_genes_transposons.201606.gff
Araport11_GFF3_genes_transposons.201606.Gp
genePredToBed Araport11_GFF3_genes_transposons.201606.Gp Ara-
port11_GFF3_genes_transposons.201606.bed
```

4. Download and prepare the raw reads. The read data listed in Table 1 can be downloaded from NCBI Short Read Archive (SRA) and converted to compressed fastq files using the NCBI SRA Toolkit (https://github.com/ncbi/sra-tools):

```
cd ../reads
prefetch -v SRR2850572
fastq-dump -outdir . -gzip /home/[User]/ncbi/public/sra/
SRR2850572.sra
```

5. Here, as well as in many places throughout the protocol, processing of one sample is shown. The same analysis can be applied to other samples by replacing the run IDs (SRR accession numbers listed in Table 1) in the command lines accordingly. One of the advantages of using ezRUN in SUSHI is that the analysis framework manages the meta-information. Once samples are defined in one dataset (*see* **Note 3**), same analysis can be applied to all samples within the dataset, without extra administrative tasks [7].

3.2 Quality Control and Read Alignment

1. For quality control and pre-processing of the reads, we show the usage of FastQC (http://www.bioinformatics.babraham.ac.uk/projects/fastqc/). Because sequencing adaptor

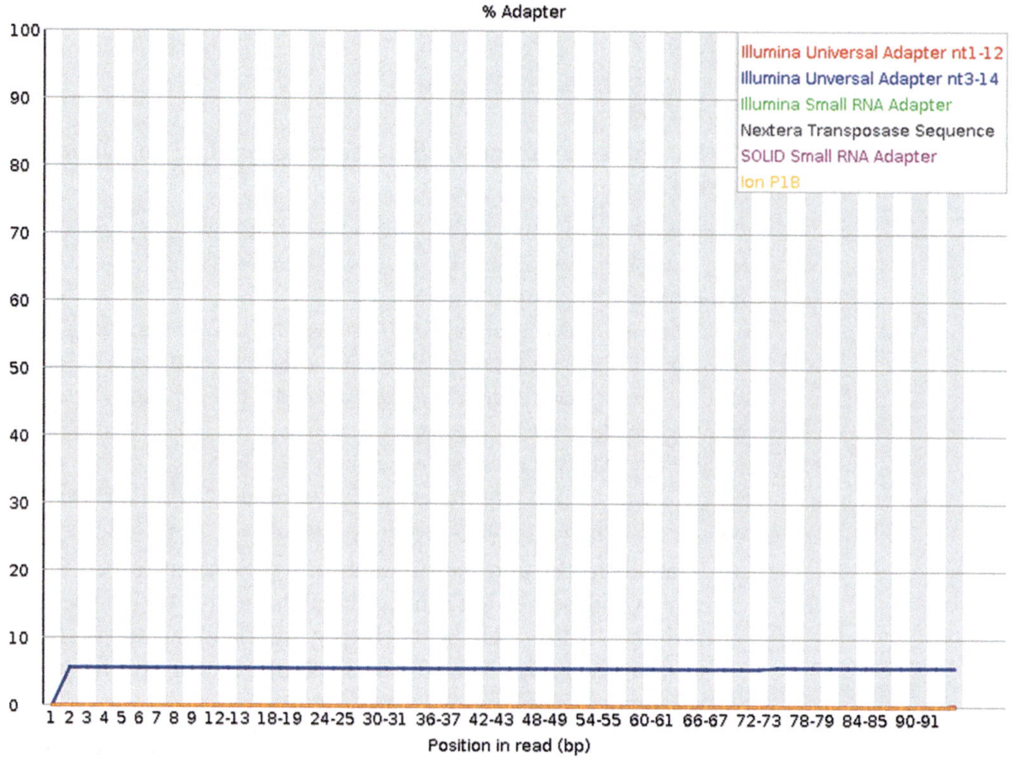

Fig. 1 Adaptor content in sequencing reads

contamination can be found in all six samples (Fig. 1) (*see* **Note 4**), Trimmomatic [9] are applied to generate quality controlled reads (adaptor trimmed, hard trimming of the first 5 bases, minimum length 50, average read quality 20).

```
cd ../
mkdir fastqc
fastqc -t 8 -o fastqc reads/*.fastq.gz
cd reads
trimmomatic-0.36.jar SE -threads 8 -phred33 SRR2850572.fastq.
gz SRR2850572.tr.fastq
ILLUMINACLIP:/usr/local/ngseq/src/Trimmomatic-0.36/adapters/
TruSeq3-SE.fa:2:30:10 HEADCROP:5 MINLEN:50 AVGQUAL:20
```

2. The quality checked reads are aligned to the reference genome using the splice aligner STAR [10]. The alignment result files (bam files) are sorted and indexed using SAMtools [11] for downstream QC and counting analysis.

```
cd ../reference
mkdir STARIndex
STAR -runMode genomeGenerate -genomeDir ./STARIndex -genomeChr-
BinNbits 16 -limitGenomeGenerateRAM 30000000000 -genomeFasta-
Files TAIR10_Chr.all.fasta -sjdbGTFtagExonParentTranscript
```

```
Parent -sjdbGTFfile Araport11_GFF3_genes_transposons.201606.
gff -sjdbOverhang 124 -runThreadN 8
cd ../
mkdir Map_STAR
cd Map_STAR
STAR -genomeDir ../reference/STARIndex/ -sjdbOverhang 124 -read-
FilesIn SRR2850572.tr.fastq -runThreadN 8 -outFilterType BySJout
-outFilterMatchNmin 30 -outFilterMismatchNmax 10 -outFilterMis-
matchNoverLmax 0.05 -alignSJDBoverhangMin 1 -alignSJoverhangMin 8
-alignIntronMax 1000000 -alignMatesGapMax 1000000 -outFilterMul-
timapNmax 50 -chimSegmentMin 15 -chimJunctionOverhangMin 15 -
chimScoreMin 15 -chimScoreSeparation 10 -outSAMstrandField in-
tronMotif -outStd BAM_Unsorted -outSAMtype BAM Unsorted >
SRR2850572.bam
samtools sort -m 3000M -@ 8 SRR2850572.bam > SRR2850572.sorted.bam
samtools index SRR2850572.sorted.bam
rm SRR2850572.bam
```

3.3 Read Alignment Quality Control and Gene Expression Quantification

1. In the following example, we use the mapping QC tool RSeQC [12] to check the mapping rate of the reads, the mapped strand, and the uniformity of read coverage on transcripts (*see* **Note 5**):

```
bam_stat.py -i Map_STAR/SRR2850572.sorted.bam
infer_experiment.py -r reference/Araport11_GFF3_genes_tran-
sposons.201606.bed -i Map_STAR/SRR2850572.sorted.bam
geneBody_coverage.py -r reference/Araport11_GFF3_genes_tran-
sposons.201606.bed -i Map_STAR/ -o BamQC
```

2. The summary statistic showed all reads from SRR2850572 can be mapped, with most being mapped uniquely:

```
Load BAM file... Done
#==================================================
#All numbers are READ count
# ================================================-
=Total records: 13312878
QC failed: 0
Optical/PCR duplicate: 0
Non primary hits 1943411
Unmapped reads: 0
mapq < mapq_cut (non-unique): 1913376

mapq >= mapq_cut (unique): 9456091
Read-1: 0
Read-2: 0
Reads map to '+': 4540183
Reads map to '-': 4915908
Non-splice reads: 6179787
```

```
Splice reads: 3276304
Reads mapped in proper pairs: 0
Proper-paired reads map to different chrom:0
```

3. The mapped strand is "reversely stranded":

```
Reading reference gene model reference/Araport11_GFF3_genes_-
transposons.201606.bed ... Done
Loading SAM/BAM file ... Total 200000 usable reads were sampled
This is SingleEnd Data
Fraction of reads failed to determine: 0.0089
Fraction of reads explained by "++,--": 0.0079
Fraction of reads explained by "+-,-+": 0.9832
```

4. The transcript overage plot (Fig. 2) suggests the transcripts are uniformly covered in all six samples.
5. For gene expression quantification we use featureCounts [13] to count the mapped reads for annotated genes, reversely stranded, allowing both multi-mapping and multi-overlapping reads:

```
cd ../
mkdir Count_featureCounts
featureCounts -T 8 -a reference/Araport11_GFF3_genes_transpo-
sons.201606.gtf -minOverlap 10 -primary -O -M -t exon -g
gene_id -s 2 -o Count_featureCounts/counts.txt Map_STAR/*.bam
```

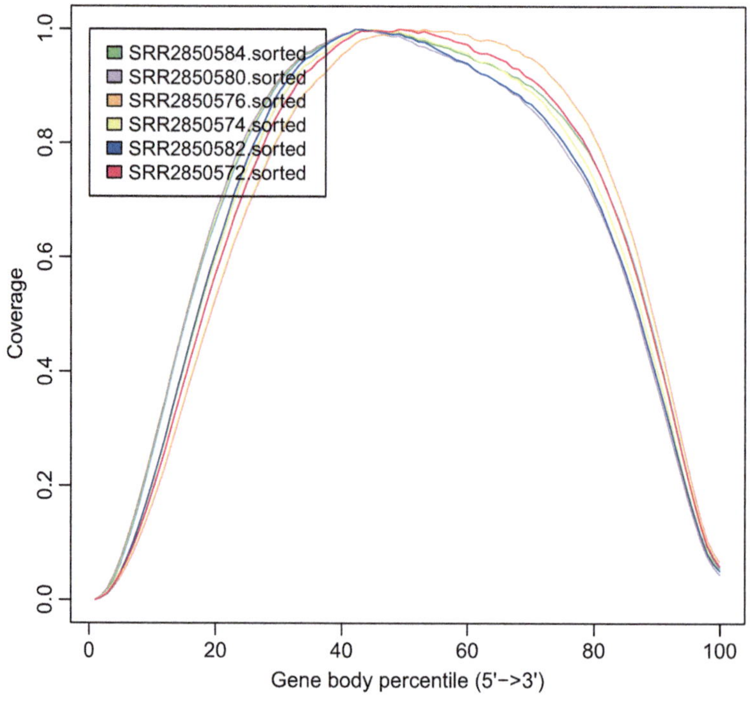

Fig. 2 Gene body coverage in all six samples

3.4 Count QC

We show the example using the R bioconductor package NOISeq [14] to check the length bias and reproducibility among biological replicates (*see* **Note 5**):

1. We first start R in our working directory:

```
R
```

2. After starting R, we install and load NOISeq package:

```
source("https://bioconductor.org/biocLite.R")
biocLite("NOISeq")
library(NOISeq)
```

3. Read in the count table, and convert it to a NOISeq object:

```
mytable<-read.table("Count_featureCounts/counts.txt", skip=1,
header=TRUE)
mycounts<-mytable[, c(7:12)]
row.names(mycounts)<-mytable[, 1]
colnames(mycounts)<-c("YFP-NEG1", "YFP-NEG2", "YFP-NEG3",
"YFP-POS1", "YFP-POS2", "YFP-POS3")
myfactors = data.frame(condition=c("YFP-NEG", "YFP-NEG",
"YFP-NEG", "YFP-POS", "YFP-POS", "YFP-POS"), conditionrun=c
("YFP-NEG-run1", "YFP-NEG-run1", "YFP-NEG-run1", "YFP-POS-
run1", "YFP-POS-run1", "YFP-POS-run1"), run=c("run1",
"run1","run1","run1","run1","run1"))
mydata <- readData(data = mycounts, factors=myfactors, length
= mytable[, c("Geneid", "Length")], chromosome = mytable[, c
("Chr", "Start", "End")])
```

4. Estimate the length bias:

```
mylenbias = dat(mydata, type = "lengthbias")
explo.plot(mylenbias)
```

5. Addressing the reproducibility among biological replicates:

```
mypca=dat(mydata, type="PCA")
explo.plot(mypca)
```

6. As expected, length bias of counts (longer genes have more counts) can be observed in all samples (Fig. 3). Biological replicates of the same condition do cluster together in the Principal Component Analysis (PCA) plot (Fig. 4).

3.5 Differential Expression Analysis

We show the example using the R bioconductor package edgeR [15]:

1. We first start R in our working directory:

```
R
```

2. After starting R, we install and load edgeR package:

```
source("https://bioconductor.org/biocLite.R")
biocLite("edgeR")
library(edgeR)
```

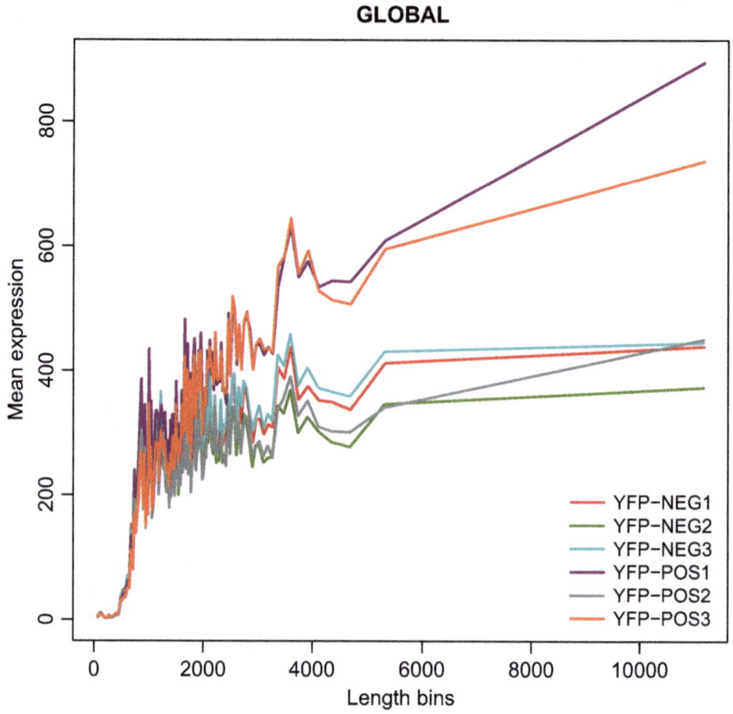

Fig. 3 Gene length versus expression. The genes are divided into bins according to gene length. Each bin contains 200 genes and the middle point of each bin is depicted in *X* axis. For each bin, the 5% trimmed mean of the corresponding expression values is computed and depicted in *Y* axis

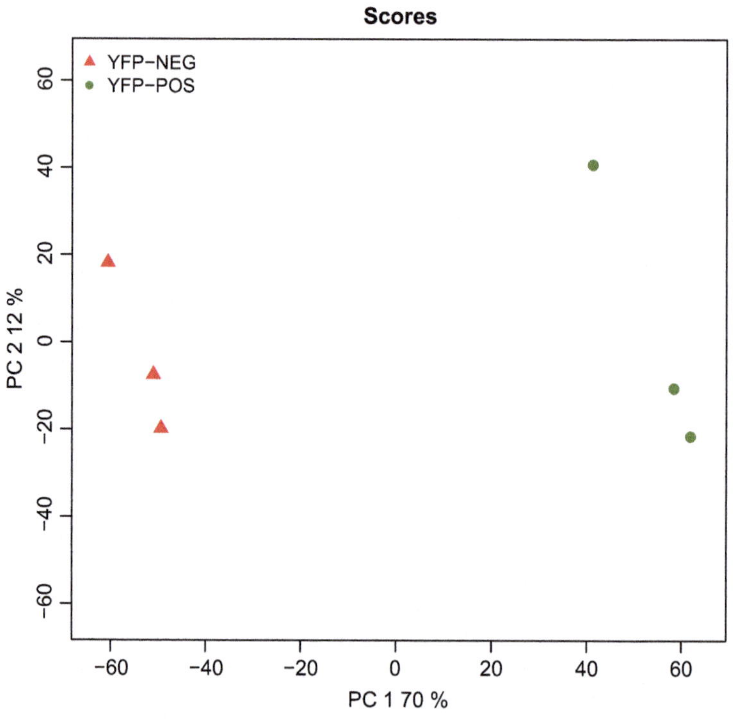

Fig. 4 Principal Component Analysis (PCA) plot colored by conditions

3. Read in the count table, and convert it to an edgeR object:

```
mytable<-read.table("Count_featureCounts/counts.txt", skip=1,
header=TRUE)
mycounts<-mytable[, c(7:12)]
row.names(mycounts)<-mytable[, 1]
colnames(mycounts)<-c("YFP-NEG1", "YFP-NEG2", "YFP-NEG3",
"YFP-POS1", "YFP-POS2", "YFP-POS3")
Condition<-c("YFP-NEG", "YFP-NEG", "YFP-NEG", "YFP-POS",
"YFP-POS", "YFP-POS")
y<-DGEList(counts=mycounts, group=Condition, genes=mytable[,
c("Geneid", "Length")])
```

4. Perform the integrated normalization method trimmed mean of *M* values (TMM) [16]:

```
y<-calcNormFactors(y, method="TMM")
```

5. Estimate common dispersion:

```
y<-estimateCommonDisp(y)
```

6. Estimate tagwise dispersion:

```
y<-estimateTagwiseDisp(y)
```

7. Perform differential expression analysis:

```
et<-exactTest(y, pair=c("YFP-NEG", "YFP-POS"))
res<-as.data.frame(topTags(et, n=37813))
```

8. Generate list of differentially expressed genes:

```
   rese2fold<-res(res$logFC>=1 | res$logFC<=-1, ]
res2foldpadj<-res2fold[res2fold$FDR<=0.01, ]
```

4 Notes

1. This protocol assumes users have basic knowledge about the Unix operation system. There are many books on this topic, as well as free tutorials on the Internet: http://korflab.ucdavis.edu/Unix_and_Perl/
2. The command line examples in this protocol assume the user install the software needed (Table 2) and make all excitable available in his/her PATH environment. Please follow the installation instruction for each software package.
3. A dataset file in SUSHI is a tab-delimited file that records meta-information about the experiment and data [7]. The dataset to initiate the analysis described here using ezRUN in SUSHI can be found here:

   http://fgcz-sushi-demo.uzh.ch/projects/p1000/GSE74458/dataset.tsv

4. Adaptor content analysis is only one of the many analysis modules available in FastQC. Detailed documentation can be found at:

   http://www.bioinformatics.babraham.ac.uk/projects/fastqc/Help/

   Here one can access the summary FastQC report on all six samples, generated by ezRUN in SUSHI: http://fgcz-sushi-demo.uzh.ch/projects/p1000/QC_Fastqc_287_2016-10-31–09-29-17/FastQC_Result/00index.html

5. There are many tools available for QC of read alignment results and count data, each with a slightly different set of QC metrics. The feasibility of some QC parameters also depends on the availability of the corresponding annotation data. The ezRun package in SUSHI provides an extensive collection of the QC metrics. For the tutorial dataset, the mapping QC summary report generated by ezRun in SUSHI can be accessed here:

   http://fgcz-sushi-demo.uzh.ch/projects/p1000/QC_RNABamStats_289_2016-10-31–11-41-37/RNA_BAM_Statistics/00index.html

   The link to the count QC summary report generated by ezRun in SUSHI:

   http://fgcz-sushi-demo.uzh.ch/projects/p1000/QC_CountQC_291_2016-10-31–12-21-18/Count_QC/00index.html

References

1. Wang Z, Gerstein M, Snyder M (2009) RNA-Seq: a revolutionary tool for transcriptomics. Nat Rev Genet 10:57–63. doi:10.1038/nrg2484
2. Conesa A, Madrigal P, Tarazona S et al (2016) A survey of best practices for RNA-seq data analysis. Genome Biol 17:13. doi:10.1186/s13059-016-0881-8
3. Rehrauer H, Opitz L, Tan G et al (2013) Blind spots of quantitative RNA-seq: the limits for assessing abundance, differential expression, and isoform switching. BMC Bioinformatics 14:370. doi:10.1186/1471-2105-14-370
4. Li W, Freudenberg J (2014) Mappability and rea d length. Front Genet 5:381. doi:10.3389/fgene.2014.00381
5. Zhao S, Zhang Y, Gordon W et al (2015) Comparison of stranded and non-stranded RNA-seq transcriptome profiling and investigation of gene overlap. BMC Genomics 16(1):675. doi:10.1186/s12864-015-1876-7
6. Li S, Labaj PP, Zumbo P et al (2014) Detecting and correcting systematic variation in large-scale RNA sequencing data. Nat Biotechnol 32:888–895
7. Hatakeyama M, Opitz L, Russo G et al (2016) SUSHI: an exquisite recipe for fully documented, reproducible and reusable NGS data analysis. BMC Bioinformatics 17:228. doi:10.1186/s12859-016-1104-8
8. Villarino GH, Hu Q, Manrique S et al (2016) Transcriptomic signature of the SHATTERPROOF2 expression domain reveals the meristematic nature of Arabidopsis gynoecial medial domain. Plant Physiol 171:42–61. doi:10.1104/pp.15.01845
9. Bolger AM, Lohse M, Usadel B (2014) Trimmomatic: a flexible trimmer for Illumina sequence data. Bioinformatics 30:2114–2120. doi:10.1093/bioinformatics/btu170
10. Dobin A, Davis CA, Schlesinger F et al (2013) STAR: ultrafast universal RNA-seq aligner.

Bioinformatics 29:15–21. doi:10.1093/bioinformatics/bts635

11. Li H, Handsaker B, Wysoker A et al (2009) The sequence alignment/map format and SAMtools. Bioinformatics 25:2078–2079. doi:10.1093/bioinformatics/btp352
12. Wang L, Wang S, Li W (2012) RSeQC: quality control of RNA-seq experiments. Bioinformatics 28:2184–2185. doi:10.1093/bioinformatics/bts356
13. Liao Y, Smyth GK, Shi W (2014) featureCounts: an efficient general purpose program for assigning sequence reads to genomic features. Bioinformatics 30:923–930. doi:10.1093/bioinformatics/btt656
14. Tarazona S, Furió-Tarí P, Turrà D et al (2015) Data quality aware analysis of differential expression in RNA-seq with NOISeq R/Bioc package. Nucleic Acids Res 43(21):e140. doi:10.1093/nar/gkv711. gkv711
15. Robinson MD, McCarthy DJ, Smyth GK (2010) edgeR: a Bioconductor package for differential expression analysis of digital gene expression data. Bioinformatics 26(1):139–140. doi:10.1093/bioinformatics/btp616
16. Robinson MD, Oshlack A (2010) A scaling normalization method for differential expression analysis of RNA-seq data. Genome Biol 11:R25. doi:10.1186/gb-2010-11-3-r25

Chapter 24

RNA-Seq Data Analysis Protocol: Combining In-House and Publicly Available Data

Marc W. Schmid

Abstract

Comparing gene expression profiles measured in a wide range of different tissue types, at different developmental stages, or under different environmental conditions can yield valuable insights into the mechanisms of cell/tissue specification and differentiation, or identify cell/tissue-type specific responses to environmental stimuli. Critical for such comparisons is the identical processing of data from different sources. This may also include the integration of a novel data set into an existing collection of data sets (e.g., in-house and publicly available data). Here, I describe a complete workflow for RNA-Seq data, from data processing steps to the comparison of gene expression profiles measured with RNA-Seq. I use publicly available data for demonstration purposes, but I also describe how to integrate your own data sets. The workflow runs on all three major operating systems (Linux, MacOS, and Windows). The scripts and the tutorial can be accessed on github.com/MWSchmid/RNAseq_protocol.

Key words RNA-Seq, Public data, Data integration, Analysis, Differential expression, Multigroup comparisons, Gene expression, Transcriptome, Workflow

1 Introduction

Many aspects in biology are controlled by the establishment of specific gene expression patterns or the activation of a specific set of genes upon an environmental stimulus. An example is the development of a multicellular organism controlled by distinct cell fate decisions, which are largely mediated through the establishment of cell- or tissue-specific gene expression patterns (transcriptomes). Spatially and temporally resolved transcriptome data thereby facilitates the identification of cell-type-specific genes or entire regulatory gene networks underlying cell differentiation and functional specification.

In an individual study, researchers frequently focus on measuring one, two, or only a few specific cell/tissue types, developmental stages, or environmental stimuli (conditions). However, aside the "planned" direct comparison between the two conditions (e.g., egg

Anja Schmidt (ed.), *Plant Germline Development: Methods and Protocols*, Methods in Molecular Biology, vol. 1669, DOI 10.1007/978-1-4939-7286-9_24, © Springer Science+Business Media LLC 2017

cells versus sperm cells), it can be interesting to include more data from public resources to, for example, identify genes, which are specific to a certain condition compared to a wider range of other conditions (e.g., egg cells compared to a variety of somatic tissues and cell types). Similarly, it may also offer the possibility to identify genes, which are generally specific to a certain condition but are activated upon perturbation (e.g., environmental stimulus or genetic mutation).

In this chapter I describe a workflow which starts with handling raw RNA-Seq data, i.e., reads, from publicly available experiments and finishes with the identification of differentially expressed genes using a variety of state-of-the-art tools for differential expression. It is important to note that this workflow represents just one specific approach and that there are many alternatives available for each step. However, there is currently no single one-fits-all solution or consensus on a best approach (*see* [1] for a recent review on RNA-Seq and en.wikipedia.org/wiki/List_of_RNA-Seq_bioinformatics_tools for a list of RNA-Seq tools). Nonetheless, it is generally worthwhile to test alternative tools and carefully select those, which seem to be most suitable for a specific analysis.

2 Software Dependencies and Genome Data

The workflow presented here runs on all major operating systems (MacOS, Linux, and Windows). It requires the installation of R [2] and certain dependencies, which are specific to the operating system. Software packages change frequently and the instructions given here may soon be outdated. Please visit github.com/MWSchmid/RNAseq_protocol to get the newest version of the installation instructions or to report problems and errors.

2.1 Installing R and Dependencies on Linux (Ubuntu-Like)

Visit cran.r-project.org and follow the instructions on the website to get the newest stable version of R (the default version in the package manager is frequently outdated). In brief, open a terminal and type (note that lines starting with "#" are comments, which are not executed):

```
## add the repository to your software sources
# sudo add-apt-repository "deb http://<MIRR>/bin/linux/ubuntu <VERS>/"
# <VERS>: Ubuntu version (code name; e.g. "trusty" for Ubuntu 14.04)
# <MIRR>: A mirror listed on https://cran.r-project.org/mirrors.html
sudo add-apt-repository "deb http://cran.rstudio.com/bin/linux/ubuntu xenial/"

## add the authentication key
sudo apt-key adv -keyserver keyserver.ubuntu.com -recv-keys E084DAB9

## update the software package list
sudo apt-get update
```

```
## install R
sudo apt-get install r-base r-base-dev r-cran-rjava

## install all other dependencies
sudo apt-get install libxml2-dev libssl-dev libcurl3-openssl-dev qt5-default
```

2.2 Installing R and Dependencies on MacOS

Install XCode (AppStore) and XQuartz (www.xquartz.org). Then download the R-installer from cran.r-project.org and install R according to the installation instructions.

2.3 Installing R and Dependencies on Windows

Rsubread [3] and many other NGS tools do not run natively on Windows. An option is to use Cygwin, a Unix-like environment and command-line interface for Windows. Download the most recent installer from (cygwin.com) and start the setup. Go through the dialog until you can select the packages you would like to install. Select the following packages in addition to what is per default selected (for the libraries ending on "-devel", double check if the version without "-devel" is selected as well).

- Archive:
  - libbz2-devel: BZip file de/compressor
- Devel:
  - gcc-core: GNU Compiler Collection (C, OpenMP)
  - gcc-g++: GNU Compiler Collection (C++)
  - gcc-fortran: GNU Compiler Collection (Fortran)
  - make: The GNU version of the 'make' utility
- Libs:
  - libcurl-devel: Multi-protocol file transfer library (development)
  - libiconv-devel: Unicode iconv() implementation
  - libicu-devel: IBM Internationalization Component for Unicode
  - libintl-devel: GNU Internationalization runtime library
  - liblzma-devel: LZMA de/compressor library (development)
  - libpcre-devel: Perl Compatible Regular Expressions library development
  - libtirpc-devel: A port of Sun's Transport-Independent RPC library
  - libxml2-devel: GNOME XML library (development)
  - zlib-devel: Gzip de/compression library (development)
- Net:
  - openssl-devel: A general purpose cryptography toolkit with TLS implementation (development)

- Science:
  - R: R Statistical computing language

  After completing the Cygwin installation, start Cygwin, type "R" and press enter to start an R console. At least for the alignment of the short reads (i.e., usage of Rsubread), you need to use this console. Note that packages installed in the Cygwin R console are not available to any other native R installation (i.e., R installed with the installer available on cran.r-project.org). Aside the R within Cygwin, I recommend to install the "native Windows R" as well.

2.4 Other Software and Packages (All Platforms)

1. Bioconductor packages and other dependencies (basic installation, SRAdb, ShortRead, Rsamtools, Rsubread [3], biomaRt [4], DESeq2 [5], edgeR [6], and limma [7]). Start R and type:

```
source("http://bioconductor.org/biocLite.R")
biocLite()
biocLite(c("SRAdb", "Rsamtools", "Rsubread",
  "biomaRt", "DESeq2", "edgeR",
  "limma", "XLConnect", "gplots",
  "colorRamps", "ShortRead"))
# Note for Windows users:
# I recommend using the "Cygwin R" only for
# the data pre-processing. Only SRAdb, Rsamtools
# and Rsubread are required for this part.
# All the other libraries should be installed
# on the "native Windows R".
# Important for Windows users:
# Current versions of Rsubread may fail to compile
# on Cygwin and multithreading may be limited as well.
# It is unclear if and when this will be fixed.
# You can download version 1.16.1 from
# bioconductor.org/packages/3.0/bioc/
# src/contrib/Rsubread_1.16.1.tar.gz
# and install it in R with:
install.packages("/path/to/Rsubread_1.16.1.tar.gz")
```

2. Download the "RNAseqWrapper_0.99.0.tar.gz" package from github.com/MWSchmid/RNAseq_protocol and install it in R with:

```
install.packages("/path/to/RNAseqWrapper_0.99.0.tar.gz",
 repos = NULL, type = "source")
# Note for Windows users:
# Install this package only in the "native Windows R".
```

3. Download the archive "workFlowData.zip" from github.com/MWSchmid/RNAseq_protocol and unpack it.
4. The workflow includes featureCounts [8] to count the number of reads per gene. However, if you would like to specifically

address the problem of reads aligning with multiple locations in the genome (multireads) or reads aligning with positions where two or more genes overlap (ambiguous reads), you can use Rcount [9]. Download the archive matching your operating system from github.com/MWSchmid/Rcount and unpack it.

2.5 Reference Sequence and Annotation

The reference sequence and annotation can be either downloaded manually or with the R script given below. Ensembl (ensembl.org) provides standardized and (mostly) up-to-date reference genomes for many organisms, but any alternative resource meeting the format requirements can be used. For the workflow presented here (and most alternative tools), the reference sequence and annotation need to be in "fasta" and "gff/gtf" format, respectively. Even though these formats are relatively well specified, incompatibilities can occur. If you encounter problems or observe an unlikely result, consult the manual of the specific software or contact the developers.

```
# Specify the local file paths and the links.
gSeqFileGZ <- "/path/to/genome/sequence.fasta(.gz)"
gAnnFileGZ <- "/path/to/genome/annotation.gtf|gff|bed(.gz)"
gSeqLink <- "http:|ftp://link/to/genome/sequence.fasta"
gAnnLink <- "http:|ftp://link/to/genome/annotation.gtf"

# Download the file.
download.file(gSeqLink, gSeqFileGZ, "auto")
download.file(gAnnLink, gAnnFileGZ, "auto")

# If the files are compressed, unpack them either manually
# or with a system command (Linux, MacOS and Cygwin). Note
# that this will fail on a native Windows R installation.
system(paste0("gunzip ", gSeqFileGZ))
system(paste0("gunzip ", gAnnFileGZ))
```

3 Workflow

The workflow comprises (a) the download of short reads and sample annotation, (b) the alignment of short reads with the reference genome, (c) the assignment of alignments to genes, and (d) the identification of differentially expressed genes (Fig. 1). Most steps are highly automated and do not require extensive user input. However, the selection of the publicly available data and the annotation of the samples is critical to the success of the analysis and thus requires the user to invest time browsing through the literature and verifying whether the automatically retrieved annotation is "correct" and complete (the latter is frequently not the case).

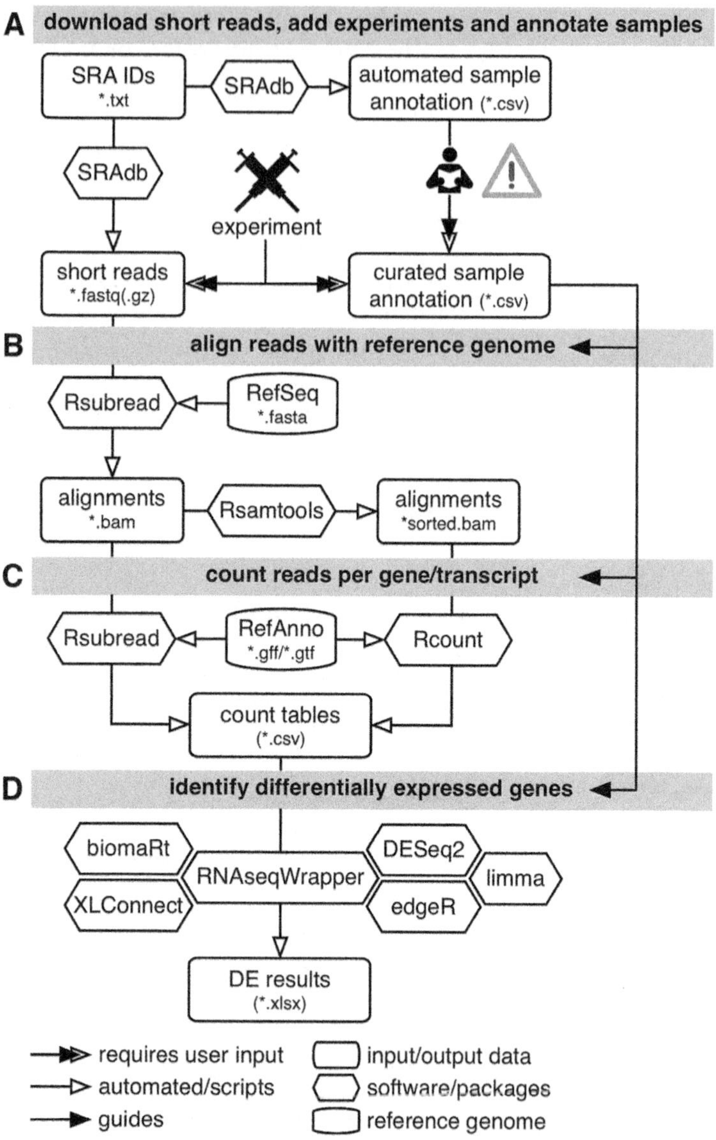

Fig. 1 Schematic representation of the workflow for the identification of differentially expressed genes using RNA-Seq data

3.1 Download Publicly Available Data

The composition of the data set is critical for the analysis. In principle, one could simply retrieve all experiments matching a certain search pattern from SRA and "blindly" process these data. However, it is more beneficial to manually curate a specific data set for which the underlying experimental procedures and the quality of the data are known (*see* [10, 11] for details on quality controls). In the following, I assume that there is a predefined collection of experiments/samples of interest, for which all SRA accession numbers are listed in a file called "mySamples.txt" (like given in the example data). Additional samples, which are not available on the

SRA repository, may be added at a later step. The following script will parse your SRA IDs of interest, download all necessary short read files containing the raw RNA-Seq data and create a table with the annotation retrieved from the SRA repository (see "autoGenAnnotation.csv" from the example data). This table will then be manually refined and extended at the next step.

```
library("SRAdb")

# Specify several file paths and directories in which
# you would like to work in.
readsDir <- "/path/to/the/raw/read/files"
metaDB <- "/path/to/metadata/database/SRAmetadb.sqlite"
mySamplesFile <- "/path/to/the/file/with/all/mySamples.txt"
autoAnnotation <- "/path/to/autoGenAnnotation.csv"

# Check for a local copy of the database and
# download it if it does not exist.
if (!file.exists(metaDB)) { getSRAdbFile(dirname(metaDB)) }

# Windows users: if the download fails while using the
# Cygwin-R, try:
if (!file.exists(metaDB)) {
 getSRAdbFile(dirname(metaDB), method = "libcurl")
}

# Open a connection to the database
sra_con <- dbConnect(SQLite(), metaDB)

# Here we assume that you have a precompiled list of
# samples/experiments/studies which are of interest to you.
# Read in the list of experiments you are interested in.
samples <- scan(mySamplesFile, what = "character")

# You may have a collection of experiments and individual runs.
# However, only individual runs can be downloaded. Note that
# the conversion of a certain SRA ID type to another requires
# the different query types not to be mixed.
sampleTypes <- unique(substr(samples, 1, 3))
splitSamples <- lapply(sampleTypes, function(x)
 grep(paste0("^", x), samples, value = TRUE))
conversions <- lapply(splitSamples, function(x)
 sraConvert(x, sra_con = sra_con ))
allSRAids <- do.call("rbind", conversions)

# Retrieve the remaining metadata to determine
# the sequencing platform, the read length, etc.
SRAdata <- lapply(allSRAids$run, function(x)
 getSRA(search_terms = x, out_types = "sra", sra_con))
SRAdata <- do.call("rbind", SRAdata)
rownames(SRAdata) <- SRAdata$run # for data access
SRAdata$approxRL <- SRAdata$bases/SRAdata$spots
summary(SRAdata$spots) # the total number of reads
summary(SRAdata$approxRL) # the average read lengths
```

```
table(SRAdata$platform) # the sequencing platform
SRAdata$sample_name[1:10] # the names of the first 10 samples
# The sample names are not always too informative
# and I recommend renaming them afterwards.

# Search for strand-specific samples with regular
# expressions. Note that this gives no guarantee
# for strand-specificity - in any case, this
# should be verified manually.
posExp <- "strand[[:space:]{0,1}|-]specific"
negExp <- paste0("no[[:alpha:]{0,1}][[:space:]{0,1}|-]", posExp)
strSpec <- apply(SRAdata, 1, function(x)
 length(grep(posExp, x)) > 0)
notStSp <- apply(SRAdata, 1, function(x)
 length(grep(negExp, x)) > 0)
strSpec[notStSp] <- FALSE
sum(strSpec) # the number of strand-specific libraries
SRAdata$strSpec <- strSpec

# Check if there are FASTQ files available for the samples.
# For the samples without FASTQ file, you will download the
# SRA archives and convert them into FASTQ (this is a lot
# slower than downloading FASTQ files).
fastqInfo <- getFASTQinfo(allSRAids$run, sra_con, 'ftp')

# Check if the paths were retrieved correctly. If there is
# no column called "fastq_ID" (e.g., FASTQ_FILES with all
# entries being "1"), the database may be broken.
head(fastqInfo)

samplesWithFastq <- unique(with(fastqInfo, run[!is.na(fastq_ID)]))
noFastqSamples <- unique(with(fastqInfo, run[is.na(fastq_ID)]))
if (length(noFastqSamples) > 0) {
 sraInfo <- getSRAinfo(noFastqSamples, sra_con)
 noFastqButSRA <- unique(with(sraInfo, run[!is.na(ftp)]))
 noFiles <- unique(with(sraInfo, run[is.na(ftp)]))
} else {
 noFiles <- c()
}
length(noFiles) # number of samples without files
# if there are samples without FASTQ or SRA file,
# go to http://www.ncbi.nlm.nih.gov/, under "Databases"
# select "SRA" and check the status of the IDs for which
# there are no files available.

# Download the FASTQ files.
# Note that an error like "no such table: fastq" is likely
# due to a corrupt SRAdb file. Delete SRAmetadb.sqlite
# and download it again. If the error persists, contact
# the SRAdb developers.
getFASTQfile(samplesWithFastq, sra_con, readsDir, 'ftp')
```

```
# If necessary, download the SRA archives of the remaining
# samples and convert them to FASTQ. Note that the conversion
# requires the "SRA toolkit" to be installed:
# www.ncbi.nlm.nih.gov/Traces/sra/sra.cgi?view=software
getSRAfile(noFastqButSRA, sra_con, readsDir, 'sra' )
commands <- sapply(noFastqButSRA, function(x)
 paste0("fastq-dump -gzip -split-files -O ",
 readsDir, " ", file.path(readsDir, x), ".sra"))
sapply(commands, system)

# Get a vector of all FASTQ file names and
# remove the reverse reads of the PE samples.
allFastq <- list.files(readsDir, ".fastq.gz")
revFastq <- list.files(readsDir, "_2.fastq.gz")
onlyForwardFastq <- setdiff(allFastq, revFastq)
sampleIDs <- gsub("_1|.fastq.gz", "", onlyForwardFastq)

# Samples with paired-end sequencing have twice the
# coverage - therefore adjust the approximated
# read length of these samples.
PES <- gsub("_2|.fastq.gz", "", revFastq)
SRAdata[PES, "approxRL"] <- SRAdata[PES, "approxRL"]/2

# Create a table with information for the alignment
# and the preprocessing. Store the sampleID, read length,
# platform type, FASTQ file names, sample/library names,
# strand-specificity and additional attributes in a csv file.
# Even though the sample/library names and the attributes
# are not complete for each sample, they may help renaming
# all samples.
sortedSRAdata <- SRAdata[sampleIDs,] # this ensures that the
tabForProcessing <- data.frame( # sample has a FASTQ file
 sampleID = sampleIDs,
 readLen = round(sortedSRAdata$approxRL, 0),
 numReads = sortedSRAdata$spots,
 platform = sortedSRAdata$platform,
 fastqFile = onlyForwardFastq,
 phredOffset = 33,
 useMulti = "true",
 useStrand = as.logical(sortedSRAdata$strSpec),
 minReads = 5,
 maxDist = 250,
 minBelowMaxDist = 1,
 processedReads = 0,
 alignedReads = 0,
 processedAlignmentsFC = 0,
 assignedAlignmentsFC = 0,
 TISSUE = "toBeProvided", # these columns depend on the
 CELLTYPE = "toBeProvided", # analysis. In another case
 STAGE = "toBeProvided", # it may be GENOTYPE, TREATMENT
 BLOCK = "toBeProvided", # or PERTURBATION.
```

```
  mySampleName = "toBeProvided",
  SRAsamSRS = sortedSRAdata$sample,
  SRAexpSRX = sortedSRAdata$experiment,
  SRAsamName = sortedSRAdata$sample_name,
  SRAlibName = sortedSRAdata$library_name,
  SRAattrib = sortedSRAdata$sample_attribute,
  stringsAsFactors = FALSE
)

# Optional: Verify that the PHRED offset is 33 and not 64.
# This is not strictly necessary as the files from SRA are
# generally having an offset of 33 (they otherwise convert
# the encoding). However, there are still some cases with
# an offset of 64.
library("RNAseqWrapper")
rownames(tabForProcessing) <- tabForProcessing$sampleID
for (cSam in rownames(tabForProcessing)) {
  fastqFile <- file.path(readsDir,
   tabForProcessing[cSam, "fastqFile"])
  phredOffset <- f.check.phred.offset(fastqFile)
  tabForProcessing[cSam, "phredOffset"] <- phredOffset
}

# Save the table.
write.csv(tabForProcessing, autoAnnotation, row.names = FALSE)
```

3.2 Manual Curation of the Sample Annotation

The sample annotation based on the information stored at the SRA repository is frequently incomplete and requires additional information for the next steps in the workflow. Open the file with the automatically generated annotation ("autoGenAnnotation.csv" from the example data). The table contains several columns and some of them need to be modified (see "curatedAnnotation.csv" from the example data). To add your own experiment, just write a new entry below the existing ones.

Columns with parameters for the RNA-Seq read counting tools:

- *phredOffset* specifies the offset for the PHRED quality scores. It can either be 33 or 64. Generally, all files deposited at SRA have a PHRED offset of 33. However, there are exceptions and your own experiment may use a PHRED offset of 64. For the data downloaded from the repository, we already verified the encoding. If you are unsure about the encoding in your own data, you can use the function provided in the RNAseqWrapper package (*see* **Note 1**).
- *useMulti* specifies whether you would like to include multireads in the analysis using Rcount [9]. For samples with a high number of multireads compared to unique reads (e.g., if more than 30% of all reads are multireads), this option should be disabled

(type "FALSE" within the respective fields; without the quotes). To get the number of unique and multireads, you can leave this parameter enabled and check the read statistics returned by Rcount-distribute. However, you may *ab initio* disable multireads for samples processed with random primers as these frequently have a high number of multireads. Alternatively, one may use a "repeat-masked" reference genome as for example provided by Ensembl (ensembl.org).

- *useStrand* specifies if the library is strand-specific. Search entries marked with "TRUE" and check whether the reads are in sense or antisense orientation to the transcripts (for this, one eventually needs to go back to the original publication and verify the library preparation protocol). Change the value "TRUE" to "sense" or "antisense" (without the quotes).
- *minReads*, *minBelowMaxDist*, and *maxDist* are parameters for Rcount. Genes must have at least *minReads* reads in total and *minBelowMaxDist* reads within the first *maxDist* bps at the 3′ end (*see* [9] for details). The parameters may be chosen according to the total number of reads and the genome/transcriptome size (note that I did not adjust these parameters in the table from the example data).
- *TISSUE*, *CELLTYPE*, and *STAGE* describe the tissue and cell type and the developmental stage of a sample. *BLOCK* identifies the experiment the sample comes from. These four columns will be concatenated into a single variable which specifies the group a sample belongs to. *BLOCK* therefore serves to separate identical sample types, e.g., seedlings, from different experiments/laboratories into separate groups to reduce the within-group variation. Note that these columns depend on the type of the analysis. You may have a single-celled organism for which you analyze the effect of different perturbations in several genotypes (e.g., *PERTURBATION* and *GENOTYPE*). The information to complete these columns may either be inferred from the columns *SRAsamName*, *SRAlibName*, and *SRAattrib* or the original publication.
- *mySampleName* holds the names of the samples. The names will be used in the count tables and figures and can be chosen freely. Rows with identical sample names will be treated as one sample. This can be useful if a sample was sequenced several times (technical replicates). It is important to identify these cases because independent sequencing runs of one biological sample should not be considered as biological replicates (e.g., SRR210764, SRR210765, SRR210766, and SRR210767 is such a case). Such cases can normally be identified if the SRA experiment and/or sample accession numbers are the same (columns *SRAexpSRX* and *SRAsamSRS*).

- **Important**:
  - Do not use special characters, including hyphen and space, in any of the fields. Use either underscores or camelCase to separate individual words.
  - Avoid numbers at the beginning of your sample identifiers (column *mySampleName*). Always prepend your sample identifiers with a letter.

3.3 Alignment of Short Reads with the Reference Genome

Reads are aligned with the reference genome with Rsubread [3]. It is a fast and accurate aligner, which is capable of spliced read alignments and also supports reads encoded in color space (SOLiD platform). The following script shows how to build an index of the reference genome, which is then used to align the reads with the reference genome.

```
library("Rsubread")

# Specify several file paths and directories in which
# you would like to work in.
gSeqFile <- "/path/to/genome/sequence.fasta"
nucIndex <- "/base/name/of/the/genome/nucIndex"
colIndex <- "/base/name/of/the/genome/colIndex"
myAnnotation <- "/path/to/the/curatedAnnotation.csv"
readsDir <- "/path/to/the/raw/read/files"
bamDir <- "/path/to/aligned/reads/files"

# Build the indices Rsubread. Check the documentation
# for the details on the arguments (?buildIndex).
# The index for the Illumina (nucleotide-encoded) reads:
buildindex(nucIndex, gSeqFile,
  gappedIndex = TRUE, # FALSE on high-RAM machines
  memory = 8000, # max 80 % of all your RAM
  colorspace = FALSE)

# The index for the SOLiD (color-encoded) reads:
buildindex(colIndex, gSeqFile,
  gappedIndex = TRUE,
  memory = 8000,
  colorspace = TRUE)

# Import the sample annotation.
sampleTab <- read.csv(myAnnotation, stringsAsFactors = FALSE)
rownames(sampleTab) <- sampleTab$sampleID

# Align the reads - allow up to 10 alignments per read.
# Note that we redirect the messages from the aligner
# into a character vector to extract the number of
# mapped reads. However, the drawback is that we
# will not see the progress messages from the aligner
# in real-time.
```

```
for (cSam in rownames(sampleTab)) {
 cat("processing", cSam, "...\n")
 isSOLiD <- (sampleTab[cSam, "platform"] == "ABI_SOLID")
 pOff <- sampleTab[cSam, "phredOffset"]
 cIdx <- ifelse(isSOLiD, colIndex, nucIndex)
 inFile <- file.path(readsDir, sampleTab[cSam, "fastqFile"])
 outFile <- file.path(bamDir, paste0(cSam, ".bam"))
 if (file.exists(outFile)&(sampleTab[cSam,"processedReads"]>0)){next}
 stdout <- capture.output( # delete this line to disable redirecting
 subjunc(cIdx, inFile, output_file = outFile,
 nthreads = 4, # adjust the number of CPUs
 unique = FALSE, # set to TRUE to avoid multireads
nBestLocations = 10, # max number of multireads
phredOffset = pOff # PHRED offset is either 33 or 64
)
) # delete this line to disable redirecting
# extract the number of processed and mapped reads
readPattern <- "(Processed|Mapped).*reads"
readMatches <- regexpr(readPattern, stdout)
readCountsStr <- regmatches(stdout, readMatches)
numberPattern <- "[,0123456789]+"
numberMatches <- regexpr(numberPattern, readCountsStr)
readCounts <- regmatches(readCountsStr, numberMatches)
sampleTab[cSam, "processedReads"] <- gsub(",", "", readCounts[1])
sampleTab[cSam, "alignedReads"] <- gsub(",", "", readCounts[2])
write.csv(sampleTab, myAnnotation, row.names = FALSE)
}

# Save the table again to store the number of processed
# and mapped reads as well. "processedReads" should be
# equal to "numReads", and the number of "alignedReads"
# may give you an indication whether the samples are
# fine (note that alignment rates around 40 % are not
# necessarily a problem - it depends on the protocol).
write.csv(sampleTab, myAnnotation, row.names = FALSE)

# Optional: Draw a histogram with the alignment percentages.
sampleTab$alPerc <- 100*with(sampleTab,
 as.numeric(alignedReads)/as.numeric(processedReads))
png("/path/to/the/figure.png")
hist(sampleTab$alPerc, xlab = "alignment rate (in %)",
 ylab = "no. samples", main = "", las = 1)
dev.off()

# Optional: Display the samples with a low alignment rate.
colsOfInt <- c("readLen", "platform", "mySampleName", "alPerc")
subset(sampleTab, alPerc < 20)[,colsOfInt] # e.g., less than 20%
```

3.4 Counting Unique Reads per Gene with featureCounts

After the reads were aligned with a reference genome, the next step is to count the number of reads/alignments per gene (or gene isoform). The next script illustrates how to count the number of unique reads per gene with featureCounts [8]. It is a fast and easy-to-use algorithm, which is very flexible (can also be used to count the number of reads per ChIP-Seq peak).

```
library("Rsubread")

# Specify several file paths and directories in which
# you would like to work in.
myAnnotation <- "/path/to/the/curatedAnnotation.csv"
gAnnFile <- "/path/to/genome/annotation.gtf|gff|saf"
bamDir <- "/path/to/aligned/reads/files"
countTabsFC <- "/path/to/featureCounts/tables"
countTabFC <- "/path/to/the/final/featureCounts_table.csv"

# NOTE: the directory with the count tables from featureCounts
# (variable countTabsFC) should not contain anything else
# aside the count tables (at least no other *.csv files).

# Import the sample annotation.
sampleTab <- read.csv(myAnnotation, stringsAsFactors = FALSE)
rownames(sampleTab) <- sampleTab$sampleID

# Count the number of (unique!) reads with featureCounts().
# I strongly recommend to read the documentation of this
# function (type ?featureCounts).
# Note that we redirect the messages from featureCounts
# into a character vector to extract the number of
# assigned reads. However, the drawback is that we
# will not see the progress messages in real-time.
sT <- c("false" = 0, "sense" = 1, "antisense" = 2)
for (cSam in rownames(sampleTab)) {
 cat("processing", cSam, "...\n")
 bamFile <- file.path(bamDir, paste0(cSam, ".bam"))
 outFile <- file.path(countTabsFC, paste0(cSam, ".txt"))
 SSmode <- tolower(sampleTab[cSam, "useStrand"])
 SSmode <- sT[SSmode]
 minBPinExon <- ceiling(.7*sampleTab[cSam, "readLen"])
 stdout <- capture.output( # delete this line to disable redirecting
 countRes <- featureCounts(bamFile,
 annot.ext = gAnnFile, # reference annotation
 isGTFAnnotationFile = TRUE, # TRUE if ref. anno. is GTF or GFF
 GTF.featureType = "exon", # map alignments to exons
 GTF.attrType = "gene_id", # use geneIDs to group exons
 useMetaFeatures = TRUE, # yes, summarize counts per gene
 allowMultiOverlap = TRUE, # count ambiguous reads
 nthreads = 4, # adjust the number of CPUs
 strandSpecific = SSmode, # set strand-specificity
 # Windows (Rsubread 1.16.1): minReadOverlap instead of minOverlap
```

```
 minOverlap = minBPinExon, # min num of BPs within exons
 # Windows (Rsubread 1.16.1): remove largestOverlap argument
 largestOverlap = TRUE) # see documentation
 )  # delete this line to disable redirecting
 # Save the count table for the current sample.
 colnames(countRes$counts) <- cSam
 write.table(countRes$counts, outFile, sep = '\t', quote = FALSE)
 # extract the number of processed and assigned reads
 readPattern <- "(Total|Successfully).*reads"
 readCountsStr <- grep(readPattern, stdout, value = TRUE)
 numberPattern <- "[0123456789]+"
 numberMatches <- regexpr(numberPattern, readCountsStr)
 readCounts <- regmatches(readCountsStr, numberMatches)
 sampleTab[cSam, "processedAlignmentsFC"] <- readCounts[1]
 sampleTab[cSam, "assignedAlignmentsFC"] <- readCounts[2]
}

# Save the sample annotation table again to store the
# number of processed and successfully assigned reads.
# The number of "processedAlignmentsFC" should equal to
# "alignedReads", and the number of "assignedAlignmentsFC"
# may give you an indication whether a sample has
# too many reads in intergenic regions.
write.csv(sampleTab, myAnnotation, row.names = FALSE)

# Read-in all the tables and merge the files which
# correspond to only one sample (some samples were
# sequenced on multiple lanes). Windows users must
# switch to the "native Windows R" to run the code
# below.
library("RNAseqWrapper")
counts_FC <- f.read.featureCounts(countTabsFC)
counts_FC <- f.summarize.columns(
 counts_FC,
 data.frame(sample = rownames(sampleTab),
 group = sampleTab$mySampleName),
 sum)
write.csv(counts_FC, countTabFC)

# Optional: Draw a histogram with the assignment percentages.
sampleTab$asPerc <- 100*with(sampleTab,
 as.numeric(assignedAlignmentsFC)/as.numeric(processedAlignmentsFC))
png("/path/to/the/figure.png")
hist(sampleTab$asPerc, xlab = "assignment rate (in %)",
 ylab = "no. samples", main = "", las = 1)
dev.off()

# Optional: Display the samples with a low assignmend rate.
colsOfInt <- c("readLen", "platform", "mySampleName", "asPerc")
subset(sampleTab, asPerc < 50)[,colsOfInt] # e.g., less than 50 %
```

3.5 Counting Weighted Alignments with Rcount

Summarizing the number of reads per gene is not trivial. First, a read may align multiple times with the genome (multireads). Second, several genes may overlap at a given position within the genome. Reads aligning with those positions are ambiguous with respect to the gene they originate from (ambiguous reads). featureCounts either counts these read types multiple times or discards them. In the previous script, multireads were discarded and ambiguous reads were counted repeatedly. An alternative, which addresses both issues, is Rcount [9]. The following script shows how to weight and summarize the alignments with Rcount.

```
library("Rsamtools")
library("RNAseqWrapper")

# Specify several file paths and directories in which
# you would like to work in.
myAnnotation <- "/path/to/the/curatedAnnotation.csv"
bamDir <- "/path/to/aligned/reads/files"
countTabsRC <- "/path/to/Rcount/tables"
countTabRC <- "/path/to/the/final/Rcount_table.csv"
RcountFormat <- "/path/to/Rcount-format(.exe)"
RcountMultir <- "/path/to/Rcount-multireads(.exe)"
RcountDistri <- "/path/to/Rcount-distribute(.exe)"
RcountProDir <- "/path/to/directory/with/Rcount/project/files"

# NOTE: the directory with the count tables from Rcount
# (variable countTabsRC) should not contain anything else
# aside the count tables (at least no other *.txt files).

# Create a database for Rcount with Rcount-format.
# See the Rcount user guide for details.
system(RcountFormat, wait = FALSE)

# Set the path for the annotation you just created.
RcountAnnota <- "/path/to/Rcount/annotation.xml"

# Import the sample annotation.
sampleTab <- read.csv(myAnnotation, stringsAsFactors = FALSE)
rownames(sampleTab) <- sampleTab$sampleID

# Sort the reads in the bam files according to
# chromosome and position, and create an index.
# Note that this can take quite some time and
# you may parallelize this task with the function
# f.sort.bam.parallel provided with the RNAseqWrapper
# package. Adjust the number of cores and maximal
# amounts of RAM used by each core according to your
# system. Windows users must set numCores to 1.
bamFiles <- paste0(rownames(sampleTab), ".bam")
sorted <- f.sort.bam.parallel(bamFiles, bamDir, bamDir,
                    numCores = 4, maxMemPerCore = 3600)
```

```
# Check if all files were sorted and indexed successfully.
if (length(sorted$fail) > 0) {
 cat("some files were not sorted or indexed.\n")
 cat(paste0(sorted$fail, collapse = '\n'), '\n')
}

# Distribute multireads based on the number of
# unique reads with Rcount-multireads
for (cSam in rownames(sampleTab)) {
 cat("processing", cSam, "...\n")
 inFile <- file.path(bamDir, paste0(cSam, "_srt.bam"))
 outFile <- file.path(bamDir, paste0(cSam, "_wts.bam"))
 calcWeights <- tolower(sampleTab[cSam, "useMulti"])
 calcWeights <- ifelse(calcWeights == "true", "y", "n")
 alloDist <- sampleTab[cSam, "readLen"]
 command <- paste0(RcountMultir, ' -c "', inFile, '","',
  outFile, '",', calcWeights, ',', alloDist)
 system(command)
}

# Create Rcount project files
for (cSam in rownames(sampleTab)) {
 cat("processing", cSam, "...\n")
 f.write.Rcount.project.file(
 sampleName = cSam,
 annotationXML = RcountAnnota,
 readsInfile = file.path(bamDir, paste0(cSam, "_wts.bam")),
 readsOutfile = file.path(bamDir, paste0(cSam, "_mpd.bam")),
 countTableFile = file.path(countTabsRC, paste0(cSam, ".txt")),
 OUTprojectXML = file.path(RcountProDir, paste0(cSam, ".xml")),
 useMulti = tolower(sampleTab[cSam, "useMulti"]),
 useStrand = tolower(sampleTab[cSam, "useStrand"]),
 minReads = sampleTab[cSam, "minReads"],
 maxDist = sampleTab[cSam, "maxDist"],
 minBelowMaxDist = sampleTab[cSam, "minBelowMaxDist"])
}

# Distribute and count alignments. Note that this
# may be done with the GUI instead of the command
# line prompt. The advantage is that the read
# statistics are added to the project files and
# browsable within the GUI.

# GUI - start it with a double click or the system
# command below, open all the projects and press
# "run all".
system(RcountDistri, wait = FALSE)

# Or run it via the command line:
proFiles <- sapply(rownames(sampleTab), function(x)
 file.path(RcountProDir, paste0(x, ".xml")))
```

```
proFiles <- paste(proFiles, collapse = ',')
command <- paste0(RcountDistri, " -c ", proFiles)
system(command)

# Read-in all the tables and merge the files which
# correspond to only one sample (some samples were
# sequenced on multiple lanes). Windows users must
# switch to the "native Windows R" to run the code
# below.
counts_RC <- f.read.Rcount(countTabsRC)
counts_RC <- f.summarize.columns(
  counts_RC,
  data.frame(sample = rownames(sampleTab),
  group = sampleTab$mySampleName),
  sum)
write.csv(counts_RC, countTabRC)
```

3.6 Identification of Differentially Expressed Genes

The last step of the workflow is the identification of differentially expressed genes with DESeq2 [5], edgeR [6], or limma [7]. The script below can be used independently of the alignment and read counting software used above (more examples with other experimental designs are available on github.com/MWSchmid/RNAseq_protocol). It can be applied to any count table and may also be used for quantitative comparisons of other data types (e.g., ChIP-Seq or sequencing of microbial communities).

The analysis is based on a model with a single factor with multiple levels. Given a simple crossed factorial design with two factors, e.g., genetic background (wild type versus mutant) and drug treatment (mock versus drug), the approach presented here would fit one coefficient for each of the four factor level combinations (Fig. 2). The comparisons of interest can then be specified as contrasts. This is on the one hand more flexible and powerful than the approach based on multiple pairwise comparisons

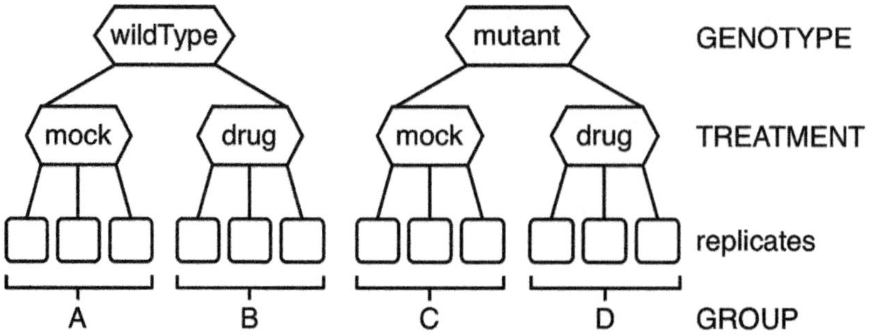

Fig. 2 Example illustrating how a simple crossed factorial design (*GENOTYPE* and *TREATMENT*) can also be analyzed using a single factor (*GROUP*). The full model using the crossed factorial design would be *(0+) GENOTYPE + TREATMENT + GENOTYPE:TREATMENT*. The model with the single factor is reduced to *(0+) GROUP*. The main effect of the genotype can then for example be tested with the contrast (A + B)/2 − (C + D)/2

followed by combining gene lists. On the other hand, it is more intuitive and easier than a model with several factors and interaction terms (*see* also section 9.5 in the limma user guide [7]). In the example data and the script below, the factors *TISSUE*, *CELLTYPE*, and *STAGE*, together with the blocking variable *BLOCK*, are combined into one single factor termed _SFAC__. We then define four contrasts to identify genes differentially expressed between (1) root tissues and green tissues, (2) female and male gametophytes, (3) gametophytes and sporophyte, and (4) gametophyte and embryo. The formula for the second contrast will for example be *(egg cell + synergids + central cell)/3 – pollen*. The results are finally stored in MS Excel workbooks with an overview sheet and one sheet for each contrast.

```
# Writing the MS Excel workbooks needs quite some RAM.
# On machines with less than 8GB of RAM, you can either
# set the ensembl variable below to NA (ensembl <- NA)
# or write csv files instead of workbooks.
options(java.parameters = "-Xmx8g")
library("biomaRt")
library("edgeR")
library("DESeq2")
library("limma")
library("XLConnect")
library("RNAseqWrapper")

# Specify several file paths and directories in which
# you would like to work in.
myAnnotation <- "/path/to/the/curatedAnnotation.csv"
countTabFC <- "/path/to/featureCounts_table.csv"
countTabRC <- "/path/to/Rcount_table.csv"
rDir <- "/directory/where/results/are/stored"

# Import the sample annotation.
sampleTab <- read.csv(myAnnotation, stringsAsFactors = FALSE)
rownames(sampleTab) <- sampleTab$sampleID

# Currently there are still some samples with
# several runs in the annotation. We need to remove
# them latest now.
colsToKeep <- c("TISSUE", "CELLTYPE", "STAGE", "BLOCK",
 "mySampleName")
sampleTab <- unique(sampleTab[,colsToKeep])
rownames(sampleTab) <- sampleTab$mySampleName

# Choose a biomart database (only if you were using
# the corresponding reference genome and annotation).
# To display available marts and datasets for animals
# and plants (there may also be other hosts):
listMarts(host = "www.ensembl.org") # animals
listMarts(host = "plants.ensembl.org") # plants
```

```
# Connect to a database and check whether there
# is a dataset for Arabidopsis available:
ensembl <- useMart("plants_mart", host = "plants.ensembl.org")
ensemblDatasets <- listDatasets(ensembl)
ensemblDatasets[grep("Arabidopsis", ensemblDatasets$description),]

# Finally connect to the database for A. thaliana:
ensembl <- useDataset("athaliana_eg_gene", mart = ensembl)

# Set the class of the biomart DB to "ensembl".
# Note that this is a simple error-workaround
# which is only necessary for some hosts.
biomaRt:::martBM(ensembl) <- "ensembl"

# Create a single factor which summarizes all the
# information given in the annotation table. Note
# that TISSUE, CELLTYPE, STAGE and BLOCK can be
# deleted/replaced/extended as required. See the
# paragraph on manual curation of the sample
# annotation above.
sampleTab$SFAC__ <- with(sampleTab,
  paste(TISSUE, CELLTYPE,
  STAGE, BLOCK,
  sep = '_'))

# Convert the new variable to a factor.
sampleTab$SFAC__ <- factor(sampleTab$SFAC__)

# Specify a generic model and the design matrix.
formulaString <- "~0+SFAC__"
design <- model.matrix(formula(formulaString),
  data = sampleTab,
  contrasts.arg = NULL)

# Specify the contrasts (comparisons) you are
# interested in. Instead of writing the entire
# formula by hand, one can define the formulas
# using regular expressions and the function
# f.formulate.simple.contrast. See ?regex for
# more details on regular expressions. However,
# you can also specify the names directly without
# using regular expressions (in this case you need
# to use fixed=TRUE in f.formulate.simple.contrast).
# See ?f.formulate.simple.contrast for more details.
# The function will specify the formula that it equals
# to the samples in the first argument minus the samples
# in the second argument (each scaled to the number of
# samples). In the example contrast "femVSmal" below,
# a positive log fold-change therefore means that the
# gene is expressed at a higher level in the female
# tissues compared to the male tissues.
# Define groups of samples based on their names
# and/or name patterns. The available names are
```

```
# given by the design matrix:
colnames(design)

# Based on the names, you can define an expression
# which specifies all the "green" tissues of a plant:
RE_green <- "__flo|__inflo|__leaf|__SAM|__sdl|__sil"

# Define some more groups:
RE_root <- "__root" # all root tissues
RE_emb <- "__emb" # all embryonic tissues
RE_femGam <- "__femGam" # all cells of the female gametophyte
RE_malGam <- "__malGam" # all samples from the male gametophyte

# And a group which includes any gametophytic tissue/cell type.
# Note that there are several ways to achieve the same:
RE_anyGam <- c("__malGam", "__femGam") # two exact words
RE_anyGam <- "__[[:alpha:]]{3}Gam" # __<three characters>Gam
RE_anyGam <- "__femGam|__malGam" # __femGam OR __malGam
myCont <- makeContrasts(
 f.formulate.simple.contrast( # rootVSgreen
 RE_root, RE_green, design),
 f.formulate.simple.contrast( # femVSmal
 RE_femGam, RE_malGam, design),
 f.formulate.simple.contrast( # gamVSspo
 RE_anyGam, RE_anyGam, design,
 invertMinus = TRUE),
 f.formulate.simple.contrast( # gamVSemb
 RE_anyGam, RE_emb, design),
 levels = design)

# Give short names to each contrast.
colnames(myCont) <- c("rootVSgreen", "femVSmal",
  "gamVSspo", "gamVSemb")

# Load the featureCounts and Rcount data sets.
myDataList <- list(
 feaCou = read.csv(countTabFC, row.names = 1),
 Rcount = read.csv(countTabRC, row.names = 1)
)

# Remove genes with less than five reads in all
# samples.
myDataList <- lapply(myDataList, function(x)
 f.strip.data(x, minVal = 5, minTimes = 1))

# Draw a sample correlation matrix. Given the size
# the data set, skip drawing all scatter plots at once.
for (dT in names(myDataList)) {
 logData <- log2(myDataList[[dT]]+1)
 f.do.some.overview(logData, rDir, dT,
 skipScatters = TRUE)
}
```

```
# Test for differential expression.
results <- lapply(myDataList, function(x)
 f.multiple.multi.level.comparisons(x, sampleTab,
  formulaString,
  myCont, design))

# Write the results into MS Excel workbooks.
for (dT in names(results)) {
 for (dM in names(results[[dT]])) {
 f.write.DEGtabs.to.workbook(results[[dT]][[dM]], rDir,
  paste0(dT, '_', dM), ensembl)
 }
}
# Note that warning messages containing
# is.na() applied to non-(list or vector)...
# can be ignored.

# Optional: Normalize the data and calculate the
# average expression levels across the specific
# conditions (average across replicates). Store
# the results for later use in .csv tables.

# Run all the normalizations.
myNormData <- lapply(myDataList, function(x)
 f.all.normalizations(x, sampleTab, formulaString, design))

# Define how the samples should be averaged.
byTab <- data.frame(sample = rownames(sampleTab),
  group = sampleTab$SFAC__,
  stringsAsFactors = FALSE)

# Average the normalized data.
meanTabs <- lapply(myNormData, function(x)
 lapply(x, function(y) f.summarize.columns(y, byTab, mean)))

# Store the tables in .csv format.
for (dT in names(myNormData)) {
 for (nM in names(myNormData[[dT]])) {
 write.csv(myNormData[[dT]][[nM]], file.path(rDir,
  paste0(dT, '_', nM, "_normalized.csv")))
 write.csv(meanTabs[[dT]][[nM]], file.path(rDir,
  paste0(dT, '_', nM, "_normalizedMeans.csv")))
 }
}
```

4 Notes

1. To determine the PHRED offset of a FASTQ file, you can use the function f.check.phred.offset() in the RNAseqWrapper package:

```
library("RNAseqWrapper")

# Guess the PHRED offset in a single file:
myPhredOffset <- f.check.phred.offset(
 "/path/to/a/fastq(.gz)/file")

# or for several files:
myFiles <- c("/path/to/a/fastq(.gz)/file",
 "/path/to/a/fastq(.gz)/file",
 "/path/to/a/fastq(.gz)/file")
myPhredOffsets <- sapply(myFiles, function(x)
  f.check.phred.offset(x),
  USE.NAMES = FALSE)
```

2. R version and packages tested on Ubuntu 16.04.02:

```
## R version 3.3.3 (2017-03-06)
## Platform: x86_64-pc-linux-gnu (64-bit)
## Running under: Ubuntu 16.04.2 LTS
##
## locale:
## [1] LC_CTYPE=en_US.UTF-8 LC_NUMERIC=C
## [3] LC_TIME=C LC_COLLATE=en_US.UTF-8
## [5] LC_MONETARY=C  LC_MESSAGES=en_US.UTF-8
## [7] LC_PAPER=C  LC_NAME=C
## [9] LC_ADDRESS=C LC_TELEPHONE=C
## [11] LC_MEASUREMENT=C LC_IDENTIFICATION=C
##
## attached base packages:
## [1] stats4 parallel stats graphics grDevices utils datasets
## [8] methods base
##
## other attached packages:
## [1] ShortRead_1.32.0  GenomicAlignments_1.10.0
## [3] BiocParallel_1.8.1 gplots_3.0.1
## [5] XLConnect_0.2-12  XLConnectJars_0.2-12
## [7] DESeq2_1.14.1  SummarizedExperiment_1.4.0
## [9] Biobase_2.34.0 edgeR_3.16.5
## [11] limma_3.30.12  biomaRt_2.30.0
## [13] Rsamtools_1.26.1  Biostrings_2.42.1
## [15] XVector_0.14.0 GenomicRanges_1.26.3
## [17] GenomeInfoDb_1.10.3 IRanges_2.8.1
## [19] S4Vectors_0.12.1  Rsubread_1.24.1
## [21] SRAdb_1.32.0 RCurl_1.95-4.8
## [23] bitops_1.0-6 graph_1.52.0
## [25] BiocGenerics_0.20.0 RSQLite_1.1-2
## [27] RNAseqWrapper_0.99.0
##
## loaded via a namespace (and not attached):
## [1] httr_1.2.1  splines_3.3.3 gtools_3.5.0
## [4] Formula_1.2-1 assertthat_0.1 latticeExtra_0.6-28
## [7] backports_1.0.5 lattice_0.20-34 digest_0.6.12
```

```
## [10] RColorBrewer_1.1-2 checkmate_1.8.2 colorspace_1.3-2
## [13] htmltools_0.3.5 Matrix_1.2-8 plyr_1.8.4
## [16] GEOquery_2.40.0 XML_3.98-1.5 genefilter_1.56.0
## [19] zlibbioc_1.20.0 xtable_1.8-2 scales_0.4.1
## [22] gdata_2.17.0 htmlTable_1.9 tibble_1.2
## [25] annotate_1.52.1 ggplot2_2.2.1 nnet_7.3-12
## [28] lazyeval_0.2.0 survival_2.40-1 magrittr_1.5
## [31] memoise_1.0.0 hwriter_1.3.2 foreign_0.8-67
## [34] tools_3.3.3 data.table_1.10.4 stringr_1.2.0
## [37] munsell_0.4.3 locfit_1.5-9.1 cluster_2.0.5
## [40] AnnotationDbi_1.36.2 caTools_1.17.1 grid_3.3.3
## [43] htmlwidgets_0.8 base64enc_0.1-3 gtable_0.2.0
## [46] DBI_0.5-1 R6_2.2.0 gridExtra_2.2.1
## [49] knitr_1.15.1 Hmisc_4.0-2 KernSmooth_2.23-15
## [52] rJava_0.9-8 stringi_1.1.2 Rcpp_0.12.9
## [55] geneplotter_1.52.0 rpart_4.1-10 acepack_1.4.1
```

3. R version and packages tested on MacOS X 10.10.5:

```
## R version 3.3.3 (2017-03-06)
## Platform: x86_64-pc-linux-gnu (64-bit)
## Running under: Ubuntu 16.04.2 LTS
##
## locale:
## [1] LC_CTYPE=en_US.UTF-8 LC_NUMERIC=C
## [3] LC_TIME=C LC_COLLATE=en_US.UTF-8
## [5] LC_MONETARY=C  LC_MESSAGES=en_US.UTF-8
## [7] LC_PAPER=C  LC_NAME=C
## [9] LC_ADDRESS=C LC_TELEPHONE=C
## [11] LC_MEASUREMENT=C LC_IDENTIFICATION=C
##
## attached base packages:
## [1] stats4 parallel stats graphics grDevices utils datasets
## [8] methods base
##
## other attached packages:
## [1] ShortRead_1.32.0  GenomicAlignments_1.10.0
## [3] BiocParallel_1.8.1 gplots_3.0.1
## [5] XLConnect_0.2-12  XLConnectJars_0.2-12
## [7] DESeq2_1.14.1  SummarizedExperiment_1.4.0
## [9] Biobase_2.34.0 edgeR_3.16.5
## [11] limma_3.30.12  biomaRt_2.30.0
## [13] Rsamtools_1.26.1  Biostrings_2.42.1
## [15] XVector_0.14.0 GenomicRanges_1.26.3
## [17] GenomeInfoDb_1.10.3 IRanges_2.8.1
## [19] S4Vectors_0.12.1  Rsubread_1.24.1
## [21] SRAdb_1.32.0 RCurl_1.95-4.8
## [23] bitops_1.0-6 graph_1.52.0
## [25] BiocGenerics_0.20.0 RSQLite_1.1-2
## [27] RNAseqWrapper_0.99.0
```

```
##
## loaded via a namespace (and not attached):
## [1] httr_1.2.1  splines_3.3.3 gtools_3.5.0
## [4] Formula_1.2-1 assertthat_0.1 latticeExtra_0.6-28
## [7] backports_1.0.5 lattice_0.20-34 digest_0.6.12
## [10] RColorBrewer_1.1-2 checkmate_1.8.2 colorspace_1.3-2
## [13] htmltools_0.3.5 Matrix_1.2-8 plyr_1.8.4
## [16] GEOquery_2.40.0 XML_3.98-1.5 genefilter_1.56.0
## [19] zlibbioc_1.20.0 xtable_1.8-2 scales_0.4.1
## [22] gdata_2.17.0 htmlTable_1.9 tibble_1.2
## [25] annotate_1.52.1 ggplot2_2.2.1 nnet_7.3-12
## [28] lazyeval_0.2.0 survival_2.40-1 magrittr_1.5
## [31] memoise_1.0.0 hwriter_1.3.2 foreign_0.8-67
## [34] tools_3.3.3 data.table_1.10.4 stringr_1.2.0
## [37] munsell_0.4.3 locfit_1.5-9.1 cluster_2.0.5
## [40] AnnotationDbi_1.36.2 caTools_1.17.1 grid_3.3.3
## [43] htmlwidgets_0.8 base64enc_0.1-3 gtable_0.2.0
## [46] DBI_0.5-1 R6_2.2.0 gridExtra_2.2.1
## [49] knitr_1.15.1 Hmisc_4.0-2 KernSmooth_2.23-15
## [52] rJava_0.9-8 stringi_1.1.2 Rcpp_0.12.9
## [55] geneplotter_1.52.0 rpart_4.1-10 acepack_1.4.1
```

4. R version and packages tested on Windows Server 2012 R2 (Cygwin R):

```
## R version 3.3.2 (2016-10-31)
## Platform: x86_64-unknown-cygwin (64-bit)
## Running under: CYGWIN_NT-6.3 WINDOWS-P8HLEBT 2.7.0(0.306/5/3) 2017-02-12 13:18
x86_64 Cygwin
##
## locale:
## [1] en_US.UTF-8/en_US.UTF-8/en_US.UTF-8/C/en_US.UTF-8/en_US.UTF-8
##
## attached base packages:
## [1] stats4 parallel stats graphics grDevices utils datasets
## [8] methods base
##
## other attached packages:
## [1] SRAdb_1.32.0 RCurl_1.95-4.8 bitops_1.0-6
## [4] graph_1.52.0 RSQLite_1.1-2 Rsamtools_1.26.1
## [7] Biostrings_2.42.1 XVector_0.14.0 GenomicRanges_1.26.3
## [10] GenomeInfoDb_1.10.3 IRanges_2.8.1 S4Vectors_0.12.1
## [13] BiocGenerics_0.20.0 Rsubread_1.16.1
##
## loaded via a namespace (and not attached):
## [1] Rcpp_0.12.9 zlibbioc_1.20.0 BiocParallel_1.8.1
## [4] R6_2.2.0  GEOquery_2.40.0 httr_1.2.1
## [7] Biobase_2.34.0 DBI_0.5-1 digest_0.6.12
## [10] memoise_1.0.0 XML_3.98-1.5
```

5. R version and packages tested on Windows Server 2012 R2 (native R):

```
## R version 3.2.5 (2016-04-14)
## Platform: x86_64-w64-mingw32/x64 (64-bit)
## Running under: Windows Server 2012 R2 x64 (build 9600)
##
## locale:
## [1] LC_COLLATE=English_United States.1252
## [2] LC_CTYPE=English_United States.1252
## [3] LC_MONETARY=English_United States.1252
## [4] LC_NUMERIC=C
## [5] LC_TIME=English_United States.1252
##
## attached base packages:
## [1] parallel stats4 stats graphics grDevices utils datasets
## [8] methods base
##
## other attached packages:
## [1] gplots_3.0.1 XLConnect_0.2-11
## [3] XLConnectJars_0.2-9 DESeq2_1.10.1
## [5] RcppArmadillo_0.6.700.3.0 Rcpp_0.12.4
## [7] SummarizedExperiment_1.0.2 Biobase_2.30.0
## [9] GenomicRanges_1.22.4 GenomeInfoDb_1.6.3
## [11] IRanges_2.4.8  S4Vectors_0.8.11
## [13] BiocGenerics_0.16.1 edgeR_3.12.1
## [15] limma_3.26.9 biomaRt_2.26.1
## [17] RNAseqWrapper_0.99.0 RevoUtilsMath_3.2.5
##
## loaded via a namespace (and not attached):
## [1] genefilter_1.52.1 gtools_3.5.0 locfit_1.5-9.1
## [4] splines_3.2.5 rJava_0.9-8 lattice_0.20-33
## [7] colorspace_1.2-6 survival_2.39-2 XML_3.98-1.4
## [10] foreign_0.8-66 DBI_0.3.1 BiocParallel_1.4.3
## [13] RColorBrewer_1.1-2 lambda.r_1.1.7 plyr_1.8.3
## [16] zlibbioc_1.16.0 munsell_0.4.3 gtable_0.2.0
## [19] futile.logger_1.4.1 caTools_1.17.1 latticeExtra_0.6-28
## [22] geneplotter_1.48.0 AnnotationDbi_1.32.3 KernSmooth_2.23-15
## [25] acepack_1.3-3.3 xtable_1.8-2 scales_0.4.0
## [28] gdata_2.17.0 Hmisc_3.17-3 annotate_1.48.0
## [31] XVector_0.10.0 gridExtra_2.2.1 ggplot2_2.1.0
## [34] grid_3.2.5  bitops_1.0-6 RCurl_1.95-4.8
## [37] RSQLite_1.0.0 Formula_1.2-1 cluster_2.0.4
## [40] futile.options_1.0.0 Matrix_1.2-5 rpart_4.1-10
## [43] nnet_7.3-12
```

References

1. Conesa A, Madrigal P, Tarazona S et al (2016) A survey of best practices for RNA-Seq data analysis. Genome Biol 17:13
2. R Core Team (2015) R: a language and environment for statistical computing. R Foundation for Statistical Computing, Vienna. https://www.R-project.org
3. Liao Y, Smyth GK, Shi W (2013) The Subread aligner: fast, accurate and scalable read mapping by seed-and-vote. Nucleic Acids Res 41:e108
4. Durinck S, Spellman P, Birney E et al (2009) Mapping identifiers for the integration of genomic datasets with the R/Bioconductor package biomaRt. Nat Protoc 4:1184–1191
5. Love MI, Huber W, Anders S (2014) Moderated estimation of fold change and dispersion for RNA-seq data with DESeq2. Genome Biol 15:550
6. Robinson MD, Oshlack A (2010) A scaling normalization method for differential expression analysis of RNA-Seq data. Genome Biol 11:R25
7. Ritchie ME, Phipson B, Wu D et al (2015) *limma* powers differential expression analyses for RNA-sequencing and microarray studies. Nucleic Acids Res 43:e47
8. Liao Y, Smyth GK, Shi W (2014) featureCounts: an efficient general purpose program for assigning sequence reads to genomic features. Bioinformatics 30:923–930
9. Schmid MW, Grossniklaus U (2015) Rcount: simple and flexible RNA-Seq read counting. Bioinformatics 31:436–437
10. Li X, Nair A, Wang S et al (2015) Quality control of RNA-Seq experiments. Methods Mol Biol 1269:137–146
11. Qi W, Schlapbach R, Rehrauer H (2017) RNA-seq data analysis: from raw data quality control to differential expression analysis. In: Schmidt A (ed) Plant germline development. Methods in molecular biology. Springer, Dordrecht

Part VI

Technologies with Outreach for Plant Manipulation, Studies of Pathways, and Applications in Crop Species

Chapter 25

Impedance Flow Cytometry as a Tool to Analyze Microspore and Pollen Quality

Iris Heidmann and Marco Di Berardino

Abstract

Analyzing pollen quality in an efficient and reliable manner is of great importance to the industries involved in seed and fruit production, plant breeding, and plant research. Pollen quality parameters, viability and germination capacity, are analyzed by various staining methods or by in vitro germination assays, respectively. These methods are time-consuming, species-dependent, and require a lab environment. Furthermore, the obtained viability data are often poorly related to in vivo pollen germination and seed set. Here, we describe a quick, label-free method to analyze pollen using microfluidic chips inserted into an impedance flow cytometer (IFC). Using this approach, pollen quality parameters are determined by a single measurement in a species-independent manner. The advantage of this protocol is that pollen viability and germination can be analyzed quickly by a reliable and standardized method.

Key words Impedance flow cytometry, IFC, Microfluidic chip, Pollen viability, Developmental stages, Pollen germination, Pollen size, Label-free

1 Introduction

Sexual plant reproduction either for maintaining genetic diversity, seed or fruit production relies on viable and functional pollen grains [1]. Off-season crossing or germplasm preservation programs benefit from conserved and stored mature pollen that retain their germination capacity [2]. Immature pollen of specific stages (microspores) are used to produce homozygous parental lines for breeding purposes [3]. However, genetic and environmental factors affect pollen development and viability [4, 5] and consequently reproduction [6–9]. Therefore, an estimation of pollen quality parameters like developmental stage, viability, and germination capacity is crucial for successful pollen-based processes.

Currently, various methods like staining with fluorescent/non-fluorescent dyes (stage and viability) [10, 11], size measurements (ploidy) [12], or in vitro pollen germination are used to determine pollen quality. However, these methods have disadvantages with

Anja Schmidt (ed.), *Plant Germline Development: Methods and Protocols*, Methods in Molecular Biology, vol. 1669, DOI 10.1007/978-1-4939-7286-9_25, © Springer Science+Business Media LLC 2017

respect to reliability, analysis time, labor intensity, genotype and species dependency. Moreover, the estimated pollen viability does often not correspond to pollen germination even when the correct protocol has been applied [13, 14]. Recently, a general, efficient, and reliable method that determines pollen viability and predicts germination using microfluidic chips in combination with an impedance flow cytometer has been described [15].

Impedance flow cytometry (IFC) is based on the electric properties of cells that react to an applied alternating current (Fig. 1). By varying the frequency (MHz) of the alternating current information about cell size, membrane integrity and cytoplasmic conductivity are simultaneously obtained. IFC has been used in microbiology and human cell cultures to characterize cell types [16] and culture conditions [17], but has not been widely used for plant cells [15].

Here we describe a detailed protocol to follow pollen development, analyze pollen viability, and predict pollen germination using IFC [15]. In the described method microspores, immature or mature pollen grains are isolated or harvested, resuspended in specific measurement buffers, and filtrated to remove particles that are larger than the actual pollen grains to prevent clogging of

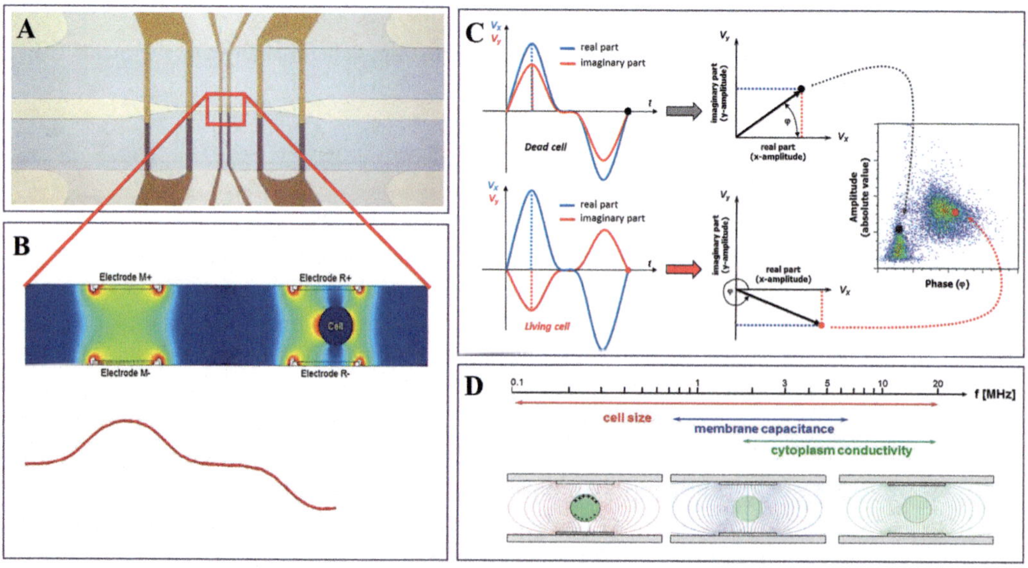

Fig. 1 Working principle of IFC. The microfluidic chip (**A**) consists of a sensing channel whose width can range from 50 to 250 μm, as well as measurement "M" and reference electrodes "R," to which an alternating current is applied (**B**). As a cell or particle crosses the electric field, it changes the impedance value of the liquid in a characteristic way. The impedance signal is split in its real and imaginary components (v_x and v_y), which are different for dead and living cells (**C**). A smaller and more compact (dead) cell gives a smaller signal in both amplitude and phase angle as a bigger and physiologically active one (**C**). According to the dielectric properties of cells these differences can be observed at higher frequencies, where membrane capacitance (integrity) and cytoplasmic conductivity are interrogated (**D**)

the microfluidic chip. The sample is then loaded into an impedance flow cytometer with default electronic settings of trigger level, amplification, and demodulation. As the measurement proceeds, the data of the analyzed cells are displayed in amplitude/phase angle dot plots and related histograms by which the cell/pollen populations are characterized. Non-developing microspores and dead pollen grains are usually reduced in amplitude value and phase angle while viable cells show a slightly higher amplitude value and a distinct higher phase angle [15]. Pollen with germination capacity are a part of the viable populations that show a higher phase angle than the non-viable ones. By adapting the electronic setting, the pollen population with germination capacity can be discriminated from the viable and non-viable ones [15]. The different populations are identified by inactivation controls; the populations that disappear after the treatment must have been the viable and/or germinating ones, as the cells are not harmed or destroyed during the measurement [15].

This simple method opens opportunities to screen for any genetic or environmental effect affecting pollen development, viability, and germination in a reliable and standardized manner. As the cells are not harmed during the passage through the IFC, the analyzed microspores or pollen can be recollected and used for further experiments or applications.

2 Materials

Most of the required equipment can be found in standard cell or molecular biology labs.

2.1 General Equipment

1. Water purification system to produce deionized water (15–18 MΩ, Millipore) (*see* **Note 1**).
2. Microwave, heater, water kettle, or water bath.
3. Sterile disposable pipettes of various volumes, micropipette and tips for small-volume dispensing, microcentrifuge tubes; 250 μl PCR-strips, fine forceps (Dumont no. 5 or 7).
4. PCR machine, 96-well with gradient option between 20 and 60 °C.
5. Vortex.
6. Table top centrifuge.
7. Cell strainers with appropriate mesh size, e.g., 30, 50, 100, and 150 μm (*see* **Note 2**).
8. Microscope with size measurement tool (*see* **Note 2**).
9. Glass slides and coverslips.
10. 2 ml rubber bulbs (grey) as used, e.g. for Pasteur pipettes.

Table 1
Standard pollen germination media

Component	Brewbaker and Kwack [23] (Solanaceae)	Vizintin and Bohanec [24] (Cucurbitaceae)
Sucrose	100 g/l	
Maltose		150 g/l
KNO_3	100 mg/l	
$Ca(NO_3)_2 \cdot 4H_2O$	300 mg/l	300 mg/l
$MgSO_4 \cdot 7H_2O$	100 mg/l	
H_3BO_3	100 mg/l	250 mg/l
Adjust to indicated pH (with 1 N KOH or HCl)	pH 5.2	pH 7.0

2.2 Special Equipment, Solutions, and Plant Material

1. Impedance Flow Cytometer Ampha Z30 (or derivatives) plus accessories including AmphaSoft software (Amphasys).
2. Microfluidic chips at 50, 120, or 250 μm channel size (Amphasys).
3. Waterpik FLW220 with tips or any other cordless electrical tooth flosser or brush that allows changing tips.
4. Micro pestles with tips fitting into 1.5 and 2 ml microcentrifuge tubes (preferentially the safe-lock type).
5. Analysis buffer AF 1–6 (Amphasys) or appropriate (*see* **Note 3**).
6. 2% (w/w) Tween 20 diluted in AF buffer.
7. 20% (w/v) potassium chloride solution (KCl).
8. Polystyrene beads of various sizes 10–100 μm (Sigma) (*see* **Note 4**).
9. Pollen germination media (examples of standard compositions are given in Table 1).
10. Gene frames (125 μl).
11. Pollen from plants either grown in the field, greenhouse, or controlled growth chambers as appropriate for the species and cultivar (*see* **Note 5**).

3 Methods

The method of pollen collection and subsequent sample preparation can vary slightly depending on the species and type of analysis. The typical workflow comprises pollen collection, extraction, filtration, and measurement (*see* Fig. 2). The identification of dead, non-germinating, and viable populations occurs by the disappearance of certain populations, the viable or germinating one, after inactivation (inactivated controls).

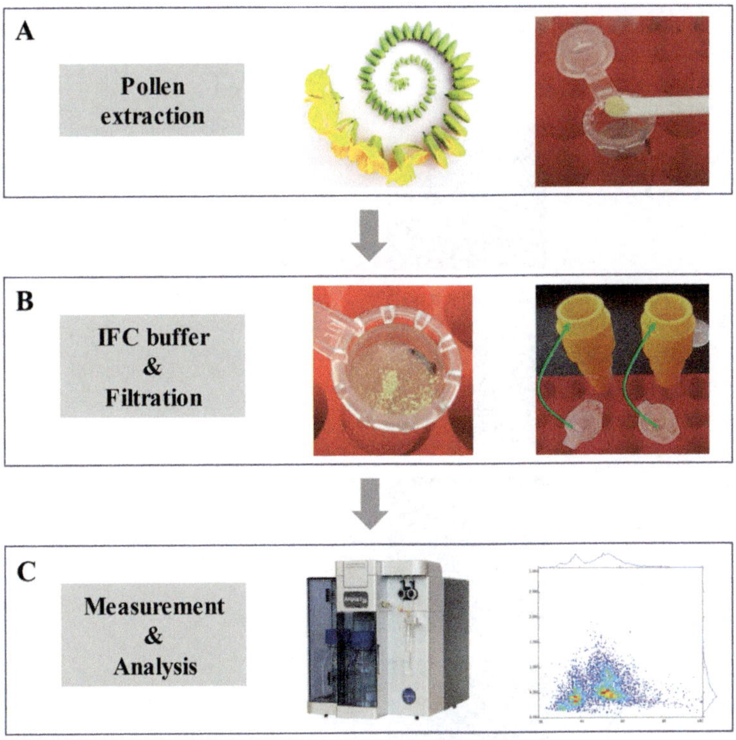

Fig. 2 Workflow. Microspores or pollen are isolated or harvested from closed buds or open flowers, respectively (**A**). Microspores or pollen resuspended in AF buffer are passed through cell strainers of appropriate size (**B**), loaded onto Ampha Z30 and analyzed (**C**)

3.1 Pollen Collection and Sample Preparation

3.1.1 Estimation of Viability and Developmental Stage

1. Collect a range of single, fresh flower buds with increasing sizes and sort them from small to big accordingly (*see* **Note 6**, Fig. 2A).
2. Place whole buds or extracted anthers into 1.5 or 2 ml microcentrifuge tubes containing 500 μl AF buffer and squeeze buds/anthers gently with a micro pestle (*see* **Note 7**).
3. Remove micro pestle, add 1 ml AF buffer, close the tube and mix the content by flicking the tube briefly or using a vortex.
4. Prepare fresh microcentrifuge tubes, insert cell strainer and pour the suspension through (Fig. 2B, *see* **Note 8**).
5. Remove cell strainer, close tube, and continue with the next sample or prepare appropriate control as described in Subheading 3.3.
6. Prepare for measurement as described in Subheading 3.4. (Figs. 2C and 3).

3.1.2 Mature Pollen Viability

1. Collect mature, dry pollen (e.g., tomato, sweet pepper, potato, wind pollinators) from open flowers by inserting the tip of the

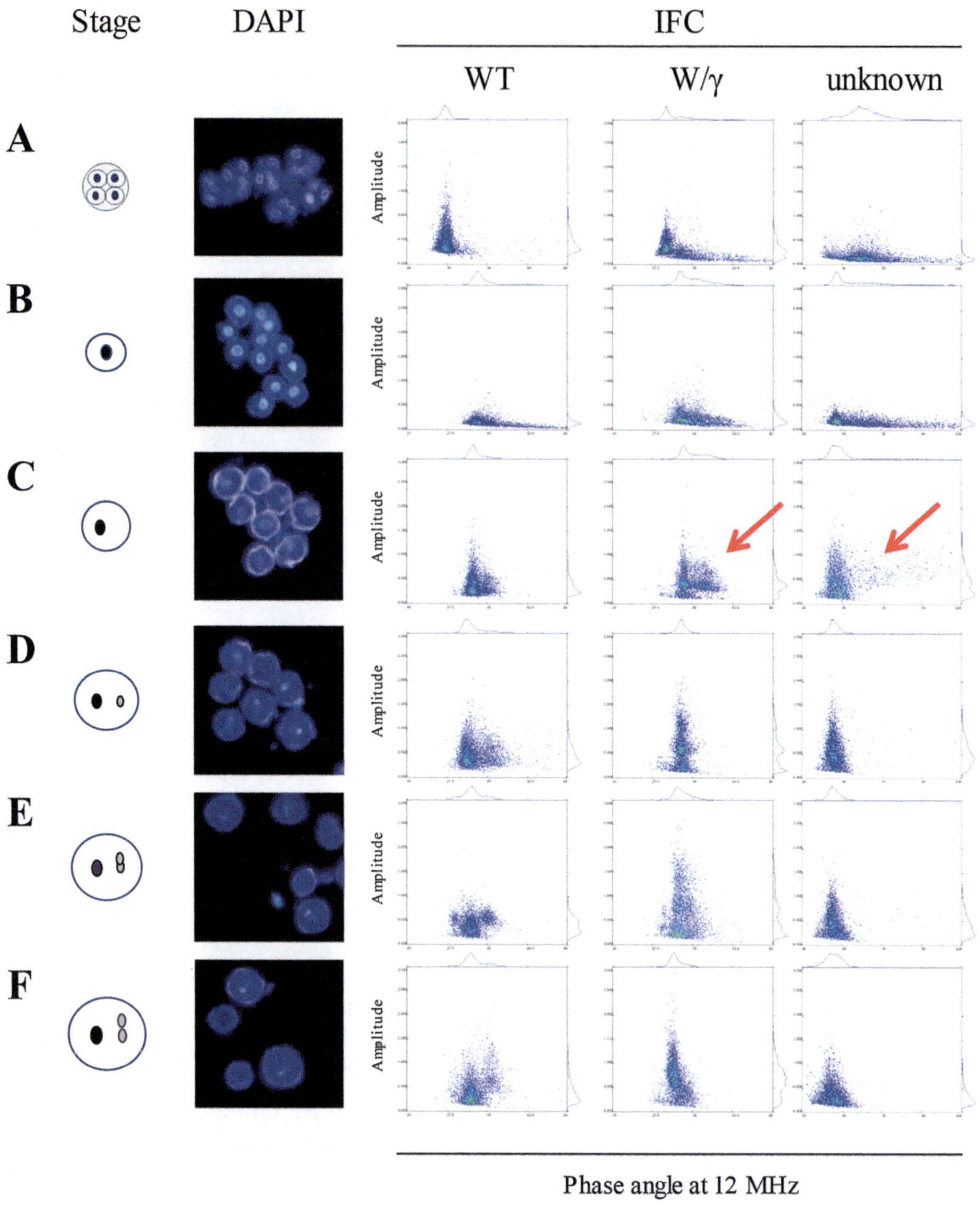

Fig. 3 Detection of developmental aberrations in potato pollen. Pollen development follows well defined steps. After release from the tetrad (**A**) the young microspore bears a central nucleus (**B**), at a later stage this nucleus migrates to the cell wall (**C**), divides to form an immature pollen grain with a generative (*gray*) and vegetative nucleus (*black*) (**D**). As the immature pollen grain matures the generative nucleus divides again (**E**) forming two sperm cells that remain attached to each other (**F**). Isolation of pollen from putative sterile plants like the potato genotype "W/y" or an undescribed sterility like in the genotype "unknown" shows that the stage of pollen abortion can be detected without DAPI staining

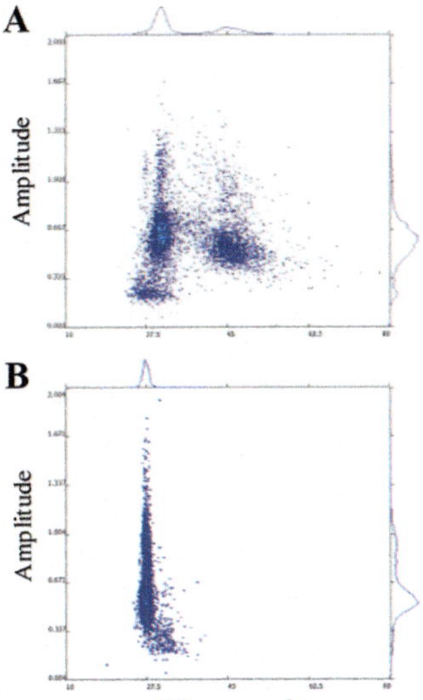

Fig. 4 Identification of dead and viable pepper pollen populations. The pollen populations are identified by heat-inactivation. Pollen populations usually display two clouds, one with a lower and one with a higher phase angle (**A**). Upon heat-inactivation the population with the higher phase angle disappears (**B**), identifying this one as the viable one that has been inactivated. The analysis was performed at 12 MHz

anthers into an 1.5 ml microcentrifuge tube and shake the flower using a waterpik, continue with **step 3** (*see* **Note 9**).

2. Collect mature, insect-transferred pollen (e.g., melon, cucumber, Brassica-species) by removing anthers from the open flowers and transfer them into 1.5 ml microcentrifuge tubes (*see* **Note 10**).
3. Add 1.5 ml of AF buffer, mix well and continue as described before (*see* Subheading 3.1.1) from **step 4** onwards (*see* **Note 11**).
4. Prepare killed pollen sample as described (*see* Subheading 3.3.1 or 3.3.2) to create a negative (dead) control sample as shown in Fig. 4.

3.1.3 Pollen Viability Combined with Germination (Tomato)

1. Collect mature pollen as described in Subheading 3.1.2.
2. Incubate pollen (fresh or dried) for 1 h in 2 ml AF6 buffer, mix from time to time.
3. Centrifuge for 3 min at 1500 × *g* in a microcentrifuge, remove liquid and resuspend in 250 μl AF6 buffer.
4. Prepare two equal aliquots of 100 μl.
5. Prepare one aliquot as germination-inactivated control as described (*see* Subheading 3.3.3).

6. Transfer the 100 μl aliquots each to 1.5 ml AF6 buffer.
7. Filtrate and measure all samples (*see* Subheading 3.4) to define the dead, viable, and germinating pollen population (Fig. 5). Verify by germination assay (*see* **Note 12**) as described below.
8. Centrifuge the remaining sample (about 250 μl) for 3 min at 1500 × *g* and remove the AF6 buffer.
9. Resuspend the pellet in 250 μl germination medium suitable for the species (*see* Table 1), spread into gene frames mounted onto glass slides and incubated for 90 min at RT in boxes with a relative humidity (RH) of about 80%.
10. Count spores as germinated when the pollen tube exceeds the diameter of the grains and calculate the percentage of germinated pollen.

3.2 Size Determination

1. Mix the original vials containing the polystyrene beads well.
2. Prepare a bead control solution for either 10, 20, or 30 μm beads by diluting the respective polystyrene suspension in AF buffer to a final concentration of approximately 5×10^4 beads/ml, and 60 or 100 μm beads to a final concentration of 1×10^4 beads/ml.
3. Add 30 μl of the diluted beads either individually or as mix (*see* **Note 4**) to each filtrated microspore/pollen sample (1.5 ml) prepared as described above (*see* Subheadings 3.1.1 and 3.1.2) and mix well (*see* **Note 13**).
4. Continue with Subheading 3.4 (measurement).

3.3 Control Sample Preparation

Dead or inactivated germination controls are essential for the identification of the analyzed populations. While temperature treatments around 40 °C inhibit pollen germination, temperatures above 54 °C completely kills pollen, at least in the tested species [15]. Based on the laboratory equipment, there are different options to create those:

3.3.1 Estimating the Killing Curve of Viable Pollen by Using a PCR Gradient

1. Prepare 1.5 ml of at least double concentrated, filtrated microspore/pollen suspension as described in Subheadings 3.1.1 and 3.1.2.
2. Fill 250 μl PCR strip vessels (12 per strip) with 100 μl of microspore/pollen suspension and close vessels.
3. Set the PCR program to a temperature gradient from 40 to 60 °C with a 15 min incubation step. Write down the different temperatures per well (*see* **Note 14**).
4. Fill 1.5 ml microcentrifuge tubes (one per treatment, tube marked) with 1 ml AF buffer.
5. Transfer the pollen suspension into the microcentrifuge tubes, close the lid and proceed to Subheading 3.4.

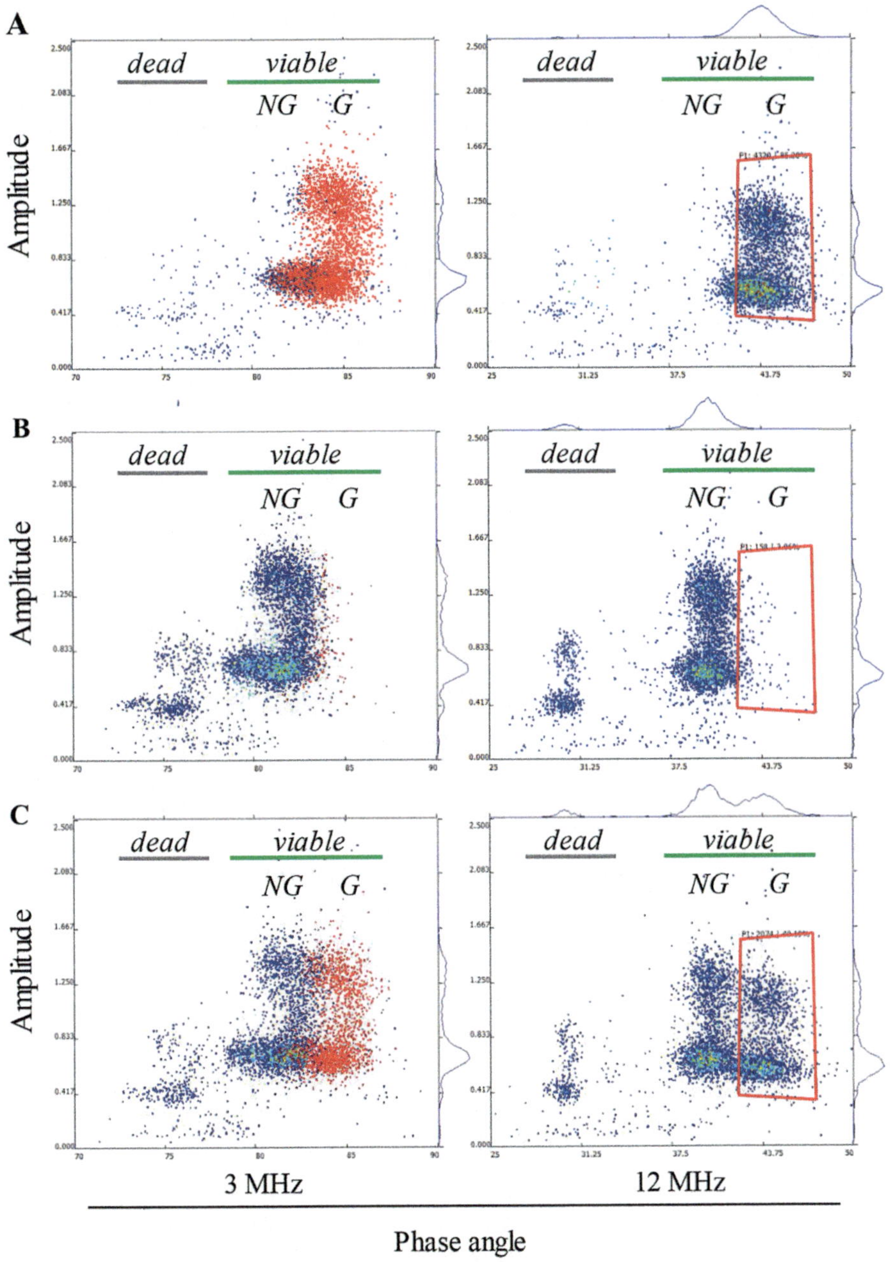

Fig. 5 Prediction of pollen germination. Pollen germination of tomato can be predicted at 3 and 12 MHz. Pollen populations usually carry dead (*grey bar*) and viable pollen grains (*green bar*) with a low and high phase angle, respectively (**A**). The viable population consists of pollen without (NG) and with germination capacity (G, marked in *red* at 3 MHz, and with a *red polygon* around the corresponding population at 12 MHz). A 15 min treatment at 40 °C inactivates in vitro tomato pollen germination but is not killing the pollen, the viable population (*blue*) is still present but the one with germination capacity is lacking (*red*) (**B**). In an equal mix of the two treatments (**C**) the different phase angle of the two populations becomes obvious

3.3.2 Killing Pollen by Hot Water Treatment

1. Place a mug (250 ml) with water into a microwave and heat up until it is cooking.
2. Remove mug and place a closed 1.5 or 2 ml safe-lock-tube containing microspore/pollen suspension into the hot water.
3. Incubate for 15 min, cool down to room temperature afterward, and proceed to Subheading 3.4 (*see* **Note 15**).

3.3.3 Inactivation of Pollen Germination

1. Resuspend a substantial amount of pollen (at least 6 × 5000 pollen grains) in AF buffer and incubate for 1 h at RT and filtrate.
2. Prepare samples and Eppendorf tubes as for the killing curve but set the PCR program to a temperature gradient from 20 to 40 °C with a 15 min incubation step (*see* **Note 16**).

3.4 Measurement

1. Start-up Ampha Z30 and software as described by the manufacturer.
2. Insert measurement chip and rinse it.
3. Choose measurement buffer, flush cycles between samples, and stop conditions (optional). Stop conditions allow terminating the measurement after a certain number of counted cells or a defined sample volume.
4. Before measurement adjust the stop condition to a higher cell count to visualize the beads along developing pollen grains and estimate the sizes (Fig. 6A), or to determine a dead population (Fig. 6B).
5. Choose frequencies (MHz), 0.5 and 12 MHz for viability, 12 MHz for germination capacity (*see* **Note 17**).
6. Prepare work space as described by the manufacturer.
7. Safe work space.
8. Shake individual samples before each measurement to create an equal distribution of pollen grains throughout suspension.
9. Insert the sample tubing tube into the microcentrifuge tube and start the measurements.
10. The machine will stop automatically if stop conditions are provided, without stop conditions the measurement has to be stopped manually (*see* **Note 18**).
11. After the flushing cycle proceed with the next sample.

3.5 Data Analysis

1. Adjust first the phase range and amplitude value for all samples by using the "zoom"- and "apply all" function (see manual of the supplier).
2. Compare the viable with the dead/inactivated control sample and place the gating accordingly (see manual of the supplier). Apply to all sets the same parameters for all measurements and

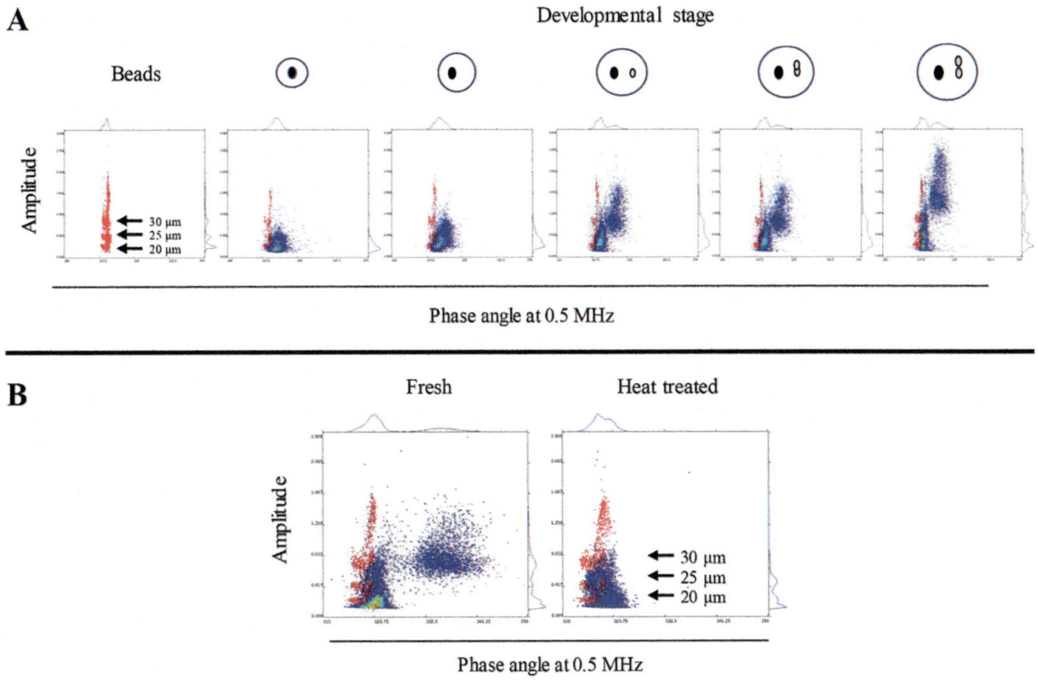

Fig. 6 Size determination by beads. In the absence of a microscope with size measurement tools a mix of polystyrene beads of different sizes (marked in *red* with size indication) can serve as marker-set to estimate the size of developing pollen (**A**) or the difference between viable and dead pollen populations (**B**) using the amplitude value at 0.5 MHz

calculates automatically the percentages of cells within the different gates (*see* **Note 19**).

3. Export collected data to excel compatible file by choosing the "cvs"- option in the report function for further statistical analysis, etc. (*see* **Note 20**).

4 Notes

1. Deionized water of 15 MΩ is the minimal required water quality to run the Ampha Z30 machine. If no system to produce water of the required quality is available, aqua dest. purchased at pharmacies or other shops can be used.
2. The use of cell strainers is essential as pollen samples usually contain debris, like trichomes, hairs, pollen aggregates, or small insects, which has to be removed prior to measurement. Omission of this step results in chip clogging and failure of the measurement. The cell strainer type is chosen in relation to the used chip and pollen size. The use of a mesh size smaller than the standard pollen size will not result in a representative sample. It is recommended to estimate the pollen size first,

either by measuring under a microscope (e.g., cell sense software Olympus), or consulting suitable websites (www.amphasys.com; http://pollen.tstebler.ch/). Alternatively, use the cell strainers to estimate pollen size by passing a suspension through a tower of different mesh sizes decreasing from 150 to 20 or 10 μm, observe the formation of fractions, and check them under the microscope. Non-developed and dead pollen is usually of a smaller size than viable pollen.

Cell strainers can be re-used after thorough cleaning and complete drying. Salt or water residues within the mesh can affect the conductivity of the buffer and interfere with the measurement.

3. The AF buffers are designed per pollen type and species. They are used as extraction and measurement buffer. As a general rule, dry pollen is best to be measured in AF4 or AF6, while AF3 or AF5 are optimal for insect-transferred, sticky pollen. A recommendation for a number of species is given at the supplier's website (www.amphasys.com).

   For the analysis of pollen development (microspore stages), the standard isolation buffer can be used. When adjusting a non-AF medium or buffer the liquid impedance has to be measured and if necessary adjusted to the optimal value designated per chip by the supplier (www.amphasys.com). It is recommended to adjust the liquid impedance with potassium chloride, e.g., the standard isolation medium for *Brassica* microspores (1/2 B5 medium [18], 13% sucrose) for an optimal separation of the dead and viable population.

   AF buffers contain sugars and should be handled under sterile conditions, or otherwise be used in small aliquots that can be used in one day.

4. The use of polystyrene beads is an option. They are delivered at specific sizes and can be mixed in certain ratios and combination, depending on the experimental design and the expected size of the pollen grain. Of course, beads can also be used as size indicator using a microscope. The beads will be visible as round spheres and with a different reflection under a light microscope. As an example, tomato pollen have a size between 20 and 25 μm, thus polystyrene beads between 15 and 30 μm are of appropriate sizes. Beads of 60 or 100 μm are suitable for bigger pollen like cucumber or corn (50–90 μm). When beads have been added to the samples, two extra rinse cycles should be applied after the measurements to thoroughly clean the fluidic tubing.

5. It should be considered that pesticide treatments, pests, or extreme temperatures might affect pollen quality [19]. For certain screenings, like for cytoplasmic male sterility (CMS), repetitive measurements are recommended, as CMS can be

influenced by environmental conditions [20] or even mimicked by those [19].

6. Remove any open flowers sheading pollen to avoid contaminations of the other samples with mature pollen. In many species, pollen development is accompanied with an increasing bud length. For a reliable analysis, it is recommended to measure the buds length by a standard ruler. For a precise stage determination, uni-nuclear to tri-cellular, a standard DAPI staining is recommended. The described technique allows a quick screening, e.g., of disrupted pollen development or a rough state determination.
7. Some flower buds are relatively sturdy or carry a big ovary, like sweet pepper or onion, which hampers the extraction of the microspores. Other flowers release quickly oxidizing compounds when wounded, like Florence fennel, which affect pollen viability. In these cases and in cases where buds carry a lot of hairs/long trichomes an anther extraction using fine forceps is recommended. Buds from *Brassica* species are soft with a small ovary and can easily be squeezed as whole.
8. Pouring the microspore suspension is the fastest option. The suspension can also be pipetted if it does not contain too much debris which hampers pipetting. Cell strainers can have some air within the mesh, a quick lift and drop of the strainer onto the microcentrifuge tube releases the captured air and allows the filtration. Filtration can further be enhanced by placing a 2 ml (grey) rubber bulb tightly onto the strainer (fits exactly onto the Celltrix cell strainers) and pressing it gently.
9. It is recommended to change the tip of the waterpik frequently or per plant to avoid pollen contamination and/or disease spreading.
10. While melon and cucumber anthers can be best collected by using a Dumont No. 7 forceps, Brassica-anthers with their relatively long filaments can be best pinched from the open flowers by placing the anthers between the opening and the lid of the microcentrifuge tube and just closing the lid.
11. Stored pollen is treated as described for mature and dry pollen. It depends on the species [21] if a hydration phase is necessary or not.
12. The in vitro germination assay in this case serves as a control when this measurement is executed for the first time. It has to be taken into consideration that not all described pollen germination protocols/media are optimal for each genotype/species, which might influence the strength of the correlation between IFC and in vitro germination data. For convenience, the compositions of two standard germination media are given in Table 1. However, additional work prior starting these

experiments is likely needed to find the optimal germination medium if the standard media are insufficient to initiate a pollen germination of at least 50%.

13. It is recommended to prepare two controls, a pollen-free bead and a bead-free pollen sample, because the beads will show no change in phase and can disguise the dead pollen population. The amount of beads added to a pollen suspension should be about equal to the amount of pollen to avoid an overrepresentation of beads.
14. Automatic gradient calculations of PCR machines do not increase the temperature/well by a fixed value. If the temperatures of two adjacent wells are too close together, like 40 and 40.5 °C, it makes sense to skip one or two wells to have a bigger temperature difference between the treatments, e.g., 40 and 42 °C.
15. This is the fastest and most easily applicable killing method. It is usually applied after the standard measurements and especially in cases when not much experience with a certain pollen type has been gained previously. The best method to fix the microcentrifuge tube in the mug is to put a piece of tape onto the lid of the closed tube and attach the ends across the open mug which holds the tube down into the hot water. Within 15 min the water temperature of the mug (at room temperature) drops from 90 to 60 °C which is sufficient to kill the pollen. A water bath or lab-heater (with microcentrifuge tube block) set to 60 °C can also be used for the 15 min treatment.
16. All pollen populations that we have analyzed so far and that were subjected to the treatment above 37–40 °C did not germinate. However, it cannot be excluded that pollen of certain species used to a climate around or above this temperature range are tolerant to these temperatures. For those species, the PCR gradient has to be adjusted accordingly.

    It is possible that germinating pollen of different species react differently and the identification of the germinating population has to be worked out by adjusting the default settings and the use of optimized AF measurement buffers for each species.
17. In most species, the discrimination between dead and viable is sufficient at 0.5 MHz but in some cases, as for the estimation the germination capacity, the measurement at 12 MHz provides more detailed information. For specific measurements, the default setting of the machine has to be adjusted in the expert window.
18. For most pollen measurements, 3000–5000 analyzed cells are sufficient for a reliable interpretation. When beads were added to the pollen sample, it is recommended to either switch off the stop condition “cells” or to set the stop condition to

8000–10,000 cells. Beads analyzed by the 120 μm chip at 0.5 MHz but not at 12 MHz can appear in a "J"-shape. This is due to the variation in size of the beads and their position within the channel as they pass the measurement electrodes, an effect described as inertial focusing [22].

19. Specific populations can be marked and named by using the option "polygon-gating" and "select cells" (see manual of the supplier).
20. Alternatively, an html-file which contains all data and the images can be created.

References

1. Heslop-Harrison JS (1992) Cytological techniques to assess pollen quality. In: Cresti M, Tiezzi A (eds) Sexual plant reproduction. Springer, Berlin, pp 41–48
2. Hanna WW, Towill LE (2010) Long-term pollen storage. In: Janick J (ed) Plant breeding reviews, vol 13. John Wiley & Sons, Oxford, pp 179–207
3. Maluszynski M (ed) (2003) Doubled haploid production in crop plants: a manual. Kluwer, Dordrecht
4. De Storme N, Geelen D (2014) The impact of environmental stress on male reproductive development in plants: biological processes and molecular mechanisms. Plant Cell Environ 37(1):1–18
5. Johnson SA, McCormick S (2001) Pollen germinates precociously in the anthers of raring-to-go, an Arabidopsis gametophytic mutant. Plant Physiol 126(2):685–695
6. Harsant J, Pavlovic L, Chiu G, Sultmanis S, Sage TL (2013) High temperature stress and its effect on pollen development and morphological components of harvest index in the C3 model grass *Brachypodium distachyon*. J Exp Bot 64(10):2971–2983
7. Hecker RJ, Stanwood PC, Soulis CA (1986) Storage of sugarbeet pollen. Euphytica 35 (3):777–783
8. Herrero MP, Johnson RR (1980) High temperature stress and pollen viability of maize. Crop Sci 20(6):796–800
9. Issarakraisila M, Considine JA (1994) Effect of temperature on pollen viability in Mango cv. "Kensington". Ann Bot 73:231–240
10. Rodriguez-Riano T, Dafni A (2000) A new prodecudre to asses pollen viability. Sex Plant Reprod 12:241–244
11. Heslop-Harrison J, Heslop-Harrison Y (1970) Evaluation of pollen viability by enzymatically induced fluorescence; intracellular hydrolysis of fluorescein diacetate. Stain Technol 45 (3):115–120
12. Kelly JK, Rasch A, Kalisz S (2002) A method to estimate pollen viability from pollen size variation. Am J Bot 89(6):1021–1023
13. Dafni A, Firmage D (2000) Pollen viability and longevity: practical, ecological and evolutionary implications. Plant Syst Evol 222(1-4):113–132
14. Johnson-Brousseau SA, McCormick S (2004) A compendium of methods useful for characterizing Arabidopsis pollen mutants and gametophytically-expressed genes. Plant J 39 (5):761–775
15. Heidmann I, Schade-Kampmann G, Lambalk J, Ottiger M, Di Berardino M (2016) Impedance flow cytometry: a novel technique in pollen analysis. PLoS One 11(11):e0165531. doi:10.1371/journal.pone.0165531
16. Crocetti S, Beyer C, Schade G, Egli M, Fröhlich J, Franco-Obregón A (2013) Low intensity and frequency pulsed electromagnetic fields selectively impair breast cancer cell viability. PLoS One 8(9):e72944
17. Pierzchalski A, Hebeisen M, Mittag A, Bocsi J, Di Berardino M, Tarnok A (2012) Label-free hybridoma cell culture quality control by a chip-based impedance flow cytometer. Lab Chip 12:4533–4543
18. Gamborg OL, Miller RA, Ojima K (1968) Nutrient Requirement of suspension cultures of soybean root cells. Exp Cell Res 50:151–158
19. Giorno F, Wolters-Arts M, Mariani C, Rieu I (2013) Ensuring reproduction at high temperatures: the heat stress response during anther and pollen development. Plants 2:489–506
20. Fan ZX, Lei WX, Hong DF, He JP, Wan LL, Xu ZH, Liu PW, Yang GS (2007) Development and primary genetic analysis of a fertility

temperature-sensitive polima cytoplasmic male sterility restorer in Brassica napus. Plant Breed 126(3):297–301

21. Franchi GG, Piotto B, Nepi M, Baskin CC, Baskin JM, Pacini E (2011) Pollen and seed desiccation tolerance in relation to degree of developmental arrest, dispersal, and survival. J Exp Bot 62(15):5267–5281
22. Hansson J, Karlsson JM, Haraldsson T, van der Wijngaart W, Russom A (2011) Inertial particle focusing in parallel microfluidic channels for high-throughput filtration. In: 2011 16th International Solid-State Sensors, Actuators and Microsystems Conference, 5–9 June 2011. IEEE, Washington, DC, pp 1777–1780. doi:10.1109/transducers.2011.5969458
23. Brewbaker JL, Kwack BH (1963) The essential role of calcium ion in pollen germination and pollen tube growth. Am J Bot 50(9):859–865
24. Vižintin L, Bohanec B (2004) In vitro manipulation of cucumber (Cucumis sativus L.) pollen and microspores: isolation procedures, viability tests, germination, maturation. Acta Biol Cracov Bot 46:177–183

Chapter 26

Agrobacterium-Mediated Sorghum Transformation

Emily Wu and Zuo-Yu Zhao

Abstract

Agrobacterium-mediated plant transformation is commonly used in crop genome modification. An optimized sorghum transformation protocol we developed is described here. Using this protocol, the transformation frequency of sorghum inbred TX430 is over 10% with *Agrobacterium* strain LBA4404 and 33% with *Agrobacterium* strain AGL1. Two different selection marker genes, moPAT and PMI, were used in this protocol.

Key words *Agrobacterium*, *Sorghum*, *Transformation*, *Plant transformation*, *Transgenic sorghum*, *Transgenic plants*, *Selection markers*, *Super-binary vector*

1 Introduction

Since the first success of *Agrobacterium*-mediated sorghum transformation published in 2000 [1], a number of protocol modifications have been reported (reviews [2–4]). Improvement of sorghum transformation frequency using microprojectile bombardment was also reported [5]. Recently, data collected from large-scale experiments demonstrated a significant enhancement of transformation frequency and the ease of transformation procedure in sorghum [6]. These improvements were largely due to the modifications of medium formulations with increased levels of copper and 6-benzylaminopurine (BAP) in the media described previously in barley transformation [7, 8]. Through these medium modifications, a comprehensive sorghum transformation protocol has been developed. Using this optimized protocol, the transformation frequency of sorghum TX430 ranged from 8 to 13.4% with *Agrobacterium* strain LBA4404. Furthermore, the transformation frequency increased to 33% using the strain AGL1. These data were collected from 848 independent transgenic events and more than 1300 transgenic T_0 plants, derived from approximately 7000 *Agrobacterium*-infected immature embryos [6].

Anja Schmidt (ed.), *Plant Germline Development: Methods and Protocols*, Methods in Molecular Biology, vol. 1669, DOI 10.1007/978-1-4939-7286-9_26, © Springer Science+Business Media LLC 2017

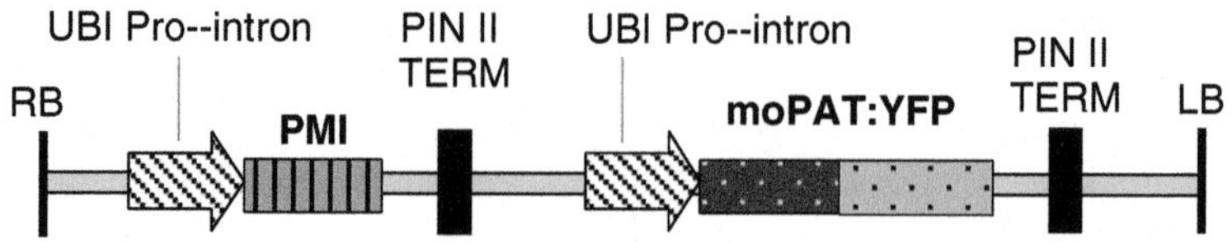

Fig. 1 Diagram of the T-DNA region of a vector used in sorghum transformation. RB: T-DNA right border, LB: T-DNA left border, UBI Pro-intron: maize ubiquitin promoter and ubiquitin intron, PMI: PMI gene, moPAT:YFP: the fusion protein gene of moPAT and YFP, PIN II TERM: terminator of PIN II gene

When evaluating a particular transformation protocol, the molecular data of those transgenic plants derived by using this protocol, including integrated T-DNA copy number, frequencies of intact or truncated T-DNA insertions, and frequencies of *Agrobacterium* vector backbone integration into the sorghum genome, are important factors for consideration (Fig. 1). The molecular data of quantitative PCR (Q-PCR) analysis of 1182 transgenic plants representing 675 independent transgenic events generated by using this optimized protocol were described in our previous report [6].

Selection marker genes used in transformation are important for sorghum, especially in the development of commercial products [9]. Sorghum can cross-pollinate with some weeds, such as johnsongrass (*Sorghum halepense* L.) [10–12] and Shattercane (*Sorghum bicolor* L.) [13] in the field. Thus, an herbicide-resistant gene may be suboptimal as the selection marker gene for sorghum transformation.

2 Materials

2.1 Plant Material, Agrobacteria Strains, Vectors, and Selection Marker Genes

1. Sorghum inbred TX430 plants.
2. *Agrobacterium tumefaciens* strains LBA4404 or AGL1 [14].
3. Super-binary vectors constructed with pSB1 and pSB11 [15, 16].

2.2 Culture Media

Unless otherwise stated, amounts of all reagents and components are given per 1 l stock solution or medium.

1. Medium PHI-L: 2.0 ml Ferrous sulfate · $7H_2O$ Stock (Stock-A), 50 ml AB buffer 20× (Stock-B), 50 ml 20× AB salts (Stock-C), 5.0 g Glucose, 1.0 ml of 50 mg/ml Spectinomycin stock, 15.0 g Bacto-agar, pH 6.8 (*see* **Note 1**).
2. Stock-A: 1.25 g Ferrous sulfate·$7H_2O$.
3. Stock-B: 60 g Potassium phosphate dibasic, 20 g Sodium phosphate monobasic anhydrous.

4. Stock-C: 20 g Ammonium chloride, 6.0 g Magnesium sulfate heptahydrate ($7H_2O$), 3.0 g Potassium chloride, 0.228 g Calcium chloride·$2H_2O$.
5. Medium PHI-M: 5.0 g Yeast extract, 10.0 g Peptone, 5.0 g Sodium chloride, 15.0 g Bacto-agar, 1.0 ml of 50 mg/ml Spectinomycin stock, pH 6.8 (*see* **Note 1**).
6. Medium PHI-I: 4.30 g MS salts, 0.5 mg Nicotinic acid, 0.5 mg Pyridoxine HCl, 1.0 mg Thiamine·HCl, 0.10 g Myo-inositol, 1.0 g Vitamin assay casamino acids, 1.5 mg 2, 4-Dichlorphenoxyacetic acid (2, 4-D), 68.50 g Sucrose, 36.0 g Glucose, pH 5.2. This medium should be filter-sterilized (No autoclave).
7. Medium PHI-T: 4.3 g MS salts, 0.5 mg Nicotinic acid, 0.5 mg Pyridoxine·HCl, 1.0 mg Thiamine·HCl, 100.0 mg Myo-inositol, 2.0 mg 2, 4-D, 20.0 g Sucrose, 10.0 g Glucose, 700.0 mg L-proline, 0.5 g 2-(*N*-Morpholino) ethanesulfonic acid (MES), 100 μM Acetosyringone, 10.0 mg Ascorbic acid, 8.0 g Agar, pH 5.8.
8. Medium DBC3-CARB: 4.3 g MS basal salt mixture, 0.25 g Myo-inositol, 1.0 g Casein hydrolysate, 1.0 mg Thiamine·HCl, 1.0 mg 2, 4-D, 30 g Maltose, 0.69 g L-proline, 1.22 ml 1 mg/ml Cupric sulfate stock, 3.5 g Phytagel, 0.5 ml of 1 mg/ml BAP, 2.0 ml of 50 mg/ml Carbenicillin stock, pH 5.8.
9. Medium DBC3-CT: 4.3 g MS basal salt mixture, 0.25 g Myo-inositol, 1.0 g Casein hydrolysate, 1.0 mg Thiamine·HCl, 1.0 mg 2, 4-D, 30 g Maltose, 0.69 g L-proline, 1.22 ml of 1 mg/ml Cupric sulfate stock, 3.5 g Phytagel, 0.5 ml of 1 mg/ml BAP stock, 0.4 ml of 250 mg/ml Cefotaxime stock, 0.6 ml of 250 mg/ml Timetin stock, pH 5.8.
10. Medium DBC3-PMIS: 4.3 g MS basal salt mixture, 5.0 g Maltose, 1.0 mg Thiamine·HCl, 0.25 g Myo-inositol, 1.0 g Casein hydrolysate, 0.69 g L-proline, 1.22 ml of 1 mg/ml Cupric sulfate stock, 1.0 mg 2, 4-D, 12.5 g Mannose, 0.5 ml of 1 mg/ml BAP stock, 3.5 g Phytagel, 2.0 ml of 50 mg/ml Carbenicillin stock, pH 5.8.
11. Medium DBC3-PMIS-CT: 4.3 g MS basal salt mixture, 5.0 g Maltose, 1.0 mg Thiamine·HCl, 0.25 g Myo-inositol, 1.0 g Casein hydrolysate, 0.69 g L-proline, 1.22 ml of 1 mg/ml Cupric sulfate stock, 1.0 mg 2, 4-D, 12.5 g Mannose, 0.5 ml of 1 mg/ml BAP stock, 3.5 g Phytagel, 0.4 ml of 250 mg/ml Cefotaxime stock, 0.6 ml of 250 mg/ml Timetin stock, pH 5.8.
12. Medium DBC3-BARS: 4.3 g MS basal salt mixture, 30 g Maltose, 1.0 mg Thiamine·HCl, 0.25 g Myo-inositol, 1.0 g Casein hydrolysate, 0.69 g L-proline, 1.22 ml of 1 mg/ml Cupric

sulfate stock, 1.0 mg 2, 4-D, 0.5 ml of 1 mg/ml BAP stock, 3.5 g Phytagel, 2.0 ml of 50 mg/ml Carbenicillin stock, 3.0 mg Bialaphos, pH 5.8.

13. Medium DBC3-BARS-CT: 4.3 g MS basal salt mixture, 30 g maltose, 1.0 mg Thiamine·HCl, 0.25 g Myo-inositol, 1.0 g Casein hydrolysate, 0.69 g L-proline, 1.22 ml of 1 mg/ml Cupric sulfate stock, 1.0 mg 2, 4-D, 0.5 ml of 1 mg/ml BAP stock, 3.5 g Phytagel, 0.4 ml of 250 mg/ml Cefotaxime stock, 0.6 ml of 250 mg/ml Timetin stock, 3.0 mg Bialaphos, pH 5.8.
14. Medium PHI-XM: 4.3 g MS basal salt mixture, 5.0 ml MS vitamin stock solution, 0.1 g Myo-inositol, 1.0 ml of 0.5 mg/ml Zeatin stock, 1.25 ml of 1 mg/ml Cupric sulfate stock, 0.7 g L-proline, 60 g Sucrose, 8.0 g Agar, 2.0 ml of 0.5 mg/ml IAA, 1.0 ml of 0.1 mM Abscisic acid (ABA) stock, 0.1 ml of 1 mg/ml Thidiazuron stock, 2.0 ml of 50 mg/ml Carbenicillin stock, pH 5.6.
15. Medium PHI-XMBARR: 4.3 g MS basal salt mixture, 5.0 ml MS vitamin stock solution, 0.1 g Myo-inositol, 1.0 ml of 0.5 mg/ml Zeatin stock, 1.25 ml of 1 mg/ml Cupric sulfate stock, 0.7 g L-proline, 60 g Sucrose, 8.0 g Agar, 2.0 ml of 0.5 mg/ml Indole acetic acid (IAA) stock, 1.0 ml of 0.1 mM ABA stock, 0.1 ml of 1 mg/ml Thidiazuron stock, 2.0 ml of 50 mg/ml Carbenicillin stock, 5.0 mg Phospinotricin (PPT, Glufosinate), pH 5.6.
16. Medium PHI-Z: 2.15 g MS basal salt mixture, 2.5 ml MS vitamin stock solution, 0.05 g Myo-inositol, 20 g Sucrose, 3.0 g Phytagel, pH 5.6.
17. Medium PHI-ZA: 2.15 g MS basal salt mixture, 2.5 ml MS vitamin stock solution, 0.05 g Myo-inositol, 20 g Sucrose, 0.5 ml of 1 mg/ml 1-Naphtypacetic acid (NAA), 0.5 ml of 1 mg/ml Indolebutyric acid (IBA), 7.0 g Sugar, pH 5.6.
18. Sterilization solution: 50% Bleach, 0.1% Tween-20.

2.3 Planting and Greenhouse Care

1. 18-well flat.
2. 1-gallon pots.
3. Soil PRO-MIX BX.
4. Pollination bags.

3 Methods

3.1 Growth of Sorghum and Preparation of Agrobacterium

1. Grow sorghum inbred TX430 plants under preferred greenhouse conditions. Healthy sorghum plants are critical for success of transformation (*see* **Note 2**). Greenhouse conditions for growing sorghum plants are: average temperatures 29 °C

during the day and 20 °C at night with a 12 h day/night photoperiod and supplemental lighting with a 3:1 ratio of metal halide and high pressure sodium lamps.

2. Use *Agrobacterium tumefaciens* strains LBA4404 or AGL1 [14] and super-binary vectors constructed with pSB1 and pSB11 [15, 16] (*see* **Notes 3** and **4**). The super-binary vector contains two gene cassettes as follows: (1) the maize UBI promoter and UBI intron driving a fusion gene of moPAT [17] (*see* **Note 5**) and YFP gene [18] and (2) the maize UBI promoter and UBI intron driving a PMI gene [19].
3. Make a master plate from *Agrobacterium* stored as a glycerol stock at −80 °C: Streak a loop-full of bacteria to produce single colonies on medium PHI-L in a Petri dish and incubate the plate inverted, at 28 °C, in the dark, for 2–3 days. Bacteria on a master plate are usable for up to 4 weeks if the plates are sealed with Parafilm and refrigerated (4 °C).
4. Make a working plate: Pick 2–3 colonies from a master plate and streak a fresh plate of medium PHI-M. Incubate bacterial plate inverted, at 28 °C, in the dark, for 1–2 days.

3.1.1 Make Agrobacterium Suspension for Immature Embryo Infection (See Note 6)

1. Add acetosyringone to 2–4 ml medium PHI-I (pre-warmed to room temperature) in a 14 ml tube to make PHI-I with 100 μM acetosyringone.
2. Scrape bacteria off working plate with a sterile *bacterium* loop and place in the tube containing PHI-I with 100 μM acetosyringone.
3. Vigorously vortex the tube to break clumps and form a uniform suspension as determined by visual inspection.
4. Take 1 ml of the *Agrobacterium* suspension to measure optical density at 550 nm. Dilute the suspension with medium PHI-I plus 100 μM acetosyringone to an OD of 0.7 (~10^9 cfu/ml).

3.2 Preparation of Immature Embryos

1. Harvest immature panicles 11–14 days after pollination depending on the growing conditions. The immature zygotic embryos used in transformation may be 1.5–2.5 mm in length and the optimal size for TX430 is 1.9–2.2 mm (*see* **Note** 7).
2. Take off immature kernels from the panicles, sterilize with 50% bleach and 0.1% Tween-20 for 30 min under vacuum condition and then rinse with sterile water three times. Keep the kernels in sterile water before isolating the embryos.
3. Aseptically dissect embryos from each sterilized sorghum kernel and place in a 2 ml microcentrifuge containing 2 ml medium PHI-I with 100 μM acetosyringone. Place 100 embryos per tube, but the embryo number can be varied (*see* **Note 8**).

3.3 *Agrobacterium* Infection, Co-cultivation, and Resting

1. Remove PHI-I liquid medium from the tube with a micropipette tip and add 1 ml *Agrobacterium* suspension.
2. Gently invert the tube a few times to mix well and then vortex the tube for 10–30 s at speed setting between 4 and 5.
3. Incubate the embryos in *Agrobacterium* suspension for 5 min at room temperature (stand the tube in the hood) and make sure all embryos are in the suspension solution during incubation.
4. Pour the *Agrobacterium* suspension with the embryos into the side of a co-cultivation PHI-T medium plate to avoid the *Agrobacterium* suspension distributing to the medium plate and remove all suspension with a micropipette tip and spread the embryos evenly on the medium plate with scutellum side up using a scalpel blade (*see* **Note 8**).
5. Culture the plate at 25 °C, in the dark, for 4–7 days for co-cultivation step.
6. Transfer the embryos to medium DBC3-CARB (if using LBA4404) or DBC3-CT (if using AGL1) with the same orientation as co-cultivation (scutellum side up) and culture for 7–10 days for resting step at 28 °C, in the dark, in an incubator or culture room.

3.4 Selection of Putative Transgenic Events

1. Start the selection at 14 days from *Agrobacterium* infection of embryos.
2. Transfer embryos to selection medium DBC3-PMIS (if using LBA4404) or DBC3-PMIS-CT (if using AGL1) with PMI as the selection marker or to DBC3-BARS (if using LBA4404) or DBC3-BARS-CT (if using AGL1) with moPAT as the selection marker (*see* **Notes 5, 8, 9–11**). Culture 20–25 embryos per plate at first round selection and 10 embryos per plate at second and third round selection and incubate at 28 °C, in the dark and subculture interval is 2–3 weeks.
3. Starting at the second month on selection, for each subculture, only keep the healthily growing callus (embryogenic callus) and discard the non-healthy tissue (non-embryogenic) as much as possible. This will help rapid plant regeneration later. Keep the tissues derived from each embryo separately. Total selection time is about 2–3 months based on the size of the freshly growing callus tissues.

3.5 Regeneration of T_0 Plants

1. Transfer the freshly growing calli on medium PHI-XM for PMI as the selection marker (no selection agent is needed in the medium during regeneration) or PHI-XM or PHI-XMBARR (selection agent included in the medium) for moPAT as the selection marker and incubate at 28 °C, in the dark, for approximately 2–3 weeks to develop shoots. When shoots appear,

move these cultures to a lighting culture room. Cover the plate top with either a screen or non-transparent board to block the strong lights during the first week in the lighting room. When the leaves turn to green, remove the screen or the board and expose the plants to the full lights.

2. Transfer the shoots (about 2–4 cm tall) to glass tubes (one shoot per tube) or plastic boxes (10 × 9 × 10 cm; 2–4 shoots per box, make sure these shoots derived from the same embryo) containing PHI-Z medium (without selection agent) if the shoots have healthy roots. If the shoots have no root or have poor roots, transfer the shoots to PHI-ZA medium for a maximum of 1 week and if the shoots still do not produce healthy roots, move the shoots back to PHI-Z medium till good roots produced.
3. Culture these shoots under the same lighting and temperature conditions for 1–2 weeks. If there are still significant amount of phenolic produced (*see* **Note 12**), transfer the shoots to a fresh medium.
4. When the plantlets reach about 5–7 cm height with healthy roots, transfer these plantlets (T_0 plants) to greenhouse.

3.6 Greenhouse Care of Transgenic Sorghum Plants

1. Greenhouse care of transgenic T_0 plants is an important part of the sorghum transformation process. Appropriate growth conditions optimize health and fertility of T_0 plants.
2. Take the plantlets out from the culture boxes or tubes and wash the roots of the plantlets with tap water to remove medium residues.
3. Transfer the plantlets into an 18-well flat filled with soil PRO-MIX BX and place the flat in a temperature (26 °C with lights and 21 °C without lights) and lighting (16 h lighting and 8 h dark) controlled growth chamber or greenhouse for 2 weeks to grow a healthy root system.
4. Transfer the plants to 1-gallon pots containing soil PRO-MIX BX, one plant per pot, and grow in the greenhouse with 16 h light and 8 h dark at 18–30 °C. T_0 sorghum plants usually require about 5.5 months in the greenhouse from transplant to harvest.
5. Place a pollination bag to cover a new emerged panicle prior flowering. Keep the bag on the panicle till harvesting.
6. Harvest and dry the panicles and take off seeds from the panicles. Usually 50–200 kernels per panicle can be obtained from each T_0 plant.
7. Label the seeds harvested from each transgenic plant with a unique ID (*see* **Notes 10** and **11**).

4 Notes

1. Adjust to pH 6.8 with 1 M NaOH for PHI-M. No adjustment is required for PHI-L.
2. Healthy embryo-donor sorghum plants are critical for optimal transformation efficiency. Healthy sorghum plants from optimal growing conditions and cares cannot be emphasized enough.
3. Other *Agrobacterium* strains with appropriate antibiotics to inhibit *Agrobacterium* growth during resting and selection can be used.
4. Other binary vectors with good T-DNA delivery capability to sorghum immature embryos can also be used. The promotor(s) and gene(s) used on the transformation vectors are based on the needs.
5. Other selection marker genes, such as pat [20], bar, and NPTII [21], can also be used for sorghum transformation.
6. *Agrobacterium* suspension for embryo infection can also be prepared in liquid culture. One day prior to transformation, about 30 ml of minimal AB medium in a 30 ml baffle flask containing 50 μg/ml spectinomycin is inoculated with a 1/8 loop-full of *Agrobacterium* from a 1- to 2-day-old working plate. The *Agrobacterium* is grown at 28 °C, in the dark, at 200 rpm overnight (about 14 h). In mid-log phase, the *Agrobacterium* cells are harvested and resuspended at 10^9 cfu/ml in PHI-I medium plus 100 μM acetosyringone using standard microbial techniques and standard curves.
7. We mainly described sorghum inbred TX430 transformation protocol here. We also tested this protocol for other sorghum varieties such as BTx2757 and African varieties: Tegemeo, Macia and Malisor and it worked very well. Some minor modifications, such as embryo size, concentration of *Agrobacterium* suspension, co-cultivation temperature, and culture length in some steps, may be applied empirically.
8. Keeping embryo intact without physical damage during isolation, infection and co-cultivation is critical for successful transformation.
9. Removing coleoptiles when embryos are transferred to selection media helps callus growth and results in better transformation (higher transformation frequency and/or less process time).
10. Molecular analysis to confirm T-DNA integration into the sorghum genome is needed. The choice of specific analytical methods performed on any transgenic event is dependent on

the transgenes. In general, transgenic events can be assayed for the presence of the transgene(s) by polymerase chain reaction (PCR), Southern blotting, and/or Southern-by-Sequencing (SbS) [22]. To demonstrate stable transformation, confirmation of the presence of the transgene(s) in the sorghum genome should be at least in two generations, T_0 and T_1, by Southern blot or SbS.

11. With *Agrobacterium*-mediated transformation, one of the important issues is the integration of the T-DNA vector backbone sequence into the sorghum genome [6]. To elucidate the integration of the backbone sequence, Southern blot, SbS, or PCR analyses against the vector backbone sequence should be performed.
12. Phenolic production during sorghum tissue culture and transformation is a common phenomenon. When tissue and media are turning dark brown, it indicates too much phenolic produced which may negatively affect root development. Transferring the shoots to fresh PHI-Z media will help root development.

Acknowledgement

The authors thank our co-workers, Brian Lenderts, Kimberly Glassman, Maya Berezowska-Kaniewska, Heather Christensen, Tracy Asmus, Shifu Zhen, Uyen Chu, and Myeong-Je Cho for their contributions of the transformation and analysis work. This manuscript was critically reviewed by Dr. Doane Chilcoat and Dr. Todd Jones. Sorghum transformation with the PMI selection marker gene published in 2014 [6] was performed as part of a humanitarian project.

References

1. Zhao ZY, Cai T, Tagliani L et al (2000) *Agrobacterium* mediated sorghum transformation. Plant Mol Biol 44:789–798
2. Pandey AK, Bhat BV, Balakrishna D et al (2010) Genetic transformation of sorghum (Sorghum bicolor (L.), Moench). Int J Biotechnol Biochem 6:45–53
3. Kumar T, Howe A, Sato S et al (2013) Sorghum transformation: overview and utility, Chapter 10. In: Paterson AH (ed) Genomics of the *Saccharinae*, plant genetics and genomics: crops and models 11. Springer, New York, NY, pp 205–221
4. Do PT, Zhang ZJ (2015) Sorghum transformation: achievements, challenges, and perspectives. In: Recent advancements in gene expression and enabling technologies in crop plants. Springer, New York, NY, pp 291–312
5. Liu G, Ian GD (2012) Highly efficient sorghum transformation. Plant Cell Rep 31:999–1007
6. Wu E, Lenderts B, Glassman K et al (2014) Optimized *Agrobacterium*-mediated sorghum transformation protocol and molecular data of transgenic sorghum plants. In Vitro Cell Dev Biol Plant 50:9–18
7. Cho MJ, Jiang W, Lemaux PG (1998) Transformation of recalcitrant barley cultivars through improvement of regenerability and decreased albinism. Plant Sci 138:229–244

8. Bregitzer P, Campbell RD, Dahleen LS et al (2000) Development of transformation systems for elite barley cultivars. Barley Genet Newsl 30:10
9. Zhao ZY (2006) Chapter 19. Sorghum (*Sorghum bicolor L.*) In: Wang K (ed) Methods in molecular biology, 343, Agrobacterium protocols, vol 1, 2nd edn. Humana Press Inc., Totowa, NJ, pp 233–244
10. Hadley HH (1953) Cytological relationships between *Sorghum vulgare* and *S. halepense*. Agron J 45:139–143
11. Baker HG (1972) Migrations of weeds. In: Valentine DH (ed) Taxonomy, phytogeography and evolution. Academic Press, London, pp 327–347
12. Doggett H (1988) Sorghum. In: Tropical agricultural series, 2nd edn. Longman Scientific, Essex
13. Defelice SM (2006) Shattercane, Sorghum Bicolor (L.) Moench Ssp. Drummondii (Nees ex Steud.) De Wet ex Davidse—Black sheep of the family. Weed Technol 20(4):1076–1083. doi:10.1614/WT-06-051.1
14. Lazo GR, Stein PA, Ludwig RA (1991) A DNA transformation-competent *Arabidopsis* genomic library in *Agrobacterium*. Biotechnology 9:963–967
15. Komari T (1990) Transformation of cultured cells of Chenopodium quinoa by binary vectors that carry a fragment of DNA from the virulence region of pTiBo542. Plant Cell Rep 9:303–306
16. Komari T, Hiei Y, Saito Y et al (1996) Vectors carrying two separate T-DNAs for co-transformation of higher plants mediated by Agrobacterium tumefaciens and segregation of transformants free from selection markers. Plant J 10:165–174
17. US patent no. 6,096,947
18. Nagai T, Ibata K, Park ES et al (2002) A variant of yellow fluorescent protein with fast and efficient maturation for cell biological applications. Nat Biotechnol 20:87–90
19. Miles JS, Guest JR (1984) Nucleotide sequence and transcriptional start point of the phosphomannose isomerase (manA) gene of Escherichia coli. Gene 32:41–48
20. Botterman J, Gosselé V, Thoen C et al (1991) Characterization of phosphinothricin acetyltransferase and C-terminal enzymatically active fusion proteins. Gene 102:33–37
21. Fraley RT, Rogers SG, Horsch RB et al (1983) Expression of bacterial genes in plant cells. Proc Natl Acad Sci U S A 80:4803–4807
22. Zastrow-Hayes GM, Lin HN, Sigmund AL et al (2015) Southern-by-sequencing: a robust screening approach for molecular characterization of genetically modified crops. Plant Genome 8(1):1–15. doi:10.3835/plantgenome2014.08.0037.

Chapter 27

Use of the Cas9 Orthologs from *Streptococcus thermophilus* and *Staphylococcus aureus* for Non-Homologous End-Joining Mediated Site-Specific Mutagenesis in *Arabidopsis thaliana*

Jeannette Steinert, Carla Schmidt, and Holger Puchta

Abstract

Since the discovery of the CRISPR/Cas system and its in vivo application for site-specific targeted mutagenesis, this technique is wildly used in a great variety of organisms, such as many plant species. Commonly used for this application is the Cas9 enzyme from *Streptococcus pyogenes*. Here, we describe the application of two Cas9 orthologs from *Streptococcus thermophilus* and *Staphylococcus aureus* for targeted non-homologous end-joining mediated mutagenesis in *Arabidopsis thaliana*. With both orthologs, we could show efficient inheritance of the induced mutations. As both Cas9 orthologs are smaller in size than the enzyme of *S. pyogenes* and as the Protospacer adjacent motifs (PAMs) differ between all orthologs, they are attractive alternative tools for genome engineering in plants.

Key words Gene editing, Double-strand break repair, CRISPR/Cas, Genome engineering, Targeted mutagenesis

1 Introduction

In 2012, the molecular mechanism of the CRISPR/Cas system (for clustered regularly interspaced short palindromic repeats; CRISPR-associated) was discovered [1] and after the transfer of the molecular mechanism of the Cas9 enzyme to human cell culture [2–4] this system is now used for targeted induction of double-strand breaks (DSBs) in all kinds of organisms including plants [5].

The components needed for site-specific target recognition and subsequent cleavage are the Cas9 enzyme and its corresponding single guide RNA (sgRNA). The sgRNA consists of a CRISPR RNA (crRNA) and a trans-activating crRNA (tracrRNA) part, fused via a linker sequence to form this single RNA component interacting with the Cas9 enzyme. Target specificity is acquired through 20 nt of the crRNA part, the so-called spacer, which

Anja Schmidt (ed.), *Plant Germline Development: Methods and Protocols*, Methods in Molecular Biology, vol. 1669, DOI 10.1007/978-1-4939-7286-9_27,

Table 1
Comparison of the Cas9 orthologsOrthogous gene. Listed are the characteristics for the Cas9 orthologs SpyCas9, Sth1Cas9, and SauCas9 regarding their PAM sequence, size, and positions for creating nickase or dCas9 constructs

	Streptococcus pyogenes (Spy)	*Streptococcus thermophilus1* (Sth1)	*Staphylococcus aureus* (Sau)
PAM sequence	NGG NAG	NNAGAA NNGGAA	NNGRRT NNGRR(N)
Cas9 size	4.1 kb	3.4 kb	3.2 kb
point mutations for nickase/dCas9	D10A, H840A	D9A, H599A	D10A, N580A

binds complementary to the target sequence, where after the Cas9 nuclease induces a DSB. Next to the genomic target sequence, which is called protospacer, a protospacer-adjacent motif (PAM) is required for correct target recognition. Not only the PAM differs between Cas9 orthologs, but also the size of the Cas9 enzyme itself (*see* Table 1). The PAMs of here described Cas9 orthologs from *Streptococcus thermophilus* (Sth1Cas9) and *Staphylococcus aureus* (SauCas9) consist of five or six nucleotides compared to the only three nucleotides required for *Streptococcus pyogenes* Cas9 (SpyCas9) target recognition. This increment of specificity should lead to less off-target effects using these orthologs compared to SpyCas9, which was already demonstrated for SauCas9 in human cell culture [6, 7].

The Cas9 orthologs used here belong to the same class II type II-A CRISPR/Cas system, but are subdivided into different subclusters, in which the composition of the involved Cas proteins and the direction of the transcription of the CRISPR locus compared to the corresponding Cas operons differ [8–10]. Also the Cas9 proteins vary in their corresponding sgRNA sequences [7, 11, 12], at which structures like hairpins or additional stem loops are necessary for interaction with the correct Cas9. The requirements regarding the interaction between crRNA and tracrRNA and also the particular PAM sequences differ between these Cas9 orthologs, wherefore the systems should be able to function independently in a single cell.

With these orthologous Cas9 enzymes, not only new targets could be aimed on due to the different PAM requirements, but also new genome engineering approaches are possible combining various catalytic activities of several orthologs. One requirement for this simultaneous application of these Cas9 orthologs is the independent and specific activity of each enzyme with each own but not a "foreign" sgRNA. This was already demonstrated in plants for SpyCas9 and SauCas9, so the simultaneous application of different

Cas9 orthologs in one single cell should be possible without any cross-interaction [13], allowing to develop applications for new genome and also transcriptome engineering approaches in plants. In human cell culture, different chromosomal loci were labeled using orthologous dead Cas9 enzymes (dCas9, catalytically inactive Cas9), leading to multicolor detection of these targeted loci [14, 15]. As these Cas9 orthologs have been shown to also work in plants [13, 16, 17], similar applications using more than one Cas9 ortholog at once should be possible in plants, too [18].

Here, we describe the design as well as the cloning procedure of the orthologous RNA-guided Cas9 nucleases from *S. thermophilus* and *S. aureus* for their application in *A. thaliana* to generate heritable homozygously mutated plant lines [13]. By adjusting 20 nt of the sgRNA by annealing two oligonucleotides and subsequent cloning into a plasmid containing the sgRNA backbone, targeting of the desired sequence can be achieved. This adapted sgRNA construct is then transferred into a binary vector containing the Cas9 expression cassette, where after the final construct can be transferred via *A. tumefaciens* transformation into plants.

2 Materials

2.1 Antibiotics and Herbicides

1. 100 mg/mL Ampicillin (stock solution).
2. 100 mg/mL Spectinomycin (stock solution).
3. 30 mg/mL Kanamycin (stock solution).

2.2 Bacterial Strains

1. NEB5α (New England Biolabs) *E. coli* strain for conventional cloning steps.
2. GV3101 pMP90 *A. tumefaciens* strain for plant transformation.
3. DB3.1 (Life Technologies Inc.) *E. coli* strain for *ccdB* encoding vectors.

2.3 Plant Material, Buffer, Media, and DNA Oligonucleotides

1. Germination medium (GM): 4.9 g/L Murashige & Skoog, 10 g/L sucrose, adjust to pH 5.7 with KOH, 8 g/L agar.
2. Lysogeny broth (LB): 10 g/L pepton, 5 g/L yeast extract, 10 g/L NaCl, (17.5 g/L agar for solid plates).
3. Yeast extract broth (YEB): 5 g/L beef extract, 5 g/L pepton, 1 g/L yeast extract, 5 g/L sucrose, 493 mg/L $MgSO_4$, (12 g/L agar for solid plates).
4. TE buffer: 10 mM Tris–HCl, 1 mM EDTA (2,2′,2″,2‴-(Ethane-1,2-diyldinitrilo)tetraacetic acid), adjust to pH 8.

Table 2
Sequences of all primers

Oligo name	Sequence 5′-3′	Orientation
CS 32	GTCATAGCTGTTTCCTTGC	rv
SS42	TCCCAGGATTAGAATGATTAGG	fw
SS102	CACCATGTTATCACATCAATCC	rv
SS61	GAGCTCCAGGCCTCCCAGCTTTCG	fw
SS72	CTTCTATCGCCTTCTTGACG	rv
SS143	CAAGAAAGCTGGGTCCTCAG	rv

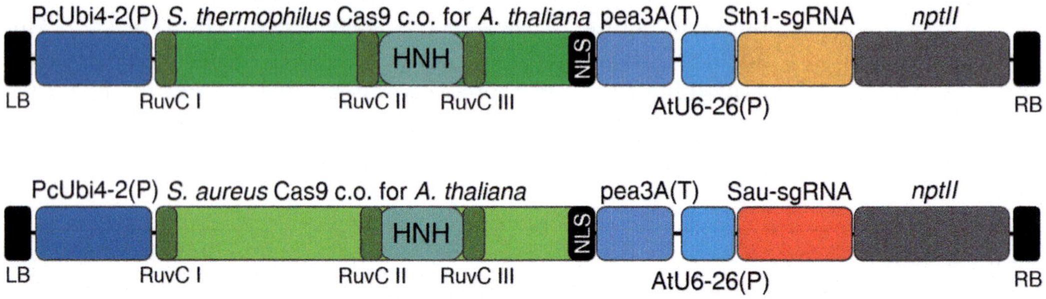

Fig. 1 Final T-DNA constructs for Sth1Cas9 nuclease and SauCas9 nuclease. The final expression cassettes include the orthologous Cas9 enzyme as well as the corresponding sgRNA on a T-DNA construct that is ready to use for plant transformation

5. 50 pmol DNA oligonucleotides, desalted purity. In Table 2, all primers described in the methods section are listed.
6. In conformity with your general transformation protocol every type of transformable plant material can be used (*see* **Note 1**).

2.4 Enzymes

1. Restriction enzymes 20,000 U/mL *AflII*, 10,000 U/mL *Bbs*I and 10,000 U/mL *NheI* with 10× buffer as supplied.
2. 1000 U/mL T4 DNA Ligase with 10× buffer as supplied.
3. LR Clonase II Enzyme mix and 800 U/mL Proteinase K.
4. Dream 5000 U/mL Taq DNA Polymerase.

2.5 Vectors

The sequence information of all described vectors can be obtained at www.botanik.kit.edu/crispr. All vectors are obtainable upon request from the authors. The final T-DNA constructs are shown in Fig. 1.

1. pEn-Sau Chimera: The sequence coding for the specific Sau sgRNA is combined with the *A. thaliana* U6-26 promotor specific for the expression of small RNAs and a termination

signal. The sgRNA is flanked by attL sites necessary for gateway cloning. Between the sgRNA and the attL sites, two recognition sites for *Bsu36I* and *MluI* have been added for inserting up to three sgRNAs according to the pEnC1.1 vector established for Spy sgRNA [19]. The vector has been synthesized by GeneArt® (Life Technologies Inc.) based on the pMA-RQ vector backbone including an ampicillin resistance cassette.

2. pEn-Sth1 Chimera: Identical to the pEn-Sau Chimera except for the exchange of Sau sgRNA into specific sgRNA coding sequence of *Streptococcus thermophilus*. This vector also features an ampicillin resistance.
3. pDe-SauCAS9: Codon optimized for *A. thaliana* SauCas9 open reading frame (ORF) synthesized by GeneArt® (Life Technologies Inc.) based on the published sequence data [7]. Afterward, the SauCas9 ORF has been integrated in the pDe-CAS9 by classical cloning using *Asc*I restriction sites. pDe-CAS9 harbors the plant-specific PcUbi4-2-promotor, pea3A-terminator and a *ccdB* gene, flanked by attR sites [20] and is located on a T-DNA-based pPZP201 backbone for *Agrobacterium*-mediated transformation [21]. Att sites are necessary for Gateway cloning for correct integration of sgRNA from pEn-Sau Chimera into the pDe-SauCAS9. The vector contains a spectinomycin resistance cassette for bacterial selection and the T-DNA includes an *nptII* gene conferring plant resistance to kanamycin.
4. pDe-Sth1CAS9: Codon optimized for *A. thaliana* Sth1Cas9 ORF synthesized by GeneArt® (Life Technologies Inc.), based on the published sequence data [22]. Beside the Cas9-ORF, pDe-Sth1CAS9 is identical to the pDe-SauCAS9 and also harbors a spectinomycin resistance cassette for bacterial selection and an *nptII* gene conferring plant resistance to kanamycin.

3 Methods

For a simple and reliable combination of Cas9 protein and specific adapted sgRNA, a cloning system containing the Gateway-based entry and destination vector has been designed for use. For both Cas9 orthologs from *S. thermophilus* and *S. aureus* the same cloning system is used, based on the previously established system designed for plant-specific expression of *S. pyogenes* Cas9 [19, 20]. Generally, this system is made up by two consecutive steps. In the first step, two complementary oligonucleotides, based on the target sequence, are integrated into the linearized entry vector, containing the sgRNA expression system. This cloning product is afterward transferred to the binary destination vector via a Gateway LR reaction (*see* Fig. 2 for an overview of the cloning procedure and Fig. 1 for the final T-DNA expression constructs).

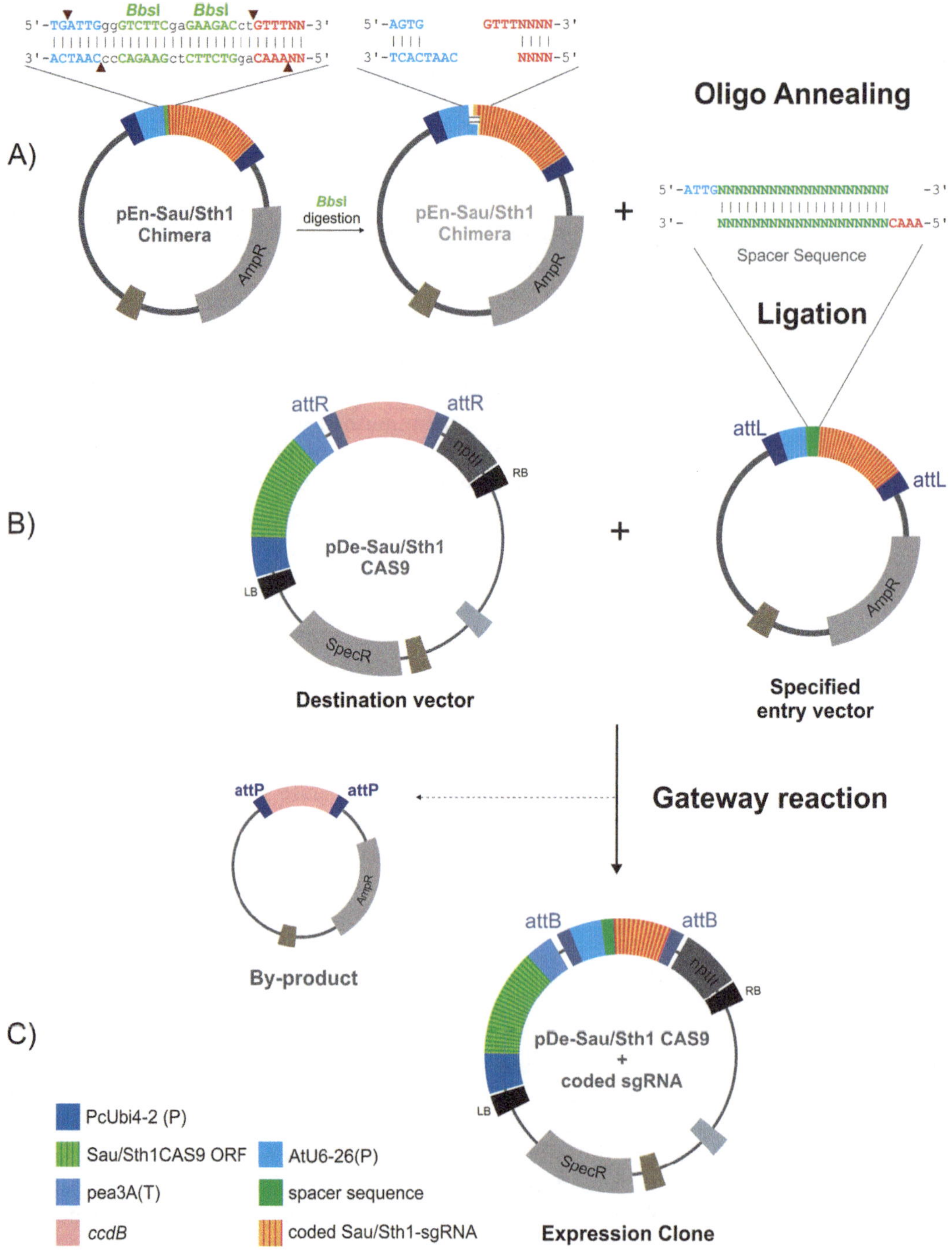

Fig. 2 Simplified overview of the general cloning procedure. (**A**) For the specification the orthologous sgRNAs, *BbsI* recognition sites are cleaved, which are located between the AtU6-26 promoter and the sgRNA sequence, leading to non-complementary sticky ends. Annealed complementary target-specific oligonucleotides (*orange*), with matching overhangs to the sticky ends created by *BbsI* can be inserted into the orthologous entry vector pEn-Sth1 Chimera or pEn-Sau Chimera by ligation. (**B**) The following gateway reaction is performed via att sites present in the specified entry vector and the destination vector (pDe-Sth1CAS9 or pDe-SauCAS9 respectively). LR Clonase II performs the exchange of the regions between att sites resulting in the corresponding expression clone and a by-product (**C**)

3.1 Determination of an Appropriate PAM Sequence for Each Cas9 Ortholog and Oligo Annealing

1. Dependent on your final application you can choose between the different orthologs and their associated PAM requirements. In contrast to the PAMs needed for SpyCas9 target recognition, variable PAM sequences are available for both described orthologs.
2. Choose your PAM sequence suitable for your experimental set up (*see* **Note 2**). For the PAM sequences SauCas9 use "NNGRR(T)"; recommended SauCas9 PAM sequences are "NNGGGT" and "NNGAA"; for PAM sequences Sth1Cas9 use "NNAGAA" or "NNGGAA."
3. Pick 20 nt protospacer, targeting your sequence of interest, upstream of the selected PAM (*see* **Notes 3** and **4**).
4. Design and order your oligonucleotides with corresponding overhangs fitting to *Bbs*I restriction sites (Fig. 2): Forward: 5′-ATTG – 20 nt protospacer sequence-3′ and Reverse: 5′-AAAC – 20 nt reverse complementary protospacer sequence-3′.
5. Adjust oligonucleotides to 50 pmol and mix 2 μL of each oligonucleotide with 46 μL ddH_2O.
6. Perform denaturation at 95 °C for 5 min. No cooling is necessary, just incubate at room temperature for 10 min.
7. The so annealed oligos can be stored at −20 °C or can be transferred directly to the entry vector.

3.2 Transfer of Target-Specific Spacer Sequence into Entry Vector

1. To integrate the annealed oligos into the entry vector (pEn-Sau Chimera, pEn-Sth1 Chimera), linearize the vector by digestion with *Bbs*I. Therefore, mix 1 μg of plasmid DNA (isolated by Miniprep with columns) with 2 μL restriction enzyme buffer, 1 μL *Bbs*I and add ddH_2O to a total volume of 20 μL. Incubate for at least 2 h by 37 °C.
2. Use a PCR purification kit to purify the digested entry vector and adjust to a final concentration of 5 ng/μL. The purified vector can be stored at −20 °C for further reactions.
3. Ligation is performed by mixing 2 μL of linearized vector with 3 μL of annealed oligos, 1 μL ligase buffer, 1 μL T4 Ligase, and 3 μL ddH_2O. Incubate at room temperature for about 2 h or at 4 °C overnight.
4. Transform 5 μL of the ligation in *E. coli* (NEB5α) according to your standard transformation protocol and plate on LB agar plates containing 100 μg/mL ampicillin.
5. Test 5–10 colonies by colony PCR, usually at least 70% of colonies should be correct (*see* **Note 5**). As forward primer the forward protospacer oligo can be used and CS 32 as reverse primer. Recommended annealing temperature is 56 °C and fragments are about 310 bp for Sau sgRNA and 390 bp for Sth1 sgRNA.

6. Inoculate LB media with 100 μg/mL ampicillin with 1 or 2 colonies and incubate overnight. Purify plasmids and sequence them with SS42 to confirm correct spacer integration.

3.3 Integration of Orthologous sgRNAs into Corresponding Destination Vectors by Gateway Cloning

1. For subsequent Gateway reactions combining Sth1 sgRNA with Sth1Cas9 or Sau sgRNA with SauCas9, adjust your destination vector (pDe-SauCAS9, pDe-Sth1CAS9) to 50 ng/μL and your previously prepared entry vector (pEn-Sth1 Chimera or pEn-Sau Chimera with integrated sgRNA) to 100 ng/μL.
2. The Gateway reaction can be performed by mixing 2 μL of target specific entry vector with 3 μL destination vector, 4 μL TE buffer and 1 μL LR Clonase II. Vortex and centrifuge the whole reaction and incubate afterward for about 2 h at room temperature.
3. To inactivate the LR Clonase II, add 1 μL Proteinase K and incubate at 37 °C for 10 min (*see* **Note 6**).
4. Transform the complete reaction into *E. coli* (NEB5α) and plate on LB agar plates containing 100 μg/mL spectinomycin.
5. Test 3–6 colonies by colony-PCR, almost all colonies should be correct. Use SS42 as forward primer and SS 102 as reverse primer. Annealing temperature is 60 °C and expected fragment lengths are about 860 bp for SauCas9 and 940 bp for Sth1Cas9.
6. Pick 1 or 2 colonies and inoculate LB medium containing spectinomycin. Incubate overnight and purify the plasmid DNA.
7. Test your plasmids by restriction digestion with *Afl*II and *Nhe*I. In case of pDe-SauCAS9 the expected bands are 5.4, 4.7,and 3.8 kb, for pDe-Sth1CAS9 the expected bands are 5.8, 4.1, 3.8, and 1 kb.
8. Sequence the plasmids with SS42 to confirm correct gateway reaction (*see* **Note 7**).
9. Verified vectors can be transformed in *A. tumefaciens* for subsequent plant transformation.

3.4 Detection of Orthologous Cas9-Induced Mutations in A. thaliana

1. Transform sequence checked vectors with specified sgRNAs into *A. tumefaciens.*
2. Stably transform *A. thaliana* plants via Agrobacterium-mediated transformation (e.g., floral dip transformation as described in [23]).
3. Perform primary transformant selection using GM plates containing 30 μg/mL kanamycin.
4. Propagate at least 20 primary transformants to acquire progeny seeds (T2 generation, *see* **Note 8**).

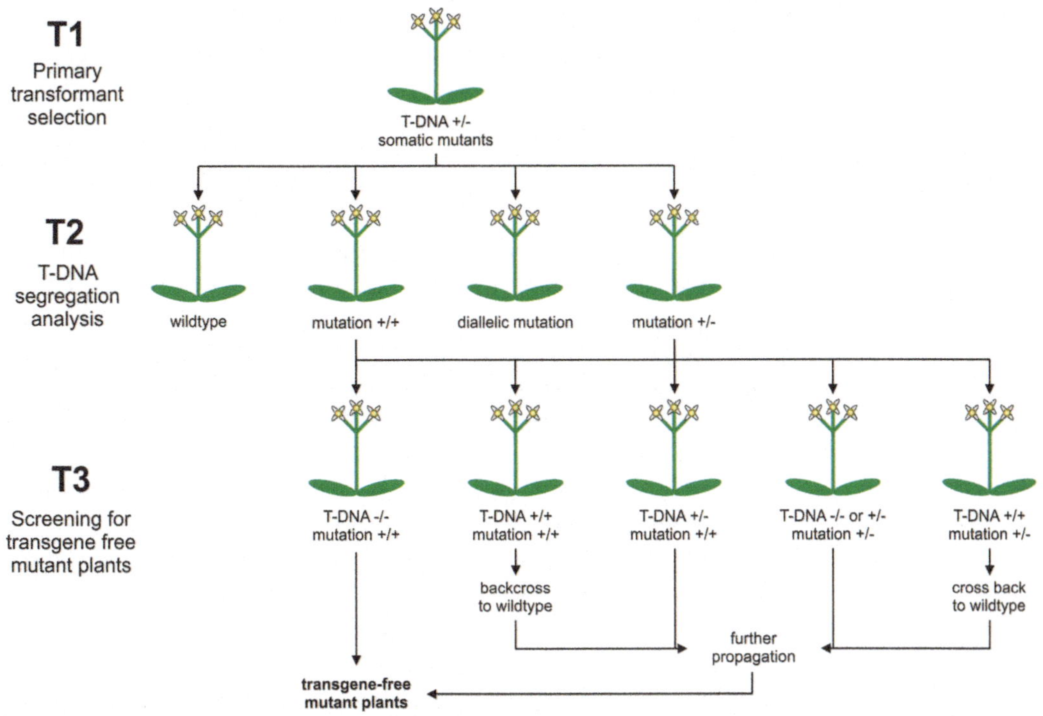

Fig. 3 Overview of the line establishment to detect transgene free mutants. In T1, transformed plants can be identified by selection. In T2, single locus lines for the T-DNA can be identified by their Mendelian segregation pattern. In this generation, heritable mutations can either be heterozygous, homozygous, or diallelic (trans-heterozygous). Already in the T2 generation, transgene-free, homozygously mutated plants might be identified. Otherwise, such plants will be found in the next generation (T3). It is important to be aware that as long as the T-DNA is expressing the nuclease, while a wild-type allele is present in the respective plants, chimerically mutated plants might arise

5. Identify correct Mendelian 3:1 segregating T2 plant lines by sowing the seeds on selection media. Correct segregation single locus lines show 75% antibiotic resistant and 25% sensitive seedlings (*see* Fig. 3).
6. Use 10 correct segregating single locus lines and sow at least 10 seeds each on GM without selection marker or on soil (*see* **Note 9**).
7. Extract DNA of 2-week-old plants. Fast protocols as described in [24] are sufficient.
8. Test your DNA samples for targeted mutagenesis events via restriction digestion or high-resolution melting analysis.
9. Confirm individuals with positive results by Sanger sequencing (*see* **Note 10**).
10. Propagate the plants showing mutations to obtain T3 seeds.
11. Sow these seeds (at least 20 seeds per line) on selection media to identify plant lines that have lost the T-DNA (*see* **Note 11**).

12. Confirm the obtained results in T2 plants by also Sanger sequencing the transgene-free T3 progeny for the presence of the mutation.

4 Notes

1. In case of *A. thaliana* transformation, the floral dip method is the recommended procedure, as previously described [23].
2. The mentioned PAM sequences were tested in *A. thaliana* for induction of targeted mutagenesis events and partially for induction of HR events [13]. For the RNA-guided Cas9 nuclease from *S. aureus*, different mutation patterns were obtained using the two here described PAM sequences. Intriguingly, the targeted mutagenesis data using the NNGGGT PAM showed a higher number of larger deletions while the data for the NNGAA PAM were leading predominantly to single nucleotide insertions. Dependent on the aim of your experiment it is recommendable to choose the PAM advisedly.
3. Of course you can always use PAM sequences as reverse complement. Therefore, your reverse-oligo should directly resemble your target sequence, downstream of the PAM, in combination with AAAC and the corresponding forward-oligo should be designed using the reverse complementary sequence.
4. On the majority, every target sequence can be used as a protospacer, there are only a few constrains you have to take heed of. Make sure that your selected sequence does not contain more than five consecutive Ts, as this can lead to an abortion of sgRNA expression. In case of the integration of more than one sgRNA into your destination vector (multiplexing) check for the absence of *Mlu*I and *Bsu36*I restriction sites.
5. Colony-PCR is an optional step to reduce the number of plasmid isolations you have to perform. For this purpose a robust *Taq*-polymerase is mandatory, that can handle inhibiting substances in the reaction. In our lab, the DreamTaq polymerase (ThermoFisher Scientific) is normally used for this application.
6. Proteinase K treatment is crucial, do not skip this step as it could inhibit the formation of correct clones.
7. Additionally, the vectors can be sequenced by SS61 binding in pea3A-terminator domain.
8. You could also check the T1 primary transformants for nuclease activity via T7 endonuclease assay or restriction digestion, but this is a mandatory step.

9. Testing 100 individual plants each should lead to the detection of a mutant. However, the exact number can vary. By testing more plants, the possibility of detecting the desired kind of mutation increases.
10. Plants carrying heterozygous mutations show double-peaked chromatograms. You can analyze the sequences using Poly Peak Parser [25]. In this case, repeat sequencing in the next generation to obtain homozygously mutated plants.
11. The optimal outcome are transgene-free T3 plants that carry the desired homozygous mutation. If the mutation or the transgene are still heterozygous in T3, the procedure can be repeated in the next generation or more plants could be tested. If no plants with outsegregated T-DNAs can be obtained (T-DNA is homozygous, all seeds germinate on selection medium), it is possible to cross these lines back with wild-type plants.

Acknowledgement

The work on our RNA-guided Cas9 vectors was funded by the European Research Council (Advanced Grant "COMREC").

References

1. Jinek M, Chylinski K, Fonfara I et al (2012) A programmable dual-RNA-guided DNA endonuclease in adaptive bacterial immunity. Science 337(6096):816–821. doi:10.1126/science.1225829
2. Cong RFA, Cox D et al (2013) Multiplex genome engineering using CRISPR/Cas systems. Science 339(6121):819–823. doi:10.1126/science.1231143
3. Mali P, Aach J, Stranges PB et al (2013) CAS9 transcriptional activators for target specificity screening and paired nickases for cooperative genome engineering. Nat Biotechnol 31(9):833–838. doi:10.1038/nbt.2675
4. Mali P, Yang L, Esvelt KM et al (2013) RNA-guided human genome engineering via Cas9. Science 339(6121):823–826. doi:10.1126/science.1232033
5. Schiml S, Puchta H (2016) Revolutionizing plant biology: multiple ways of genome engineering by CRISPR/Cas. Plant Methods 12:8. doi:10.1186/s13007-016-0103-0
6. Friedland AE, Baral R, Singhal P et al (2015) Characterization of Staphylococcus aureus Cas9: a smaller Cas9 for all-in-one adeno-associated virus delivery and paired nickase applications. Genome Biol 16:257. doi:10.1186/s13059-015-0817-8
7. Ran FA, Cong L, Yan WX et al (2015) In vivo genome editing using Staphylococcus aureus Cas9. Nature 520(7546):186–191. doi:10.1038/nature14299
8. Hsu PD, Lander ES, Zhang F (2014) Development and applications of CRISPR-Cas9 for genome engineering. Cell 157(6):1262–1278. doi:10.1016/j.cell.2014.05.010
9. Fonfara I, Le Rhun A, Chylinski K et al (2014) Phylogeny of Cas9 determines functional exchangeability of dual-RNA and Cas9 among orthologous type II CRISPR-Cas systems. Nucleic Acids Res 42(4):2577–2590. doi:10.1093/nar/gkt1074
10. Louwen R, Staals RHJ, Endtz HP et al (2014) The role of CRISPR-Cas systems in virulence of pathogenic bacteria. Microbiol Mol Biol Rev 78(1):74–88. doi:10.1128/MMBR.00039-13
11. Haeussler M, Concordet J (2016) Genome editing with CRISPR-Cas9: can it get any better? J Genet Genomics 43(5):239–250. doi:10.1016/j.jgg.2016.04.008
12. Chylinski K, Le Rhun A, Charpentier E (2013) The tracrRNA and Cas9 families of type II

CRISPR-Cas immunity systems. RNA Biol 10 (5):726–737. doi:10.4161/rna.24321

13. Steinert J, Schiml S, Fauser F et al (2015) Highly efficient heritable plant genome engineering using Cas9 orthologues from Streptococcus thermophilus and Staphylococcus aureus. Plant J 84(6):1295–1305. doi:10.1111/tpj.13078
14. Chen B, Hu J, Almeida R et al (2016) Expanding the CRISPR imaging toolset with Staphylococcus aureus Cas9 for simultaneous imaging of multiple genomic loci. Nucleic Acids Res 44 (8):e75. doi:10.1093/nar/gkv1533
15. Ma H, Naseri A, Reyes-Gutierrez P et al (2015) Multicolor CRISPR labeling of chromosomal loci in human cells. Proc Natl Acad Sci U S A 112(10):3002–3007. doi:10.1073/pnas.1420024112
16. Karvelis T, Gasiunas G, Young J et al (2015) Rapid characterization of CRISPR-Cas9 protospacer adjacent motif sequence elements. Genome Biol 16(1):253. doi:10.1186/s13059-015-0818-7
17. Kaya H, Mikami M, Endo A et al (2016) Highly specific targeted mutagenesis in plants using Staphylococcus aureus Cas9. Sci Rep 6:26871. doi:10.1038/srep26871
18. Puchta H (2015) Using CRISPR/Cas in three dimensions: towards synthetic plant genomes, transcriptomes and epigenomes. Plant J 87 (1):5–15. doi:10.1111/tpj.13100
19. Schiml S, Fauser F, Puchta H (2014) The CRISPR/Cas system can be used as nuclease for in planta gene targeting and as paired nickases for directed mutagenesis in Arabidopsis resulting in heritable progeny. Plant J 80 (6):1139–1150. doi:10.1111/tpj.12704
20. Fauser F, Schiml S, Puchta H (2014) Both CRISPR/Cas-based nucleases and nickases can be used efficiently for genome engineering in Arabidopsis thaliana. Plant J 79 (2):348–359. doi:10.1111/tpj.12554
21. Hajdukiewicz P, Svab Z, Maliga P (1994) The small, versatile pPZP family of Agrobacterium binary vectors for plant transformation. Plant Mol Biol 25(6):989–994
22. Esvelt KM, Mali P, Braff JL et al (2013) Orthogonal Cas9 proteins for RNA-guided gene regulation and editing. Nat Methods 10 (11):1116–1121. doi:10.1038/nmeth.2681
23. Clough SJ, Bent AF (1998) Floral dip: a simplified method for Agrobacterium-mediated transformation of Arabidopsis thaliana. Plant J 16(6):735–743
24. Zhang J, Stewart J (2000) Economical and rapid method for extraction cotton genomic DNA. J Cotton Sci 2000(4):193–201
25. Hill JT, Demarest BL, Bisgrove BW et al (2014) Poly peak parser: method and software for identification of unknown indels using sanger sequencing of polymerase chain reaction products. Dev Dyn 243(12):1632–1636. doi:10.1002/dvdy.24183

Chapter 28

Transgenic Reproductive Cell Ablation

Shai J. Lawit and Mark A. Chamberlin

Abstract

Numerous cell ablation technologies are available and have been used in reproductive tissues, particularly for male tissues and cells. The importance of ablation of reproductive tissues is toward a fundamental understanding reproductive tissue development and fertilization, as well as, in developing sterility lines important to breeding strategies. Here, we describe techniques for developing ablation lines for both male and female reproductive cells. Also discussed are techniques for analysis, quality control, maintenance, and the lessening of pleiotropism in such lines.

Key words Fluorescent reporter, Megagametophyte, Embryo sac, Pollen ablation, Egg cell ablation

1 Introduction

Transgenic cell ablation or targeted killing of a cell by expression of a genome-encoded gene technology uses a cell or tissue-preferred promoter to express a cytotoxic gene. This technique may be useful for both academic and applied pursuits and thus augment or even replace the use of mutants. For instance, some cell types may not be adequately disrupted by available mutants to provide an understanding of their function or developmental roles, such as the antipodal cells of the megagametophyte [1] or neighboring meiotic products for aposporous initial cells [2]. If a cell-preferred promoter is available for such cell types, then cell ablation of the cell type may help define a loss of function phenotype [3]. Such techniques may also better define the temporal limits of cell/tissue requirement for reproductive development [4].

Reproductive cell ablation may be useful to maintain a linked (trans)gene in a perpetually (hemi/hetero)zygous state, with the aid of positive selection. In the case of pollen ablation, the cell ablation prevents paternal gene transmission. Such systems can also be found in practical applications such as hybrid seed production [5–7], production of self-sterile plants [8], to prevent transgene outflow through pollen or seed [9–11], or to reduce the

Anja Schmidt (ed.), *Plant Germline Development: Methods and Protocols*, Methods in Molecular Biology, vol. 1669, DOI 10.1007/978-1-4939-7286-9_28, © Springer Science+Business Media LLC 2017

invasiveness of a species [12]. Another application of reproductive cell ablation has combined both pollen ablation and egg cell ablation [13] to produce an embryo-less population for genetic screening for parthenogenesis [14].

We discuss methods of development for a cell ablation system designed to specifically inactivate cell types of the megagametophyte. Development of such examples can be facilitated by visual tracking of each cell type [13, 15]. Variants of this system can be useful in many tissues and organs to investigate development and function of specific cell types.

2 Materials

2.1 Molecular Constructs and Transgenic Lines

1. Cell-specific promoter sequences for reproductive tissues (*see* Table 1) as in: [13, 15, 21, 23–28].
2. Coding sequences for cellular toxin (such as barnase, ADP-ribosylase, or DNA (Adenosine-N6-)-Methyltransferase

Table 1
A sampling of characterized promoters appropriate for expression of toxic gene products for cell- or tissue-specific ablation

Promoter	Tissue/Cell type	Accession	Plant type	Reference
NtTA29	Tapetum	LOC107824681	Tobacco, oilseed rape	[5, 16]
Zm5126	Anther	I75204	Maize	[17]
NtTA56	Circular cell cluster, stomium, and contiguous connective regions	U57825	Tobacco	[18]
SlLAT52	Pollen	NM_001309250	Tomato	[19]
AtMGH3	Sperm cell	AT1G19890	*Arabidopsis*	[20]
AtSPL	MMC	AT4G27330	*Hieracium*	[2]
AtSDS	Meiocytes	AT1G14750	*Arabidopsis*	[11]
AtDD45	Egg cell	AT2G21740	*Arabidopsis*, Soy	[13, 15, 21]
AtRKD2	Egg cell	AT1G74480	*Arabidopsis*	[13, 22]
ZmRKD2	Egg cell	GRMZM2G071322	Maize, Sorghum	[15]
AtDD31	Synergid	AT1G47470	*Arabidopsis*	[13, 21]
AtDD2	Synergid	AT5G43510	Soy	[15]
AtDD65	Central cell	AT3G10890	*Arabidopsis*	[13, 21]

(DAM)) with intron to prevent bacterial expression (*see* **Notes 1** and **2**).

3. Coding sequence for cognate antitoxin (such as barstar for barnase), if desired (*see* **Notes 3** and **4**).
4. Low to moderate expression constitutive promoter to drive cognate antitoxin.
5. Standard molecular cloning materials [29].
6. Transformation vectors appropriate to the species of interest. In some systems, supertransformation may be required (*see* **Note 5**), thus two transformation vectors with different selective resistance genes may be required.
7. Plant material grown to appropriate developmental stage (*see* Subheading 3.1).

3 Methods

3.1 Transgenic Lines (See Ref. 15)

1. Prepare *Arabidopsis* lines as in Lawit et al. [13].
2. Transform rapeseed lines as in Chen et al. [30].
3. Transform soybean lines as in Li et al. [31].
4. Transform maize lines as in Cho et al. [32].
5. Transform sorghum lines as in Wu et al. [33].

3.2 Male Sterility

1. Anther Tissue Ablation: Multiple approaches are available to disrupt male fertility through anther tissue ablation. Promoters expressing during early anther development, in the tapetum during early microspore development [16, 17], or in the circular cell cluster, stomium, and contiguous connective regions [18] may be used to disrupt anther development, disrupt pollen development, or prevent dehiscence, respectively (*see* Table 1). Lines must be maintained by pollinating from a male fertile source.
2. Pollen Ablation: Many pollen-preferred promoters [34] are available (Fig. 1a, b). Anecdotal evidence suggests that pollen promoters are often very strongly expressing, which may mask their low-level leaky expression in other tissues (*see* **Note 5**; Fig. 2).

3.3 Female Cell Ablation, Maintainer Constructs, and Seed Color Markers

1. Megagametophyte cell ablation requires highly specific cell expression in order to ablate specific cell types. Female-specific promoters that express prior to megagametophyte formation, such as pre-meiotic megaspore mother cell or post-meiotic megaspores, are currently unknown.
2. Transformation of megagametophyte ablation constructs using floral dip requires prior transformation with a maintainer

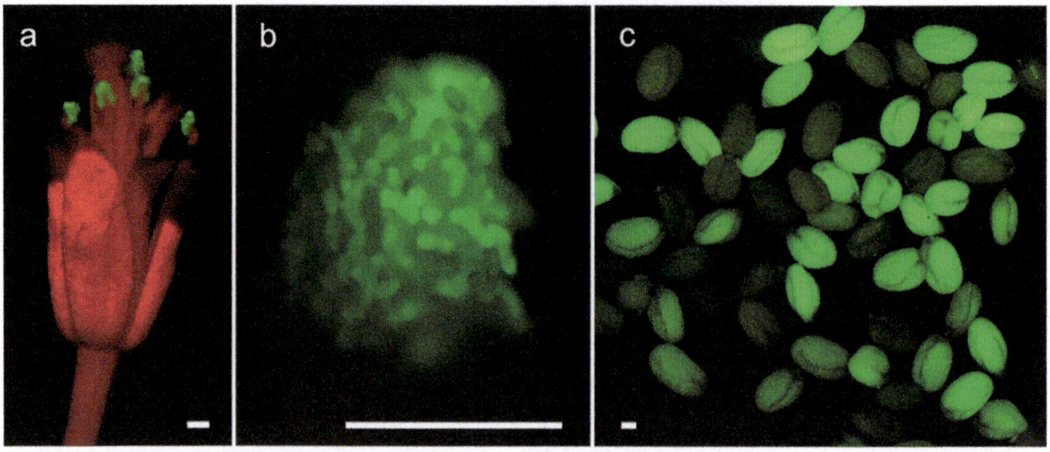

Fig. 1 Fluorescence micrographs of GFP reporters. An *Arabidopsis* line with a *LAT52-like* pollen-preferred promoter driving GFP expression seen in all the anthers of a whole flower (**a**) or just a single anther (**b**). An *Arabidopsis* line with a seed-preferred GFP reporter segregating 1:1 in the anthers due to molecular pyramiding with a pollen ablation cassette (**c**). Bar = 100 μm

Fig. 2 Plants (**a**) and flower clusters (**b**) of *Arabidopsis* expressing barnase from a *LAT52-like* promoter. Although GFP characterization suggested pollen specificity, leaky barnase expression resulted in severe pleiotropism. Flowers notably lacked sepals and petals among other developmental abnormalities

construct (*see* Subheading 3.5 and **Note 6**), or another control mechanism to quench ablation in the T_0 megagametophyte.

3. Depending on the application, it may be advantageous to create a maintainer line that quenches activity of the toxin. Maintainer constructs will typically include a strong constitutive promoter expressing the anti-toxin; a compatible selectable marker; and a method to segregate the maintainer construct from the ablation construct. One example in *Arabidopsis* [13] includes a strong constitutive promoter driving barstar; a mature seed expressed fluorescent reporter as a selectable marker (Fig. 1c); and a pollen ablation construct as a method

to maintain 1:1 segregation of the maintainer in a selfing population. Additional cytological analysis can be performed as described in Chamberlin and Lawit [15].

3.4 Antitoxin Co-expression

1. To ameliorate pleiotropisms with barnase, it is advised to co-express barstar at a low constitutive level. The high affinity of barstar for barnase will allow for nearly a 1:1 inactivation of barnase in cells with leaky barnase expression.
2. Molecular pyramiding of the barnase and barstar cassettes may lessen the time needed for evaluation of the combination, although supertransformation and crossing to bring the cassettes into the same genome may also be used.

3.5 Segregation Analysis

To understand the impact of a cell ablation line on reproductive success, segregation and transmission analysis is key. Gametophytic versus sporophytic affects are critical to consider. Maintainer constructs may alter the expected segregation when effective.

1. Pollen ablation: The effect is gametophytic, therefore segregation of 1:1 is expected in seed from selfed plants (Fig. 3a); however, this is an insensitive assay for transgene transmission. In this case, no transmission is expected from outcrossing and outcrossing to WT is the most sensitive assay for transgenic "escaped" pollen.
2. Male sterile application: The effect is sporophytic, therefore without a maintainer, transmission can only occur through the female. Pollen must be supplied from a fertile source for sexual reproduction. Segregation in seed is expected at 1:1 (Fig. 3b). Sterility should be observed without carrying in pollen to the female. Using a maintainer construct allows for selection of plants homozygous for the male sterile construct (Fig. 3c).
3. Megagametophyte cell ablation: The effect is gametophytic, therefore segregation of 1:1 is expected in seed from selfed plants (Fig. 3d), as well as seed when the plant is used as a male in a cross. No transmission is expected when the plant is crossed as a female (*see* **Note** 7). However, to date only egg cell and synergid cell ablation have been demonstrated [13]. These demonstrations utilized a maintainer construct to select for plants homozygous for the megagametophyte cell ablation construct (Fig. 3e). Although attempts have been made at antipodal cell and central cell ablation (unpublished results), the results have been inconclusive.

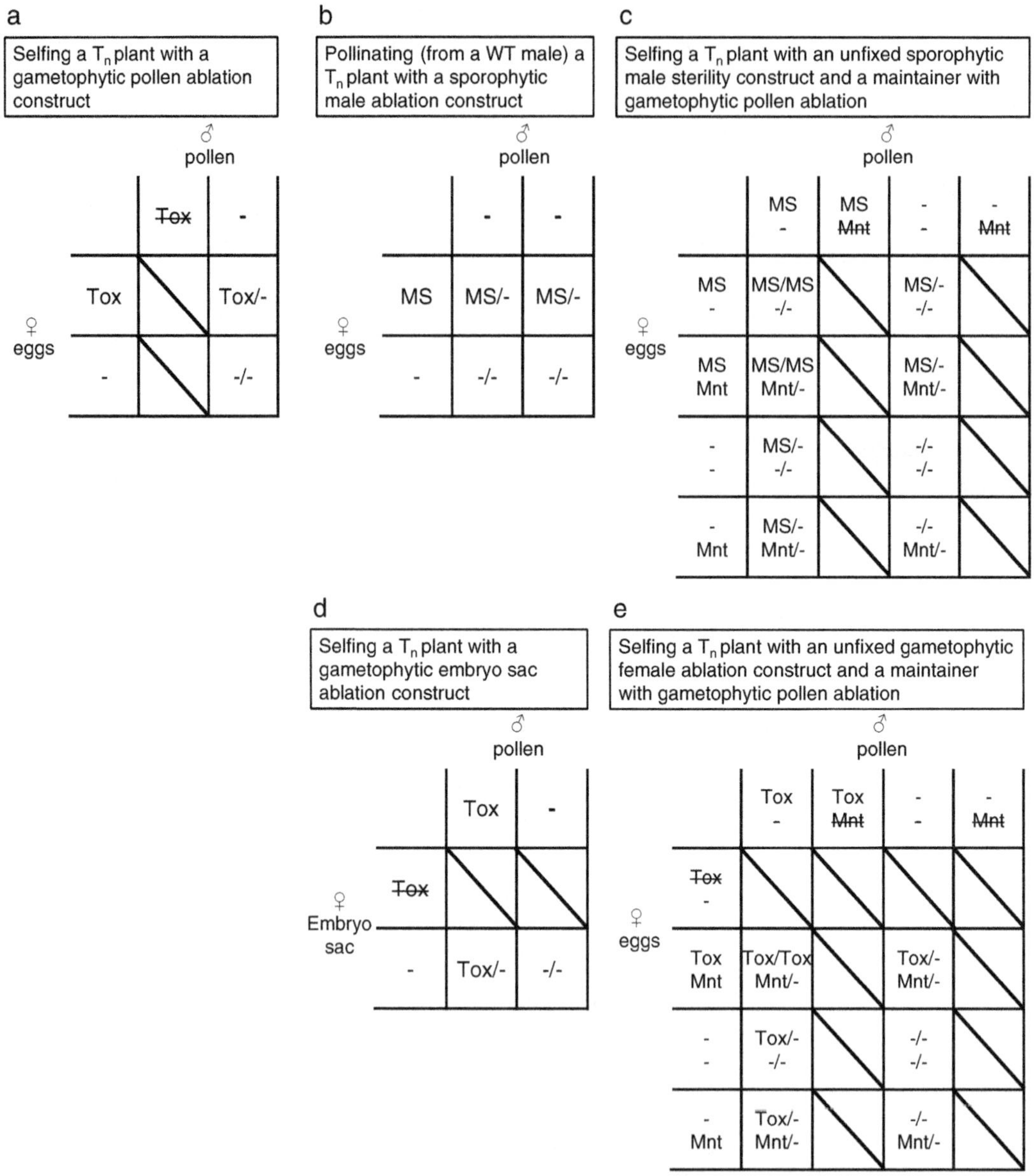

Fig. 3 Punnett squares for a number of scenarios of male or female reproductive cell ablation and how they are affected by maintainer lines. (**a**) A Punnet square for a T_n plant with gametophytically active pollen ablation, segregation is predicted to be 1:1 with wt and hemizygous progeny produced. (**b**) A Punnett squate for a T_n plant with sporophytically active male ablation polinated by a WT plant. Observed segregation is predicted to be 1:1 with wt and hemizygous (male sterile) progeny produced. (**c**) A Punnett square for a T_n plant with a hemizygous sporophytically active male ablation construct which is maintained as a male fertile plant with a maintainer construct at a second locus which is itself kept hemizygous with gametophytic pollen ablation. Segregation is complex, but one in eight progeny may carry homozygous sporophytic male sterility and a hemizygous maintainer. If the maintainer has a genetically linked detectable seed marker, seed sorting may be used to predict male sterile plants and male fertile plants. (**d**) A Punnett squate for a selfed T_n plant with a gametophytically active embryo sac cell ablation construct. (**e**) A Punnett squate for a selfed T_n plant with a gametophytically active embryo sac cell ablation construct with a maintainer construct at a second locus

4 Notes

1. A number of cytotoxic genes and strategies are available. Cytotoxic gene products such as ADP ribosylase/diphtheria toxin A chain (DTA) [35], barnase [5], and even enzymes, not typically considered as a cytotoxin, such as DNA deoxyadenosine methylase [13, 36], have been used.
2. Many toxic gene products have broad cytotoxic specificity. As such, prokaryotic cells used for cloning (*E. coli*) or transformation (*Agrobacterium tumefaciens*) may be adversely affected by the presence of an undisrupted protein toxin coding sequence. To prevent difficulties with cloning, it is recommended to disrupt the coding sequencing with an intron [37] of choice.
3. Toxic gene products, like other toxins, display a dose response in cells. In our experience with the three cytotoxins mentioned in **Note 1**, barnase is the most toxic with ADP ribosylase a close second. In our experience, DAM methylase requires a high dose for complete cell toxicity.
4. Due to dosage responses and promoter specificities, it may be advantageous to choose a toxin, such as barnase [38], which can be mediated by a cognate antitoxin (barstar). Few promoters show the extreme cell specificity required to prevent any pleotropic phenotypes from leaky expression of barnase or ADP ribosylase. Therefore, low-level, constitutive expression of barstar can greatly enhance health in transgenic plants expressing barnase in a cell-preferred fashion. Higher level barstar expression may also be used to create maintainer lines for cell ablation transgenic lines that are otherwise sterile.
5. Use of highly potent toxins such as barnase or ADP ribosylase may lead to pleotropic phenotypes with any leaky expression (Fig. 2). If using a sufficiently high expressing promoter, using a low potency toxin such as DAM methylase is advised.
6. Mature megagametophyte cell ablation in *Arabidopsis*, and other species transformed by floral dip, requires supertransformation. Due to the developmental stage at which *Arabidopsis* floral dip transformation occurs [39], transgenic cytotoxins driven by a megagametophyte cell-specific promoter may

Fig. 3 (Continued) which is itself kept hemizygous with gametophytic pollen ablation. One in six progeny may carry homozygous gametophytic female cell ablation and a hemizygous maintainer. If the maintainer has a genetically linked detectable seed marker, seed sorting may be used to predict cell ablated and female fertile plants. Tox = gametophytic toxin gene; Tox = gametophytic toxin gene eliminating the gamete; - = a null segregant; MS = a male sterility ablation transgene; Mnt = a maintainer transgene with gametophytic pollen ablation; Mnt = a maintainer transgene with gametophytic pollen ablation that has eliminated the pollen within which it has segregated; diagonal lines = no progeny due to parental gamete elimination

become active prior to fertilization or seed formation [13]. Therefore, for *Arabidopsis* it is recommended to *pre*-express a maintainer construct (*see* **Note 4**) prior to cytotoxin transformation into the plant in order to obtain functional transgenic lines with the desired cell ablation activity in the megagametophyte.

7. Significant plasticity in megagametophyte cell development leading to the switching of cell types (e.g., a synergid reprogramming to an egg with the ablation of the original egg) appears to be exaggerated with cell ablation [13] and may thus affect genetic segregation.

Acknowledgements

This work was developed at and supported by DuPont Pioneer. We thank Katherine Thilges, Eric Caswell, and others for technical support of this work.

References

1. Song X, Yuan L, Sundaresan V (2014) Antipodal cells persist through fertilization in the female gametophyte of Arabidopsis. Plant Reprod 27(4):197–203
2. Koltunow AM, Johnson SD, Rodrigues J, Okada T, Hu Y, Tsuchiya T, Wilson S, Fletcher P, Ito K, Suzuki G (2011) Sexual reproduction is the default mode in apomictic Hieracium subgenus Pilosella, in which two dominant loci function to enable apomixis. Plant J 66 (5):890–902
3. Goldberg RB, Sanders PM, Beals TP (1995) A novel cell-ablation strategy for studying plant development. Philos Trans R Soc Lond B Biol Sci 350(1331):5–17
4. Roberts MR, Boyes E, Scott RJ (1995) An investigation of the role of the anther tapetum during microspore development using genetic cell ablation. Sex Plant Reprod 8(5):299–307
5. Mariani C, Beuckeleer MD, Truettner J, Leemans J, Goldberg RB (1990) Induction of male sterility in plants by a chimaeric ribonuclease gene. Nature 347(6295):737
6. Williams ME, Leemans J, Michiels F (1997) Male sterility through recombinant DNA technology. In: Shivanna K, Sawhney V (eds) Pollen biotechnology for crops production and improvement. Cambridge University Press, Cambridge, pp 237–258
7. Wu Y, Fox TW, Trimnell MR, Wang L, Xu R, Cigan AM, Huffman GA, Garnaat CW, Hershey H, Albertsen MC (2016) Development of a novel recessive genetic male sterility system for hybrid seed production in maize and other cross-pollinating crops. Plant Biotechnol J 14(3):1046–1054
8. Thorsness MK, Kandasamy MK, Nasrallah ME, Nasrallah JB (1993) Genetic ablation of floral cells in Arabidopsis. Plant Cell 5(3):253–261
9. Gressel J (2015) Dealing with transgene flow of crop protection traits from crops to their relatives. Pest Manag Sci 71(5):658–667
10. Kobayashi K, Munemura I, Hinata K, Yamamura S (2006) Bisexual sterility conferred by the differential expression of Barnase and Barstar: a simple and efficient method of transgene containment. Plant Cell Rep 25(12):1347–1354
11. Huang J, Smith AR, Zhang T, Zhao D (2016) Creating completely both male and female sterile plants by specifically ablating microspore and megaspore mother cells. Front Plant Sci 7 (30). doi:10.3389/fpls.2016.00030
12. Gardner N, Felsheim R, Smith AG (2009) Production of male-and female-sterile plants through reproductive tissue ablation. J Plant Physiol 166(8):871–881
13. Lawit SJ, Chamberlin MA, Agee A, Caswell ES, Albertsen MC (2013) Transgenic manipulation of plant embryo sacs tracked through cell-type-specific fluorescent markers: cell labeling, cell ablation, and adventitious embryos. Plant Reprod 26(2):125–137

14. Cigan AM, Lawit SJ (2012) Method to screen plants for genetic elements inducing parthenogenesis in plants. US Patent application no. 13/445,276
15. Chamberlin MA, Lawit SJ (2017) Development and observation of mature megagametophyte cell-specific fluorescent markers. In: Schmidt A (ed) Plant Germline Development, vol. 1669, Methods in Molecular Biology. Springer, New York, NY
16. Burgess DG, Ralston EJ, Hanson WG, Heckert M, Ho M, Jenq T, Palys JM, Tang K, Gutterson N (2002) A novel, two-component system for cell lethality and its use in engineering nuclear male-sterility in plants. Plant J 31 (1):113–125
17. Unger E, Betz S, Xu R-j, Cigan AM (2001) Selection and orientation of adjacent genes influences DAM-mediated male sterility in transformed maize. Transgenic Res 10 (5):409–422
18. Beals TP, Goldberg RB (1997) A novel cell ablation strategy blocks tobacco anther dehiscence. Plant Cell 9(9):1527–1545
19. Twell D (1992) Use of a nuclear-targeted β-glucuronidase fusion protein to demonstrate vegetative cell-specific gene expression in developing pollen. Plant J 2(6):887–892
20. Brownfield L, Hafidh S, Borg M, Sidorova A, Mori T, Twell D (2009) A plant germline-specific integrator of sperm specification and cell cycle progression. PLoS Genet 5(3): e1000430
21. Steffen JG, Kang IH, Macfarlane J, Drews GN (2007) Identification of genes expressed in the Arabidopsis female gametophyte. Plant J 51 (2):281–292
22. Koszegi D, Johnston AJ, Rutten T, Czihal A, Altschmied L, Kumlehn J, Wust SE, Kirioukhova O, Gheyselinck J, Grossniklaus U, Baumlein H (2011) Members of the RKD transcription factor family induce an egg cell-like gene expression program. Plant J 67 (2):280–291
23. Johnston A, Meier P, Gheyselinck J, Wuest S, Federer M, Schlagenhauf E, Becker J, Grossniklaus U (2007) Genetic subtraction profiling identifies genes essential for Arabidopsis reproduction and reveals interaction between the female gametophyte and the maternal sporophyte. Genome Biol 8(10):R204
24. Ohnishi T, Takanashi H, Mogi M, Takahashi H, Kikuchi S, Yano K, Okamoto T, Fujita M, Kurata N, Tsutsumi N (2011) Distinct gene expression profiles in egg and synergid cells of rice as revealed by cell type-specific microarrays. Plant Physiol 155(2):881–891
25. Sánchez-León N, Arteaga-Vázquez M, Alvarez-Mejía C, Mendiola-Soto J, Durán-Figueroa N, Rodríguez-Leal D, Rodríguez-Arévalo I, García-Campayo V, García-Aguilar M, Olmedo-Monfil V, Arteaga-Sánchez M, Martínez de la Vega O, Nobuta K, Vemaraju K, Meyers BC, Vielle-Calzada J-P (2012) Transcriptional analysis of the Arabidopsis ovule by massively parallel signature sequencing. J Exp Bot 63(10):3829–3842
26. Wuest SE, Vijverberg K, Schmidt A, Weiss M, Gheyselinck J, Lohr M, Wellmer F, Rahnenführer J, von Mering C, Grossniklaus U (2010) Arabidopsis female gametophyte Gene expression map reveals similarities between plant and animal gametes. Curr Biol 20 (6):506–512
27. Yang W, Jefferson RA, Huttner E, Moore JM, Gagliano WB, Grossniklaus U (2005) An egg apparatus-specific enhancer of Arabidopsis, identified by enhancer detection. Plant Physiol 139(3):1421–1432
28. Yu H-J, Hogan P, Sundaresan V (2005) Analysis of the female gametophyte Transcriptome of Arabidopsis by comparative expression profiling. Plant Physiol 139(4):1853–1869
29. Green M, Sambrook J (2012) Molecular cloning: a laboratory manual. Cold Spring Harbor Laboratory Press, New York
30. Chen W, Tulsieram L (2015) Microprojectile bombardment transformation of Brassica. US Patent 8,993,845
31. Li Z, Xing A, Moon BP, McCardell RP, Mills K, Falco SC (2009) Site-specific integration of transgenes in soybean via recombinase-mediated DNA cassette exchange. Plant Physiol 151(3):1087–1095
32. Cho M-J, Wu E, Kwan J, Yu M, Banh J, Linn W, Anand A, Li Z, TeRonde S, Register JC III (2014) Agrobacterium-mediated high-frequency transformation of an elite commercial maize (Zea mays L.) inbred line. Plant Cell Rep 33(10):1767–1777
33. Wu E, Lenderts B, Glassman K, Berezowska-Kaniewska M, Christensen H, Asmus T, Zhen S, Chu U, Cho M-J, Zhao Z-Y (2014) Optimized Agrobacterium-mediated sorghum transformation protocol and molecular data of transgenic sorghum plants. In Vitro Cell Dev Biol Plant 50(1):9–18
34. Rutley N, Twell D (2015) A decade of pollen transcriptomics. Plant Reprod 28(2):73–89
35. Czako M, An G (1991) Expression of DNA coding for diphtheria toxin chain a is toxic to plant cells. Plant Physiol 95(3):687–692
36. Cigan AM, Albertsen MC (1997) Transgenic plants and DNA comprising anther specific

promoter 5126 and gene to achieve male sterility. US Patent 5,689,051

37. Vancanneyt G, Schmidt R, O'Connor-Sanchez A, Willmitzer L, Rocha-Sosa M (1990) Construction of an intron-containing marker gene: splicing of the intron in transgenic plants and its use in monitoring early events in Agrobacterium-mediated plant transformation. Mol Gen Genet MGG 220(2):245–250

38. Hartley RW (1988) Barnase and barstar. Expression of its cloned inhibitor permits expression of a cloned ribonuclease. J Mol Biol 202(4):913–915

39. Desfeux C, Clough SJ, Bent AF (2000) Female reproductive tissues are the primary target of Agrobacterium-mediated transformation by the Arabidopsis floral-dip method. Plant Physiol 123(3):895–904

Chapter 29

In Vivo Reporters for Protein Half-Life

Pavel Reichman and Nico Dissmeyer

Abstract

Determination of the general capacity of proteolytic activity of a certain cell or tissue type can be crucial for an assessment of various features of an organism's growth and development and also for the optimization of biotechnological applications. Here, we describe the use of chimeric protein stability reporters that can be detected by standard laboratory techniques such as histological staining, selection using selective media or fluorescence microscopy. Dependent on the expression of the reporters due to the promoters applied, cell- and tissue-specific questions can be addressed. Here, we concentrate on methods which can be used for large-scale screening for protein stability changes rather than for detailed protein stability studies.

Key words Protein degradation, Proteostasis, N-end rule, Ubiquitination, Histology

1 Introduction

Protein in vivo stability has been linked to multiple aspects. Structural determinants can be defined on every level of the protein structure ranging from primary to tertiary/quaternary level. The selectivity in proteolysis is governed by generation and recognition of specific sites on substrates, so-called degronslt-degron. In eukaryotes, degrons on short-lived proteins are recognized by ubiquitin (Ub) ligases (N-recognins), which mediate the transfer of the Ub to an internal Lys of the substrate, directing it to the 26S proteasome, where it is being degraded in an ATP-dependent manner. In bacteria, which lack Ub, specific adaptor proteins recognize and deliver protein substrates to the proteasome.

Proteasome's sequence preferences provide a second component to the degradation code to fine-tune protein half-life in vivo [1]. Manipulating the disordered initiation region by the binding of modifier molecules has been shown to be one of the efficient methods to conditionally modulate the proteasomal degradation of specific proteins [2, 3]. Focus of the chapter will be a compilation of degron sequences that can be used to fuse to reporters in order to observe protein turnover in tissues.

Anja Schmidt (ed.), *Plant Germline Development: Methods and Protocols*, Methods in Molecular Biology, vol. 1669, DOI 10.1007/978-1-4939-7286-9_29, © Springer Science+Business Media LLC 2017

1.1 N-End Rule Degrons, N-Degrons

The first degron identified in short-lived proteins is represented by a single N-terminal residue, which is targeted by the N-end rule pathway [4]. A functional N-degron is typically composed of a destabilizing N-terminal residue, internal Lys residue(s) (the site of formation of a polyubiquitin chain), and an unstructured N-terminal extension. Current knowledge about a destabilizing N-terminal residue and the processes leading to their creation is summarized in Fig. 1.

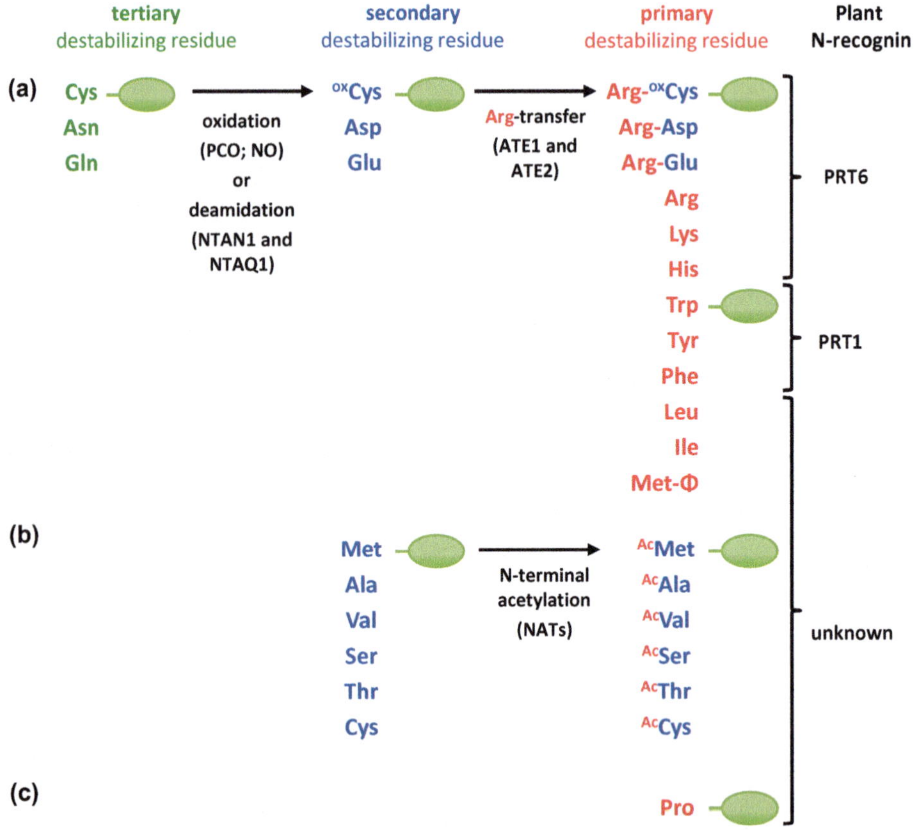

Fig. 1 N-Terminal processing of destabilizing residues. After co-translational removal of the N-terminal Met or after the proteolytic event, cytoplasmic proteins expose N-terminal amino acids which can be enzymatically modified or directly recognized by N-recognins through three different branches. (**a**) Arg/N-end rule recognizes substrates with unmodified basic or hydrophobic residues. N-terminal Cys, tertiary destabilizing residue, can be in presence of oxygen and nitric oxide enzymatically oxidized by PLANT CYSTEINE OXIDASE (PCO). Oxidized N-terminal Cys (^{OX}Cys) is then arginylated by ARGINYL tRNA TRANSFERASEs (ATEs) [5]. N-terminal Asn and Gln can be deamidated by NTAN1 or NTAQ1 enzymes to produce Asp and Glu respectively, which can be directly subjected to arginylation. Charged residues Arg, Lys and His are bound by the N-recognin (ubiquitin E3-ligase) PROTEOLYSIS6 (PRT6) [6] while hydrophobic residues Trp, Tyr, and Phe are recognized by PROTEOLYSIS1 (PRT1) [7–9] and subsequently degraded via the 26S proteasome. It has been also shown in yeast that N-terminal Met followed by one of the large hydrophobic Φ residues Leu, Phe, Tyr, Trp, or Ile (Met-Φ) can be recognized via Arg/N-end rule [10]. There is also recent evidence for a Met-Φ pathway in plants as an artificial Met-Tyr initiated fragment of BIG BROTHER is highly instable [8]. (**b**) Ac/N-end rule targets proteins bearing Nt-acetylated residues. Nt-Met or the newly exposed N-terminal residue can be acetylated (Ac) by N-terminal

All nascent proteins bear N-terminal Met (formyl-Met in prokaryotes). Methionine aminopeptidases (MetAPs) can remove N-terminal Met only if the amino acid at position 2 is small enough, but will not remove N-terminal Met if it is followed by any of the 12 destabilizing residues in the N-end rule [14–18]. The exception to this rule is Cys, whose side chain is small enough to allow the MetAPs cleave of the Met-Cys bond, and which has been shown to be a tertiary destabilizing N-terminal residue in plants and other multicellular eukaryotes [19, 20]). Several approaches to expose the destabilizing N-terminal residue have been developed, deploying different proteases [21], but the method of choice to produce an N-degron in vivo in a protein of interest is to express the protein as a ubiquitin fusion whose junctional residue becomes N-terminal on co-translational removal of the Ub moiety with deubiquitinating enzymes (DUBs). Such a system brings several advantages, the cleavage site is not, with the exception of Pro [22], influenced by the amino acid in P1′ position, and it has been found that expression of a protein as a Ub fusion can augment the protein yield in either bacterial or eukaryotic hosts. The obstacles created by Pro in the P1′ position can be overcome by using mouse Ub-specific protease Unp (or its human homolog Unph) [23].

If expressed in bacterial cells, Ub fusions can be deubiquitylated in vitro using purification system developed by Baker et al. [24, 25], but when coexpressing the DUB, they can be produced in *E. coli* with co-translational cleavage as well [26, 27]. An alternative way to generate N-degron in bacteria is using the ubiquitin-like protein SUMO and its corresponding SUMO-hydrolase instead of Ub and DUB [28].

A useful modification of Ub fusion technique (UFT) has been used for validation of the second *bona fide* plant N-end rule E3 ligase [6]. Here, a mutated stable version of Ubiquitin (UbK48R) was N-terminally tagged with a second reporter to follow fusion protein after cleavage [29, 30]. Since the tagged UbK48R contains mutations preventing the recognition by proteasomal machinery, it can serve as an internal standard. For an overview of the techniques to reveal N-degron, *see* Fig. 2.

1.1.1 e^K and Its Derivatives

The e^K is the earliest portable N-degron described and is still among the most efficacious known and most frequently used. The e^K consist of 45 amino acid sequence derived from a stretch of the *E. coli lac* repressor that was extensively characterized in *Saccharomyces cerevisiae* [4, 35]. It was subsequently shown to work as a

Fig. 1 (Continued) acetyltransferases (NATs) [10, 11]. This branch of N-end rule pathway has been confirmed to function in yeast and mammals, but is predicted to work in plants as well. (**c**) Pro/N-end rule branch recognizing N-terminal Pro as a destabilizing signal has been recently described to be functional in Yeast [12], but sequelogs [13] of the recognition component do exist in animals and plants

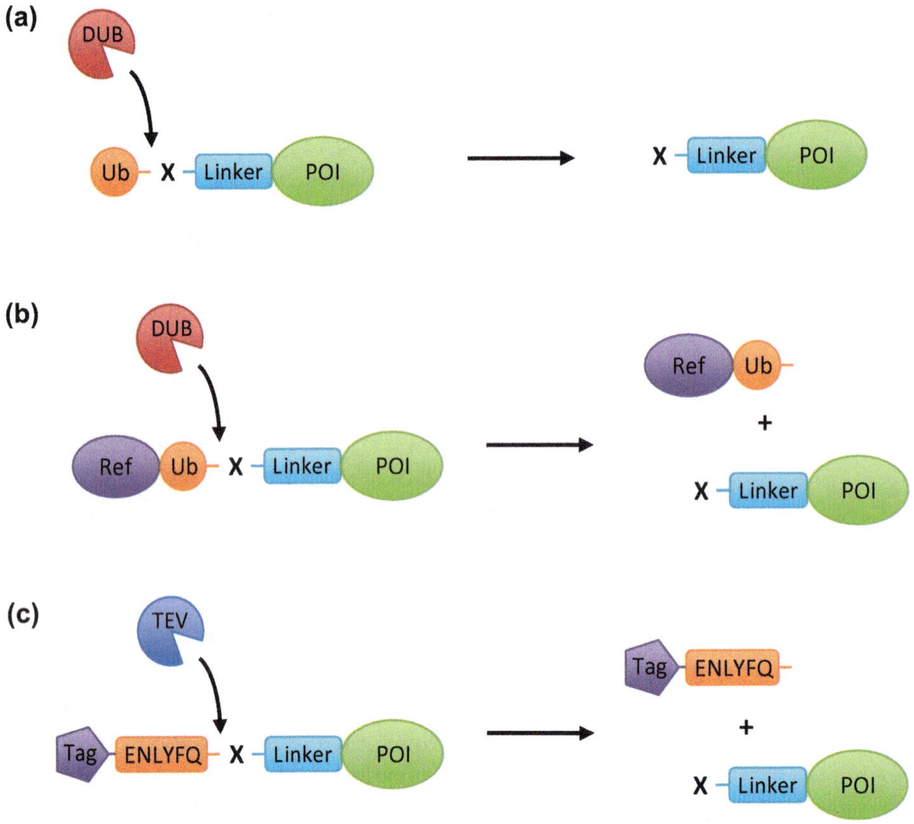

Fig. 2 Basic techniques for generation of N-Degrons. These three examples represent the most common techniques used to generate destabilizing N-terminal amino acid both in vivo and in vitro. X stands for any destabilizing amino acid with the exception of proline (see the main text for more details). Unstructured N-terminal extension (Linker) is necessary for efficient processing by the proteasome [1, 2]. (**a**) The ubiquitin fusion technique (UFT). Fusion of Ub to the protein of interest (POI) is cleaved after the last C-terminal residue of Ub Gly76 by deubiquitylating enzymes (DUBs) [4]. The in vivo cleavage of a Ub fusion is being carried out mostly co-translationally [31]. (**b**) The ubiquitin-protein-reference (UPR). A tripartite fusion containing a reference protein (Ref) fused to the N-termini of the Ub moiety. The C-terminus of the Ub is linked to a protein of interest (POI), same as in UFT [30, 32]. In vivo, this fusion is again co-translationally cleaved by deubiquitylating enzymes (DUBs) at the Ub-X junction, yielding equimolar amounts of the POI and Ref-Ub, the reference protein bearing a C-terminal Ub moiety. Mutated version of Ub is being used to prevent recognition and degradation of Ref-Ub fusion by the proteasome. If Ref-Ub is a long-lived protein, by determining its ratio to POI as a function of time we will gain an insight into relative in vivo metabolic stability of the POI. (**c**) Protease cleavage site incorporation. While this technique, using TEV protease, can be used to generate desired N-termini in vivo (add reference), it is, thanks to availability and stability of the TEV protease, a preferred technique for in vitro studies. TEV is highly sequence-specific to its recognition site ENLYFQIX, where X in the P1′ position can be any destabilizing amino acid with the exception of proline [33, 34]. The recognition sequence ENLYFQ is inserted as a linker between a tag, which usually serves for affinity purification, and the X-POI, where X stands again for the N-terminal amino acid to be exposed after the cleavage

signal for ubiquitin-dependent proteolysis in mammals [22], slime mold *Dictyostelium discoideum* [36], and most importantly, it has been used to create a reporter transgene to identify *Arabidopsis thaliana* mutants in ubiquitin-dependent protein degradation

[37]. Since then, the e^K sequence and its derivatives have been often used to generate artificial substrates and prime them for rapid degradation in vivo.

The stability of the reporter sequence is highly dependent on the N-terminal amino acid but also on the presence and the position of the KRK (Lys-Arg-Lys) motif that is available for ubiquitination in a defined distance from the N-terminal amino acid [35]. Modifying the N-terminal amino acid or the KRK motif alters general properties of the e^K sequence and might be therefore used to fine-tune the system, e.g., the addition of KRK motifs will increase the strength of an N-degron without losing its specificity [32].

In the initial work, Bachmair et al. [4] employed the ubiquitin fusion technique to generate e^K sequence fused to *E. coli* protein β-gal (β-galactosidase), and proved that it can be destabilized in *S. cerevisiae.* Dependent on its N-terminal amino acid, the protein half-life ranged from 3 min to more than 20 h, initially defining the N-end rule. Later they fused e^K, starting with destabilizing N-termini, to dihydrofolate reductase (DHFR) and used this construct in a forward genetics screen in *Arabidopsis thaliana*, which led to identification of mutants deficient in N-terminal protein degradation [37, 38].

Park et al. [39] used 31 amino acid sequence stretch of e^K to lower the basal level of ARD1 (ARrest Defective 1, a subunit of the N-terminal acetyltransferase A) expression and to ensure rapid elimination of ARD1 protein after the expression was shut-off. They used ubiquitin fusion technique again to reveal the N-terminal amino acid Arg or Met of e^K sequence followed by a hemagglutinin epitope for detection. The observed half-life of a destabilized ARD1 was under 3 min.

The adjustability of the e^K degron through positioning of KRK motif makes it especially attractive when creation of proteins with different specific protein half-lives is desirable. Using UFT, the e^K sequence has been exploited to design a family of destabilized cyan fluorescent proteins (CFP) with half-lives of 5, 50, and 70 min to be used as transcriptional reporters [40] or more recently the application of the e^K sequence was reported to generate an N-degron version of GFP (green fluorescent protein)Fluorescent marker with a half-life of 7 min [41]. High brightness and fast maturation time of this GFP variant, combined with the fast turnover rate, are properties well suited for an in vivo temporal transcriptional reporter.

1.1.2 Synthetic N-Degrons

With the aim to produce modified, enzymatically active reporter proteins with shorter in vivo half-lives for transcriptional activity studies, Worley et al. investigated N-end rule in transient tobacco protoplast expression system [42]. Ubiquitin fusion technique has been used again to produce destabilized versions of two common reporter proteins, the cytosolic version of firefly luciferase (cLUC) and the *E. coli* β-glucoronidase (GUS). The degron sequence used as the N-terminal extension of the GUS reporter was X-QIPGYGQSLM and the N-terminal extension sequence of the cLUC was X-QICRSTDLHSGTVGKM, where X stands for the N-termini M, T, R, K, F, for both and L for GUS only. Turnover rate has been measured indirectly by comparing the enzymatic activity of stable and destabilized versions. Strong dependence of activity on the N-terminus has been observed with cLUC, and milder effect with GUS, where basic amino acid at the N-terminus did not reduce the activity at all. L-GUS and F-GUS were reduced to 25% and 36% of the activity observed with M-GUS. A mutated non-cleavable version of a Ubiquitin moiety has been used as a control to produce instable protein.

Same LUC reporter with various N-termini has also been used to study the plant N-end rule by Graciet et al. [43].

1.1.3 Temperature-Inducible N-Degrons and the lt-Degron

Temperature-inducible N-degron has been developed to serve as an alternative approach to generation of heat-sensitive mutants in *S. cerevisiae* [44]. POI is fused to an N-terminal degron which targets the entire protein for proteasome-dependent degradation at a restrictive temperature, providing us with a rapid and reversible depletion or accumulation of POI based on the conditions applied. Temperature-inducible N-degron is composed of an N-terminal Ub that is co-translationally removed (UFT, see above), a destabilizing N-terminal residue after the Ub cleavage site connected by a short linker to the temperature sensitive (ts) mouse dihydrofolate reductase ($DHFR^{ts}$), containing Pro66Leu mutation, that triggers protein degradation at the restrictive (high) temperature; but remains stable at the permissive temperature.

The temperature range of the original td N-degron is 23 °C (permissive temperature) to 37 °C (restrictive temperature) [44, 45], but several variants have been developed until today, differing in the operating temperature range by introducing a point mutations to the DHFR reporter. A thoroughly tested variant for use in plant systems has been published by Faden et al., using a 13 °C as a restrictive temperature and 29 °C as a permissive temperature [46]. They used Phe as a destabilizing residue and to lower the restrictive temperature they generated a temperature-inducible N-degron variant containing a triple-mutated $DHFR^{T39A/P67L/E173D}$ with two additional substitutions (T39A/E173D). Semi-restrictive temperatures resulted in an intermediate phenotype.

1.1.4 nsP4

Another sequence established to be used as a part of the N-degron is the N-terminal part of the *Sindbis virus* RNA polymerase nsP4. It has been described as a native N-end rule substrate [47] and has been used as an adapter in peptide-based pull-downs of endogenous N-end rule E3 ligases in mice [48]. In this work, synthetic 11-mer peptides starting with varying N-termini derived from nsP4 have been used. In the same study, UPR-based fusion was used to co-translationally produce the fDHFRh-UbR48 reference protein and X-nsP4-FLAG test protein (where X is Met, Arg, or Tyr, marked with FLAG tag for detection). Thanks to the similar behavior to the e^K sequence, it is likely that the nsP4 sequence might as well be used as a constitutive N-terminal degradation signal in plants.

1.1.5 Protein Stability Determinants in Chloroplasts

Research of the protein in chloroplasts found that major stability determinants are located in the N-terminal region of proteins and revealed similarities to the bacterial system underlining three factors contributing to the stability of plastid proteins in vivo: first the action of methionine aminopeptidase that exposes the penultimate N-terminal amino acid residue, second N-end rule-like protein degradation pathway, and third additional sequence determinants in the N-terminal region [49]. Apel et al. investigated stability properties of native plastid N-termini and therefore, instead of using cleavable fusion proteins, prepared 20 protein versions by modifying the first codon after the initiator Met. They used GFP as a reporter protein and fused it to eight amino acid long N-terminal domain of a native plastid genome-encoded protein PsbE (α-subunit of cytochrome b559, a component of photosystem II [50]). Using this setup, the measured stability of the reporter protein reflects the substrate specificities of the methionine aminopeptidase revealing the N-terminal amino acid rather than the nature of the penultimate amino acid itself.

1.1.6 Other Native N-Degrons

There have been many proteins reported to follow the N-end rule, making them a possible source of the N-degron sequences to destabilize proteins. One of the promising candidates is a C-terminal cleavage product of *S. cerevisiae* SCC1p (MCD1p, Mitotic Chromosome Determinant). This 298 amino acid long protein fragment with N-terminal arginine has a half-life of ~2 min [51] and has been used as a stability and affinity reporter for the investigation structural basis of *S. cerevisiae* UBR1p substrate binding [52].

From plant field potentially applicable portable N-degrons are now emerging. Examples include the *Arabidopsis thaliana* RPM1 INTERACTING PROTEIN 4 (RIN4) [53] or N-terminal domain of the B-2 subfamily of group VII ETHYLENE RESPONSE FACTORs (ERFs), such as from RAP2.12 (RELATED TO APETALA 2.12) transcription factors [20, 54, 55]. The latter transcription factors are initiated by a highly conserved N-terminal MCGGAIL sequence, removal of N-terminal methionine by MetAPs reveals the tertiary destabilizing residue cysteine in proteins starting with Met-Cys (MC). These neo-N-terminal Cys residues were recently been shown to be specifically oxidized to cys sulfinic acid by PCOs. This partially oxidized version of Cys accepts Arg residues transferred by

ATE1 which in turn are suggested to be recognized by PRT6 of the plant N-end rule pathway [5].

The most recent example of N-end rule substrate described in plants is the central organ size regulator BIG BROTHER [8].

1.2 Cell Cycle Probes

The cell cycle requires precise spatiotemporal regulation of protein abundances. This is being achieved by regulation at transcriptional, translational and posttranslational level together. Cell cycle is being heavily controlled by ubiquitin-mediated proteolysis, providing us with numerous protein determinants which can be turned into portable degrons.

1.2.1 Destruction Boxes of Mitotic Cyclins

Cyclins are interaction partners of Cyclin-dependent kinases (Cdks) in Cdk/cyclin complexes and confer the substrate specificity, whereas the Cdk itself is the catalytically active subunit required for phosphotransfer [56, 57].

Cyclin degradation contributes to the establishment of a phase of low Cdk activity. The anaphase-promoting complex/cyclosome (APC/C), an E3 Ub ligase composed of more than ten subunits, mediates the degradation of so-called destruction box (DB) containing substrates such as cyclin B [58, 59]. The destruction box pathway is well conserved in plants. If an N-terminal fragment of a mitotic cyclin such as CYCLIN B1;1 (CYCB1;1), containing the DB, is fused to otherwise stable proteins such as β-glucuronidase (GUS) [60] or yellow fluorescent protein (YFP) [61] those become recognized as a substrate for APC/C and rapidly degraded by the 26S proteasome in late mitosis. Therefore, DBs are used as markers for G2/M transition.

For cloning, 902 bp of the genomic coding sequence of CYCB1;1 can be used including exons 1–3 and a part of exon 4, being translated into 149 amino acids of which the nine residues "30-RQVLGDIGN-38" (numbering according to the residue position in the translated CYB1;1 moiety) confers the actual DB motif. Chimeric reporter proteins using DB such as DB-GUS or YFP-DB have been used to monitor transcriptional activity and promoter dynamics with high spatiotemporal resolution, e.g., the ProCYCB1;2:DB-GUS shows the typical patchy "salt and pepper" expression pattern of a cell cycle-regulated gene due to protein degradation regulated by the cell cycle [62]. DB fusions allow the in vivo analysis of promoter activity with a high temporal resolution since the carry-over of the reporter to the daughter cells after mitosis is precluded [61–64]. To confirm that the implementation of the DB sequence does not affect the intrinsic properties the fusion protein, a mutated non-degradable variant of the DB motif (from 30-RxxLxxIxN-38 to 30-GxxVxxIxN-38) can be used as a control [65].

1.2.2 Probes for Other Cell Cycle Phases

Cyclin B fragments containing the destruction box are often used to generate G2 biosensors; however, there are more cell-cycle-specific markers which can be used as indicators for the various phases and events of the cell division cycle. Fluorescent markers can be used to discriminate between different cell cycle phases in living cells [66]. Cell-cycle marker proteins of various phases can be created by fusion to fluorescent proteins which even may be combined in cotransfection or cotransformation depending on the choice and spectral properties of the fluorophores. For example, S-phase-specific proteins, such as proliferating cell nuclear antigen 1 and 2 (PCNA1 and PCNA2), histone H4, CDT1, CDC6 or DNA ligase I, can accumulate during DNA replication due to their spatiotemporally specific expression and stability. Therefore, they specifically mark this phase of the cell cycle [67–71].

G1 phases can be marked with a fragment of human DNA helicase B [72] and Cdt1, a proteasomal substrate that accumulates in G1. This can be combined with other markers that are reciprocally stable such as geminin, a substrate of the APC which accumulates in S, G2, and M phases but is degraded in G1 [73].

DNA ligase I in combination with Dnmt1 (DNA methyltransferase) allows to discriminate between all cell-cycle phases due to different nuclear distribution during cell-cycle progression [68]. Also PCNA combined with DNA helicase B was used to monitor cell-cycle progression at the single cell level [74].

1.3 Proteasome Adaptors

This system is based on proteasome adaptors that bind to the proteasome and target protein simultaneously and feed the target to the proteasome for degradation. The adaptor itself escapes degradation and is able to recycle. The target protein is fused to an interaction domain that binds to the proteasome adaptor only in the presence of a chemical inducer of dimerization, and a disordered region that allows the proteasome to initiate degradation efficiently. This system provides a direct way to tune the abundance of a target protein concentration by rapidly depleting proteins from cells without the need for ubiquitination [75].

2 Materials

2.1 General Equipment and Infrastructure

1. Autoclave.
2. Fume hood for seed sterilization.
3. Fridge or cold room for seed stratification.
4. Plant growth chambers with controlled environment.
5. 1.5- and 2-ml microcentrifuge tubes.
6. Flat-tip forceps.
7. Pipets accurately delivering 10, 100, and 1000 μl.
8. Ultra-pure water.

2.2 Cloning and Transformation

1. Standard cloning equipment.
2. Specific oligonucleotides for PCR.
3. Proofreading polymerase such as Pfu.
4. Gateway BP and LR kits (Invitrogen) for recombination cloning.
5. Gateway-compatible Entry vector such as pDONR201 (Invitrogen).
6. *E. coli* DH5α or equivalent cloning host strain.
7. Binary Gateway-compatible Destination vector.
8. PCR templates, e.g., CYCB1;1-DB [61], DHFR of the lt-degron [46] and e^K [37] (*see* **Note 1**).

2.3 Seed Surface Sterilization and Plant Growth

1. Sodium hypochlorite 12% solution (NaOCl).
2. 37% Hydrochloric acid (HCl).
3. Plastic container with a closure.
4. Plant growth medium: 2.2 g/l MS (Murashige and Skoog) salts including vitamins, 1% (w/v) sucrose; 0.8% (w/v) plant agar (Duchefa), 0.5 g/l 2-(N-morpholino)ethanesulfonic (MES), adjust pH to 5.9 using KOH and sterilize by autoclaving for 20 min.

2.4 Quantification of β-Glucoronidase Activity

1. Stock solutions: 1 M NaH_2PO_4 (monobasic sodium phosphate), 1 M Na_2HPO_4 (dibasic sodium phosphate), 500 mM sodium phosphate buffer (pH 7.0, by mixing 39 (v/v) % of 1 M NaH_2PO_4 (monobasic) with 61 (v/v) % 1 M Na_2HPO_4 (dibasic) stock solutions of sodium phosphate, for details, *see* doi:10.1101/pdb.rec8303), 500 mM EDTA, disodium salt (pH 8.0, adjust the pH with NaOH (~20 g of NaOH pellets per l)), 10% (w/v) SDS in H_2O, 10% (v/v) Triton X-100 in H_2O.
2. GUS extraction buffer: 50 mM Na-phosphate, 10 mM EDTA, 0.1% SDS, 0.1% Triton X-100. Add fresh 10 mM β-mercaptoethanol and EDTA-free protease inhibitor cocktail (Roche).
3. GUS assay stock solution: 25 mM 4-methylumbelliferyl β-D-glucuronide (4-MUG, Santa Cruz Biotechnology) in extraction buffer. Prepare fresh and keep in dark at 4 °C.
4. GUS assay solution: 1 mM 4-MUG in extraction buffer. Prepare fresh and keep in dark at 4 °C.
5. 10 mM 4-methylumbelliferone (4-MU, Santa Cruz Biotechnology) stock solution in dimethylsulfoxide (DMSO). Dilute to working solution of 100 μM with DMSO.
6. 4-MU solutions in GUS extraction buffer in order to build a calibration curve: prepare 10, 50, 100, 150, and 200 nM fresh and keep in dark at 4 °C.

7. Protein quantification kit: BCA (bicinchoninic acid) protein assay kit or Bradford reagent.
8. Protein standard: 1 mg/ml bovine serum albumin (BSA).
9. Dewar container and liquid nitrogen.
10. Black 96-well plate (Nunc FluoroNunc; *see* **Note 2**).
11. Plate reading spectrofluorometer for quantification of β-glucoronidase activity and for obtaining the calibration curve, compatible with emission/excitation spectra of 4-methylumbelliferone (excitation maximum at 365 nm and emission maximum at 455 nm) (Tecan M1000).
12. Spectrophotometer set to measure absorbance at 595 nm (for Bradford-based method) or to 562 nm (for BCA-based method).
13. mixer mill (Retsch) with adapters for 2 ml microcentrifuge test tubes and stainless steel beads (Mühlmeier, Nirosta, 3.175 mm) or mortar with pestle.

2.5 Methotrexate and Hygromycin Selection

1. 1 mg/ml methotrexate (MTX) in dimethyl sulfoxide (DMSO, toxic).
2. Hygromycin B (ready-made solution of Duchefa; toxic).
3. Optional: anti-DHFR (human dehydrofolate reductase) antibody: mouse monoclonal DHFR A-4 (sc-74,593, Santa Cruz Biotechnology) in a 1:1000 dilution in standard Tris buffered saline with Tween 20 (TBST) containing 5% (w/v) milk.
4. Optional: anti-hygromycin phosphotransferase (hpt) antibody: mouse monoclonal anti-HPT IgG (STJ10181, St John's Laboratory Ltd) in a 1:1000 dilution in TBST containing 5% (w/v) milk (*see* **Note 3**).

3 Methods

Determining the protein stability is the first step in assessing whether or not the function of a protein is regulated by proteolysis under specific physiological conditions. Traditional and well-established methods to determine a protein's half-life include pulse-chase analysis and cycloheximide blocking. These methods, in detail described by Zhou in 2004 for use with yeast and cultured mammalian cells [76], are still among the most frequently used ones, despite their limitations. Variation of the cycloheximide assays to measure protein degradation in vivo in plants has been recently published by Gilkerson et al. [77]. Two approaches have been described, pooled-seedling degradation assays and single-seedling degradation assays using luciferase fusion protein. Other possible techniques include Fluorescent reporters (standard fluorescent proteins, photo-switchable fluorescent proteins, or fluorescent timers), chemical reporters, or selectable markers.

Depending on the selection of the reporter such as CYCB1;1-DB [61], the lt-degron [46] or e^K [37], various adaptations to the analytical approaches have to be made because they simply rely on different methods (*see* **Note 1**). We concentrate here on description of two methods which can be used for large-scale screening of plant seedlings after, e.g., mutagenesis to assess changes in protein levels.

3.1 General Remarks and Time Considerations

Use standard molecular biology methods for cloning and transformation (*see* **Note 1**). For all below-mentioned methods one should count 2–3 weeks for growing of plants.

1. The time requirement of the GUS staining depends on the strength of the used promoter (approximately 1 day).
2. Selection using methotrexate takes up to 2 weeks after germination and hygromycin selection takes approximately 10 days after germination.

3.2 Seed Surface Sterilization

For cultivation on plates, *Arabidopsis* seeds need to be surface sterilized. We recommend using chlorine dioxide gas sterilization protocol.

1. Place approximately 30 μl of seeds into a 2 ml tube.
2. Create the chlorine dioxide gas (toxic) by mixing 10 ml of sodium hypochlorite with 5 ml of 37% hydrochloric acid (37%) in a closed box under a fume hood (*see* **Note 4**).
3. Expose the seeds to chlorine dioxide gas for at least 45 min but not more than 4 h to prevent damage to the seeds. When sterilizing generation T0 seeds obtained after floral dip transformation of *Arabidopsis*, sterilize for full 3–4 h.

3.3 Plant Growth

1. If not stated otherwise, grow plants on plates with *Arabidopsis* growth media (*see* Subheading 2.3, **step 4**).
2. Transfer the seeds on plates in a density of about 200 seeds per petri dish (9 cm diameter).
3. After stratification of 3–4 days at 4 °C in the dark *Arabidopsis* seeds germinate and plants should be grown under stringently regulated conditions in a phytochamber using specific grow conditions: long day (16/8 h light/dark) 100 $\mu mol/m^2/s$, constant 21 °C and 60% humidity (*see* **Note 5**).

3.4 Quantification of β-Glucoronidase Activity

A method for quantification of β-glucoronidase activity using the fluorogenic substrate 4-MUG has been described by Jefferson et al. [78], modified by Kim et al. [79] and optimized by Faden et al. [46]. In this method the fact that the fluorogenic substrate 4-MUG is being hydrolyzed by GUS to fluorochrome 4-MU is exploited.

Using an excitation wavelength of 365 nm and measuring the emission at 455 nm, the amounts of 4-MU generated can be accurately quantified in a high-throughput plate read-based format.

1. For tissue disruption with a mixer mill place two steel beads in every 2 ml microcentrifuge tube (also *see* **step 5**).
2. Collect 100 mg plant tissue (five whole 2-week old seedlings) per sample and pool them.
3. Add the plant material to the microcentrifuge tube.
4. Snap-freeze in liquid nitrogen. Frozen material can be stored at −80 °C until assaying, *see* also **Note 6**.
5. Precool the mixer mill adapters for 2 ml microcentrifuge tubes in liquid nitrogen, place tubes into the adapters and use the mill at settings of 24 Hz two times for 90 s. Chill the samples in liquid nitrogen in between the two runs. Alternatively, use precooled mortar and pestle (omit **steps 1** and **3**) and grind the tissues properly to a fine powder.
6. Suspend the powder in 200 μl of GUS extraction buffer and store on ice while processing remaining samples.
7. Vortex each sample and centrifuge at 20,000 × *g* for 20 min at 4 °C.
8. Transfer the cleared supernatant (crude extract) to a fresh 1.5-ml microcentrifuge tube and place on ice.
9. Determine total protein concentration of each sample (method is compatible with, e.g., common Bradford- or BCA-based methods). Start with building a calibration curve using protein standard (such as BSA).
10. Transfer 20 μl of each crude extract to individual wells of the 96-well microtiter plate.
11. Prepare a reaction mixture containing 1 mM 4-MUG in GUS extraction buffer (from 25 mM stock solution), and add 980 μl aliquots into the 96-well plate wells. Work with triplicates. Keep the plate on ice. The development of the reaction can be optionally stopped, *see* **Note** 7.
12. Measure the fluorescence of the standard 4-MU solutions using excitation wavelength 365 nm and emission wavelength 455 nm; filter wavelength at 430 nm (*see* **Note 8**) and plot a standard curve of fluorescence against the 4-MU concentrations.
13. Collect the fluorometric data at 37 °C in a spectrophotometer for 60 min, taking one data point every 1–2 min (*see* **Note 9**) and normalize the data to the total protein level. Use the linear region to calculate the regression function.
14. Determine the GUS activity as pmol 4-MU per seconds and μg of total protein.

3.5 Methotrexate Selection

Follow the instructions for plant sterilization and growth (*see* Subheadings 3.2 and 3.3). Work with gloves. When germinated on agar medium containing 0.1 mg/l MTX, *Arabidopsis* seedlings die at a very early developmental stage. Most of them barely project a root from the seed coat. Introduction of a gene coding for murine DHFR with reduced affinity for MTX confers resistance to *Arabidopsis* plants.

We describe here selection using MTX resistance conferred by constructs carrying DHFR, which has been destabilized by introducing N-terminal degron sequences [37, 46]. A counteracting stabilization can be achieved by, e.g., mutation of an E3 ligase targeting the degron, in the two here-referenced cases, this is PRT1.

Rapid degradation of the DHFR should lead to a steady-state level too low for MTX resistance to emerge in transformed plants. Seedlings containing the destabilized *DHFR* transgene conferring resistance to MTX will develop to a slightly later stage than untransformed ones, but roots might not be able to penetrate into the agar containing MTX. This protocol is relevant for lt-degron [46] and e^K [37] containing reporter constructs.

3.6 Hygromycin Selection

Hygromycin selection can be used in two different contexts here. First, the hygromycin resistance gene *hygromycin phosphotransferase* (*hpt*) can be used for detection of the e^K-DHFR reporter cassette because it is contained on one transgene (transfer DNA or T-DNA, *see* **Note 10**). Second, if fused to e^K or DB, hpt can be used as instable reporter protein itself as it is immunogenic and confers antibiotic resistance.

1. Follow the instructions for plant sterilization and growth (*see* Subheadings 3.2 and 3.3). Work with gloves.
2. Use a rather smaller amount of seeds per plate (e.g., about 50), as the density of the germinating seedlings might hamper the identification of resistant individuals.
3. Always prepare a negative control, i.e., plants of the same genetic background (same ecotype) but not transformed with a transgene conferring resistance, and, if possible, use also positive control (with transgene).
4. We recommend using a concentration of 20 μg/ml Hygromycin B. For selection of primary transformants, a concentration of 25 μg/ml can be used. Growth retardation phenotype should become clearly visible 10 after germination.

4 Notes

1. For a detailed description of cloning and plant transformation techniques, we refer to Dissmeyer and Schnittger [80]. The mentioned constructs which are suggested for cloning may of

course also be used in its entity for transformation of, e.g., plant lines of interest or already published germplasm may be used in experiments of choice. From this transgenic material, the constructs can also be introgressed into plant lines of interest and the experimental approach needs to be adapted to the current needs.

2. Although not advisable, it is also possible to use the white 96-well plates, but due to the reflective properties it will result in higher background.
3. Optionally, these two antibodies can be used to track levels of DHFR or HPT when used as protein half-life reporter or when used as selective marker. We have extensively tested antibodies directed against both DHFR and HPT and want to clearly note that the anti-DHFR works very well in our hands compared to the products of other manufacturers and, more importantly, that the anti-HPT antibody was the only one giving reliable results among those few which are available from research laboratories and commercial suppliers. So far, it is tremendously difficult to find published information on this antibody that can also be used to track HPT translated from *hpt*-containing transgenes in transgenic plants.
4. Be sure to use fresh sodium hypochlorite or solution that has been stored in a fridge for not more than one year from purchasing, since the concentration of the active Chlorine drops significantly with the time.
5. Do not exceed light values over 120 μmol/m^2/s to avoid putting additional stress on plants, and carefully check the plant growth conditions.
6. For GUS quantification, the protein extracts can be stored at −80 °C for at least 2 months, but we rather recommend storing the plant material at −80 °C before extracting proteins. When doing the assay for the first time, we highly suggest storing the samples overnight at −80 °C and continuing with the assay on the next day.
7. A solution of 0.2 M sodium carbonate can be added to stop the reaction, as described in the original protocol, but when using time-course kinetics, the reaction does not need to be stopped.
8. We used 5 nm bandwidth both for the excitation and for the emission wavelength (365 and 455 nm respectively). Depending on the instrument possibilities and the strength of the GUS signal, these variables should be subjected to optimization.
9. Depending on the number of samples. When 96 samples are measured simultaneously taking of all 96 data point might take longer than 1 min, thus the time has to be increased.

10. When selecting just for the presence of the transgene, e.g., after transformation of plants, it is possible to use an alternative method described by Harrison et al. "A rapid and robust method of identifying transformed *Arabidopsis thaliana* seedlings" [81] which allows you to identify resistant plants in less than 4 days after germination.

Acknowledgements

This work was supported by a grant for setting up the junior research group of the *ScienceCampus Halle – Plant-based Bioeconomy* to N.D., a Ph.D. scholarship from the DAAD (Deutscher Akademischer Austauschdienst) to P.R., by DI 1794/3-1 of the German Research Foundation (DFG) to N.D., and by grant LSP-TP2-1 of the Research Focus Program "Molecular Biosciences as a Motor for a Knowledge-Based Economy" from the European Regional Development Fund (EFRE) to N.D. Financial support came from the Leibniz Association, the state of Saxony-Anhalt, the DFG Graduate Training Center GRK1026 "Conformational Transitions in Macromolecular Interactions" at Halle, and the Leibniz Institute of Plant Biochemistry (IPB) at Halle, Germany. N.D. lab is the participant of the European Cooperation in Science and Technology (COST) Action BM1307—"European network to integrate research on intracellular proteolysis pathways in health and disease (PROTEOSTASIS)".

References

1. Fishbain S, Inobe T, Israeli E, Chavali S, Yu H, Kago G, Babu MM, Matouschek A (2015) Sequence composition of disordered regions fine-tunes protein half-life. Nat Struct Mol Biol 22(3):214–221. https://doi.org/10.1038/nsmb.2958
2. Takahashi K, Matouschek A, Inobe T (2015) Regulation of proteasomal degradation by modulating proteasomal initiation regions. ACS Chem Biol 10(11):2537–2543. https://doi.org/10.1021/acschembio.5b00554
3. Yu H, Singh Gautam AK, Wilmington SR, Wylie D, Martinez-Fonts K, Kago G, Warburton M, Chavali S, Inobe T, Finkelstein IJ, Babu MM, Matouschek A (2016) Conserved sequence preferences contribute to substrate recognition by the proteasome. J Biol Chem 291(28):14526–14539. https://doi.org/10.1074/jbc.M116.727578
4. Bachmair A, Finley D, Varshavsky A (1986) In vivo half-life of a protein is a function of its amino-terminal residue. Science 234(4773):179–186
5. White MD, Klecker M, Hopkinson R, Weits D, Mueller C, Naumann C, O Neill R, Wickens J, Yang J, Brooks-Bartlett JC, Garman EF, Grossmann TN, Dissmeyer N, Flashman E (2017) Plant cysteine oxidases are dioxygenases that directly enable arginyl transferase-catalyzed arginylation of N-end rule targets. Nat Commun 8:14690. https://doi.org/10.1101/069336
6. Garzon M, Eifler K, Faust A, Scheel H, Hofmann K, Koncz C, Yephremov A, Bachmair A (2007) PRT6/At5g02310 encodes an Arabidopsis ubiquitin ligase of the N-end rule pathway with arginine specificity and is not the CER3 locus. FEBS Lett 581(17):3189–3196. https://doi.org/10.1016/j.febslet.2007.06.005
7. Stary S, Yin XJ, Potuschak T, Schlogelhofer P, Nizhynska V, Bachmair A (2003) PRT1 of Arabidopsis is a ubiquitin protein ligase of the

plant N-end rule pathway with specificity for aromatic amino-terminal residues. Plant Physiol 133(3):1360–1366. https://doi.org/10.1104/pp.103.029272 [pii]

8. Dong H, Dumenil J, Lu FH, Na L, Vanhaeren H, Naumann C, Klecker M, Prior R, Smith C, McKenzie N, Saalbach G, Chen L, Xia T, Gonzalez N, Seguela M, Inze D, Dissmeyer N, Li Y, Bevan MW (2017) Ubiquitylation activates a peptidase that promotes cleavage and destabilization of its activating E3 ligases and diverse growth regulatory proteins to limit cell proliferation in Arabidopsis. Genes Dev. https://doi.org/10.1101/gad.292235.116
9. Mot AC, Prell E, Klecker M, Naumann C, Faden F, Westermann B, Dissmeyer N (2017) Real-time detection of N-end rule-mediated ubiquitination via fluorescently labeled substrate probes. New Phytol. https://doi.org/10.1111/nph.14497
10. Kim HK, Kim RR, Oh JH, Cho H, Varshavsky A, Hwang CS (2014) The N-terminal methionine of cellular proteins as a degradation signal. Cell 156(1–2):158–169. https://doi.org/10.1016/j.cell.2013.11.031
11. Shemorry A, Hwang CS, Varshavsky A (2013) Control of protein quality and stoichiometries by N-terminal acetylation and the N-end rule pathway. Mol Cell 50(4):540–551. https://doi.org/10.1016/j.molcel.2013.03.018
12. Chen SJ, Wu X, Wadas B, Oh JH, Varshavsky A (2017) An N-end rule pathway that recognizes proline and destroys gluconeogenic enzymes. Science 355(6323). https://doi.org/10.1126/science.aal3655
13. Varshavsky A (2004) 'Spalog' and 'sequelog': neutral terms for spatial and sequence similarity. Curr Biol 14(5):R181–R183. https://doi.org/10.1016/j.cub.2004.02.014
14. Bradshaw RA, Brickey WW, Walker KW (1998) N-terminal processing: the methionine aminopeptidase and N alpha-acetyl transferase families. Trends Biochem Sci 23(7):263–267
15. Boissel JP, Kasper TJ, Shah SC, Malone JI, Bunn HF (1985) Amino-terminal processing of proteins: hemoglobin South Florida, a variant with retention of initiator methionine and N alpha-acetylation. Proc Natl Acad Sci U S A 82(24):8448–8452
16. Huang S, Elliott RC, Liu PS, Koduri RK, Blair LC, Bryan KM, Ghosh-Dastidar P, Einarson B, Kendall RL (1987) Specificity of cotranslational amino-terminal processing of proteins in yeast. Biochemistry 26(25):8242–8246. https://doi.org/10.1021/bi00399a033
17. Bonissone S, Gupta N, Romine M, Bradshaw RA, Pevzner PA (2013) N-Terminal protein processing: a comparative proteogenomic analysis. Mol Cell Proteomics 12(1):14–28. https://doi.org/10.1074/mcp.M112.019075
18. Sherman F, Stewart JW, Tsunasawa S (1985) Methionine or not methionine at the beginning of a protein. Bioessays 3(1):27–31. https://doi.org/10.1002/bies.950030108
19. Davydov IV, Varshavsky A (2000) RGS4 is arginylated and degraded by the N-end rule pathway in vitro. J Biol Chem 275(30):22931–22941. https://doi.org/10.1074/jbc.M001605200
20. Gibbs DJ, Lee SC, Isa NM, Gramuglia S, Fukao T, Bassel GW, Correia CS, Corbineau F, Theodoulou FL, Bailey-Serres J, Holdsworth MJ (2011) Homeostatic response to hypoxia is regulated by the N-end rule pathway in plants. Nature 479(7373):415–418. https://doi.org/10.1038/nature10534
21. Naumann C, Mot AC, Dissmeyer N (2016) Generation of artificial N-end rule substrate proteins in vivo and in vitro. Methods Mol Biol 1450:55–83. https://doi.org/10.1007/978-1-4939-3759-2_6
22. Gonda DK, Bachmair A, Wunning I, Tobias JW, Lane WS, Varshavsky A (1989) Universality and structure of the N-end rule. J Biol Chem 264(28):16700–16712
23. Gilchrist CA, Gray DA, Baker RT (1997) A ubiquitin-specific protease that efficiently cleaves the ubiquitin-proline bond. J Biol Chem 272(51):32280–32285
24. Baker RT, Catanzariti AM, Karunasekara Y, Soboleva TA, Sharwood R, Whitney S, Board PG (2005) Using deubiquitylating enzymes as research tools. Methods Enzymol 398:540–554. https://doi.org/10.1016/S0076-6879(05)98044-0
25. Catanzariti AM, Soboleva TA, Jans DA, Board PG, Baker RT (2004) An efficient system for high-level expression and easy purification of authentic recombinant proteins. Protein Sci 13(5):1331–1339. https://doi.org/10.1110/ps.04618904
26. Baker RT, Smith SA, Marano R, McKee J, Board PG (1994) Protein expression using cotranslational fusion and cleavage of ubiquitin. Mutagenesis of the glutathione-binding site of human Pi class glutathione S-transferase. J Biol Chem 269(41):25381–25386
27. Piatkov K, Graciet E, Varshavsky A (2013) Ubiquitin reference technique and its use in ubiquitin-lacking prokaryotes. PLoS One 8(6):e67952. https://doi.org/10.1371/journal.pone.0067952

28. Schmidt R, Zahn R, Bukau B, Mogk A (2009) ClpS is the recognition component for Escherichia coli substrates of the N-end rule degradation pathway. Mol Microbiol 72(2):506–517. https://doi.org/10.1111/j.1365-2958.2009.06666.x
29. Varshavsky A (2005) Ubiquitin fusion technique and related methods. Methods Enzymol 399:777–799. https://doi.org/10.1016/S0076-6879(05)99051-4
30. Levy F, Johnsson N, Rumenapf T, Varshavsky A (1996) Using ubiquitin to follow the metabolic fate of a protein. Proc Natl Acad Sci U S A 93(10):4907–4912
31. Turner GC, Varshavsky A (2000) Detecting and measuring cotranslational protein degradation in vivo. Science 289(5487):2117–2120
32. Suzuki T, Varshavsky A (1999) Degradation signals in the lysine-asparagine sequence space. EMBO J 18(21):6017–6026. https://doi.org/10.1093/emboj/18.21.6017
33. Kapust RB, Tozser J, Copeland TD, Waugh DS (2002) The P1′ specificity of tobacco etch virus protease. Biochem Biophys Res Commun 294 (5):949–955. https://doi.org/10.1016/s0006-291x(02)00574-0
34. Renicke C, Spadaccini R, Taxis C (2013) A tobacco etch virus protease with increased substrate tolerance at the P1′ position. PLoS One 8(6):e67915. https://doi.org/10.1371/journal.pone.0067915
35. Bachmair A, Varshavsky A (1989) The degradation signal in a short-lived protein. Cell 56 (6):1019–1032. 0092-8674(89)90635-1 [pii]
36. Detterbeck S, Morandini P, Wetterauer B, Bachmair A, Fischer K, MacWilliams HK (1994) The 'prespore-like cells' of Dictyostelium have ceased to express a prespore gene: analysis using short-lived beta-galactosidases as reporters. Development 120(10):2847–2855
37. Bachmair A, Becker F, Schell J (1993) Use of a reporter transgene to generate Arabidopsis mutants in ubiquitin-dependent protein degradation. Proc Natl Acad Sci U S A 90 (2):418–421
38. Potuschak T, Stary S, Schlogelhofer P, Becker F, Nejinskaia V, Bachmair A (1998) PRT1 of Arabidopsis thaliana encodes a component of the plant N-end rule pathway. Proc Natl Acad Sci U S A 95(14):7904–7908
39. Park EC, Finley D, Szostak JW (1992) A strategy for the generation of conditional mutations by protein destabilization. Proc Natl Acad Sci U S A 89(4):1249–1252
40. Hackett EA, Esch RK, Maleri S, Errede B (2006) A family of destabilized cyan fluorescent proteins as transcriptional reporters in S. cerevisiae. Yeast 23(5):333–349. https://doi.org/10.1002/yea.1358
41. Houser JR, Ford E, Chatterjea SM, Maleri S, Elston TC, Errede B (2012) An improved short-lived fluorescent protein transcriptional reporter for Saccharomyces cerevisiae. Yeast 29(12):519–530. https://doi.org/10.1002/yea.2932
42. Worley CK, Ling R, Callis J (1998) Engineering in vivo instability of firefly luciferase and Escherichia coli beta-glucuronidase in higher plants using recognition elements from the ubiquitin pathway. Plant Mol Biol 37 (2):337–347
43. Graciet E, Mesiti F, Wellmer F (2010) Structure and evolutionary conservation of the plant N-end rule pathway. Plant J 61(5):741–751. https://doi.org/10.1111/j.1365-313X.2009.04099.x
44. Dohmen RJ, Wu P, Varshavsky A (1994) Heat-inducible degron: a method for constructing temperature-sensitive mutants. Science 263 (5151):1273–1276
45. Dohmen RJ, Varshavsky A (2005) Heat-inducible degron and the making of conditional mutants. Methods Enzymol 399:799–822. https://doi.org/10.1016/S0076-6879(05)99052-6
46. Faden F, Ramezani T, Mielke S, Almudi I, Nairz K, Froehlich MS, Hockendorff J, Brandt W, Hoehenwarter W, Dohmen RJ, Schnittger A, Dissmeyer N (2016) Phenotypes on demand via switchable target protein degradation in multicellular organisms. Nat Commun 7:12202. https://doi.org/10.1038/ncomms12202
47. de Groot RJ, Rumenapf T, Kuhn RJ, Strauss EG, Strauss JH (1991) Sindbis virus RNA polymerase is degraded by the N-end rule pathway. Proc Natl Acad Sci U S A 88 (20):8967 8971
48. Tasaki T, Mulder LC, Iwamatsu A, Lee MJ, Davydov IV, Varshavsky A, Muesing M, Kwon YT (2005) A family of mammalian E3 ubiquitin ligases that contain the UBR box motif and recognize N-degrons. Mol Cell Biol 25 (16):7120–7136. https://doi.org/10.1128/MCB.25.16.7120-7136.2005
49. Apel W, Schulze WX, Bock R (2010) Identification of protein stability determinants in chloroplasts. Plant J 63(4):636–650. https://doi.org/10.1111/j.1365- 313X.2010.04268.x
50. Pakrasi HB, De Ciechi P, Whitmarsh J (1991) Site directed mutagenesis of the heme axial ligands of cytochrome b559 affects the stability of the photosystem II complex. EMBO J 10 (7):1619–1627

51. Rao H, Uhlmann F, Nasmyth K, Varshavsky A (2001) Degradation of a cohesin subunit by the N-end rule pathway is essential for chromosome stability. Nature 410 (6831):955–959. https://doi.org/10.1038/35073627
52. Choi WS, Jeong BC, Joo YJ, Lee MR, Kim J, Eck MJ, Song HK (2010) Structural basis for the recognition of N-end rule substrates by the UBR box of ubiquitin ligases. Nat Struct Mol Biol 17(10):1175–1181. https://doi.org/10.1038/nsmb.1907
53. Takemoto D, Jones DA (2005) Membrane release and destabilization of Arabidopsis RIN4 following cleavage by Pseudomonas syringae AvrRpt2. Mol Plant Microbe Interact 18 (12):1258–1268. https://doi.org/10.1094/MPMI-18-1258
54. Licausi F, Kosmacz M, Weits DA, Giuntoli B, Giorgi FM, Voesenek LA, Perata P, van Dongen JT (2011) Oxygen sensing in plants is mediated by an N-end rule pathway for protein destabilization. Nature 479(7373):419–422. https://doi.org/10.1038/nature10536
55. Gibbs DJ, Md Isa N, Movahedi M, Lozano-Juste J, Mendiondo GM, Berckhan S, Marin de la Rosa N, Vicente Conde J, Sousa Correia C, Pearce SP, Bassel GW, Hamali B, Talloji P, Tome DF, Coego A, Beynon J, Alabadi D, Bachmair A, Leon J, Gray JE, Theodoulou FL, Holdsworth MJ (2014) Nitric oxide sensing in plants is mediated by proteolytic control of group VII ERF transcription factors. Mol Cell 53(3):369–379. https://doi.org/10.1016/j.molcel.2013.12.020
56. Dissmeyer N, Weimer AK, Pusch S, De Schutter K, Alvim Kamei CL, Nowack MK, Novak B, Duan GL, Zhu YG, De Veylder L, Schnittger A (2009) Control of cell proliferation, organ growth, and DNA damage response operate independently of dephosphorylation of the Arabidopsis Cdk1 homolog CDKA;1. Plant Cell 21(11):3641–3654. https://doi.org/10.1105/tpc.109.070417
57. Dissmeyer N, Nowack MK, Pusch S, Stals H, Inze D, Grini PE, Schnittger A (2007) T-loop phosphorylation of Arabidopsis CDKA;1 is required for its function and can be partially substituted by an aspartate residue. Plant Cell 19(3):972–985. https://doi.org/10.1105/tpc.107.050401
58. Fang G, Yu H, Kirschner MW (1999) Control of mitotic transitions by the anaphase-promoting complex. Philos Trans R Soc Lond B Biol Sci 354(1389):1583–1590. https://doi.org/10.1098/rstb.1999.0502
59. Pesin JA, Orr-Weaver TL (2008) Regulation of APC/C activators in mitosis and meiosis. Annu Rev Cell Dev Biol 24:475–499. https://doi.org/10.1146/annurev.cellbio.041408.115949
60. Genschik P, Criqui MC, Parmentier Y, Derevier A, Fleck J (1998) Cell cycle-dependent proteolysis in plants. Identification of the destruction box pathway and metaphase arrest produced by the proteasome inhibitor mg132. Plant Cell 10(12):2063–2076
61. Jakoby MJ, Weinl C, Pusch S, Kuijt SJ, Merkle T, Dissmeyer N, Schnittger A (2006) Analysis of the subcellular localization, function, and proteolytic control of the Arabidopsis cyclin-dependent kinase inhibitor ICK1/KRP1. Plant Physiol 141(4):1293–1305. https://doi.org/10.1104/pp.106.081406
62. Schnittger A, Schobinger U, Stierhof YD, Hulskamp M (2002) Ectopic B-type cyclin expression induces mitotic cycles in endoreduplicating Arabidopsis trichomes. Curr Biol 12 (5):415–420
63. King RW, Deshaies RJ, Peters JM, Kirschner MW (1996) How proteolysis drives the cell cycle. Science 274(5293):1652–1659
64. Bernhardt A, Lechner E, Hano P, Schade V, Dieterle M, Anders M, Dubin MJ, Benvenuto G, Bowler C, Genschik P, Hellmann H (2006) CUL4 associates with DDB1 and DET1 and its downregulation affects diverse aspects of development in Arabidopsis thaliana. Plant J 47 (4):591–603. https://doi.org/10.1111/j.1365-313X.2006.02810.x
65. Weingartner M, Criqui MC, Meszaros T, Binarova P, Schmit AC, Helfer A, Derevier A, Erhardt M, Bogre L, Genschik P (2004) Expression of a nondegradable cyclin B1 affects plant development and leads to endomitosis by inhibiting the formation of a phragmoplast. Plant Cell 16(3):643–657. https://doi.org/10.1105/tpc.020057
66. Kurzawa L, Morris MC (2010) Cell-cycle markers and biosensors. ChemBioChem 11 (8):1037–1047. https://doi.org/10.1002/cbic.200900729
67. Menges M, Hennig L, Gruissem W, Murray JA (2002) Cell cycle-regulated gene expression in Arabidopsis. J Biol Chem 277 (44):41987–42002. https://doi.org/10.1074/jbc.M207570200
68. Easwaran HP, Leonhardt H, Cardoso MC (2005) Cell cycle markers for live cell analyses. Cell Cycle 4(3):453–455. https://doi.org/10.4161/cc.4.3.1525
69. Leonhardt H, Rahn HP, Weinzierl P, Sporbert A, Cremer T, Zink D, Cardoso MC (2000) Dynamics of DNA replication factories in living cells. J Cell Biol 149(2):271–280

70. Essers J, Theil AF, Baldeyron C, van Cappellen WA, Houtsmuller AB, Kanaar R, Vermeulen W (2005) Nuclear dynamics of PCNA in DNA replication and repair. Mol Cell Biol 25 (21):9350–9359. https://doi.org/10.1128/MCB.25.21.9350-9359.2005
71. Kisielewska J, Lu P, Whitaker M (2005) GFP-PCNA as an S-phase marker in embryos during the first and subsequent cell cycles. Biol Cell 97 (3):221–229. https://doi.org/10.1042/BC20040093
72. Gu J, Xia X, Yan P, Liu H, Podust VN, Reynolds AB, Fanning E (2004) Cell cycle-dependent regulation of a human DNA helicase that localizes in DNA damage foci. Mol Biol Cell 15 (7):3320–3332. https://doi.org/10.1091/mbc.E04-03-0227
73. Sakaue-Sawano A, Kurokawa H, Morimura T, Hanyu A, Hama H, Osawa H, Kashiwagi S, Fukami K, Miyata T, Miyoshi H, Imamura T, Ogawa M, Masai H, Miyawaki A (2008) Visualizing spatiotemporal dynamics of multicellular cell-cycle progression. Cell 132(3):487–498. https://doi.org/10.1016/j.cell.2007.12.033
74. Hahn AT, Jones JT, Meyer T (2009) Quantitative analysis of cell cycle phase durations and PC12 differentiation using fluorescent biosensors. Cell Cycle 8(7):1044–1052. https://doi.org/10.4161/cc.8.7.8042
75. Wilmington SR, Matouschek A (2016) An inducible system for rapid degradation of specific cellular proteins using proteasome adaptors. PLoS One 11(4):e0152679. https://doi.org/10.1371/journal.pone.0152679
76. Zhou P (2004) Determining protein half-lives. Methods Mol Biol 284:67–77. https://doi.org/10.1385/1-59259-816-1:067
77. Gilkerson J, Tam R, Zhang A, Dreher K, Callis J (2016) Cycloheximide assays to measure protein degradation in vivo in plants. BioProtocol 6(17). 10.21769/BioProtoc.1919
78. Jefferson RA, Kavanagh TA, Bevan MW (1987) GUS fusions: beta-glucuronidase as a sensitive and versatile gene fusion marker in higher plants. EMBO J 6(13):3901–3907
79. Kim KW, Franceschi VR, Davin LB, Lewis NG (2006) Beta-glucuronidase as reporter gene: advantages and limitations. Methods Mol Biol 323:263–273. https://doi.org/10.1385/1-59745-003-0:263
80. Dissmeyer N, Schnittger A (2011) Use of phospho-site substitutions to analyze the biological relevance of phosphorylation events in regulatory networks. Methods Mol Biol 779:93–138. https://doi.org/10.1007/978-1-61779-264-9_6
81. Harrison SJ, Mott EK, Parsley K, Aspinall S, Gray JC, Cottage A (2006) A rapid and robust method of identifying transformed Arabidopsis thaliana seedlings following floral dip transformation. Plant Methods 2:19. https://doi.org/10.1186/1746-4811-2-19

Chapter 30

Conditional Modulation of Biological Processes by Low-Temperature Degrons

Nico Dissmeyer

Abstract

Conditional modulation of biological processes plays key roles in basic and applied research and in translation. It can be achieved on various levels via a multitude of approaches. One of the directions is manipulating target protein levels and activity by transcriptional, posttranscriptional, translational, and posttranslational control. Because in most of these techniques, the synthesis of the target proteins is adjusted to the needs, they all rely on the specific half-life of the target protein and its turn-over. Therefore, their time-of-action, in direct correlation to the desired reprogramming of molecular phenotypes caused by altering the target levels, is fixed and determined by the naturally inherent properties. We have introduced the low-temperature degron (lt-degron) to various intact multicellular organisms which allows to control target protein levels and therefore function and activity directly on the level of active protein. The lt-degron uses a combination of Ubiquitin-fusion technique linking target protein degradation to the N-end rule pathway of targeted proteolysis coupled with the use of cell- and tissue-specific promoters.

Key words Conditional allele, Conditional genetics, Protein degradation, Proteostasis, N-end rule, Ubiquitination

1 Introduction

The lt-degron can be used to conditionally accumulate, deplete, or tune levels of target proteins in vivo in multicellular organisms and to conditionally complement mutants (Fig. 1) [1]. Especially the latter possibility makes it a prime tool for dissection of gene and protein function in contexts which are otherwise hard or impossible to address, for example, lethal mutations of essential genes or those genes which are expressed in a cell-type or spatiotemporally specific manner [2].

In our experience, the lt-degron worked reliably for a broad number of proteins of interest (POIs) such as transcription factors, kinases, hydrolases, decarboxylases, proteases, highly active enzymes, and very stable proteins in various host systems and multicellular organisms like plants, insects, yeast, and cell cultures [1].

Anja Schmidt (ed.), *Plant Germline Development: Methods and Protocols*, Methods in Molecular Biology, vol. 1669, DOI 10.1007/978-1-4939-7286-9_30, © Springer Science+Business Media LLC 2017

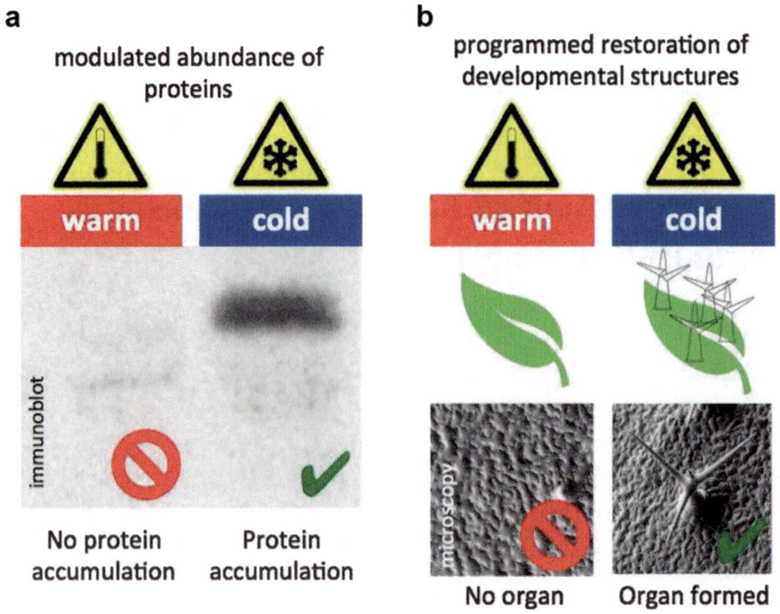

Fig. 1 Application of the lt-degron system. (**a**) Production of proteins in conditional alleles and (**b**) example for the establishment of phenotypes on demand in vivo

However, several precautions are to be taken into account in terms of target selection. The POI needs to be at least transiently located in one of the compartments with an active N-end rule pathway and Ubiquitin proteasome system, i.e., the nucleus and the cytosol.

Depending on the POI, intrinsic factors such as aggregation and oligomerization might play a role in final degradation dynamics and lead to longer protein half-lives. Highly active enzymes, cytotoxic proteins, or those with a high "biological" activity such as certain transcription factors might require special handling and tuning of the lt-degron system as a slow response of the system or a leakiness will lead to a failure of the conditionality. Last but not least, potential interferences of the relatively small N-terminal lt-degron tag which mainly consists of a thermolabile mouse dihydrofolate reductase (DHFR, 22 kDa) and linker sequences, need to be taken into account (Fig. 2). Steric hindrance might – as in each and every fusion protein – cause structural problems such as conformational errors and in the end loss of function of the lt-degron fusion. The temperature might interfere with desired effects such as protein function, enzyme activity, or development of the tissue or cell type of interest which shall be modulated via this technique.

Once a testable system is generated, that is either a stably transformed line or a transiently transformed or transfected tissue or cell culture, the restrictive and permissive temperatures need to

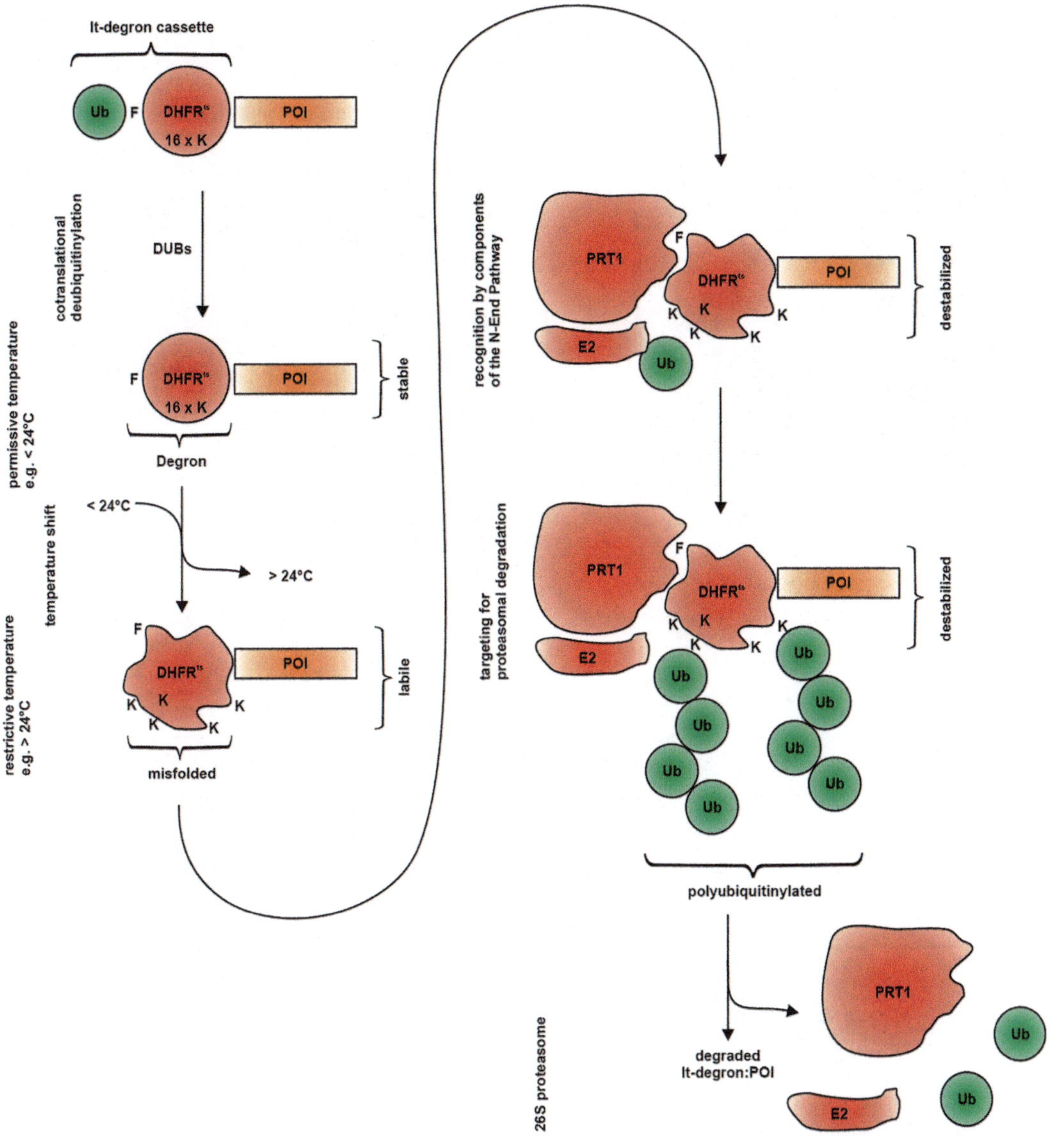

Fig. 2 The lt-degron approach. Low temperature (lt)-degron constructs contain an N-terminal degron cassette starting with a sequence encoding one single ubiquitin (Ub; Ub-fusion technique) followed by mouse dihydrofolate reductase (DHFR) with the first triplet coding for a destabilizing residue (here the bulky hydrophobic amino acid phenylalanine; F). The protein of interest (POI) is fused to the amino-terminal end of the cassette. The lt-degron fusion protein consists of an N-terminal Ub (76 amino acids, 8.5 kDa) and the destabilizing residue preceding the temperature-sensitive (ts) variant of mouse DHFR (22 kDa). DHFR contains 16 Lys (K) residues that can be partially exposed to the surface at restrictive temperatures. The Ub-fusion technique (UFT) is based on the cotranslational deubiquitylation by deubiquitylating enzymes (DUBs) and Ub-specific processing proteases (Ubps) or deubiquitinating enzymes (DUBs) revealing the actual N-degron which is (partially) inactive at permissive temperature (up to 25 °C in Arabidopsis, tobacco, Drosophila an cell cultures). A shift to restrictive temperature (above 25 °C using our low-temperature (lt-)degron) promotes DHFRts flexibility and exposure of internal Lys residues. The destabilizing N-terminal residue (F) is recognized by the N-recognin PROTEOLYSIS1 (PRT1), an E3 ligase, which represents the initial N-end rule pathway recognition component (Mot et al. 2017). E1 Ub-activating enzyme and E2 Ub-conjugating enzyme prime Ub for transfer to the DHFR moiety of the fusion protein. Polyubiquitinylation targets the entire fusion protein for degradation by the 26S proteasome

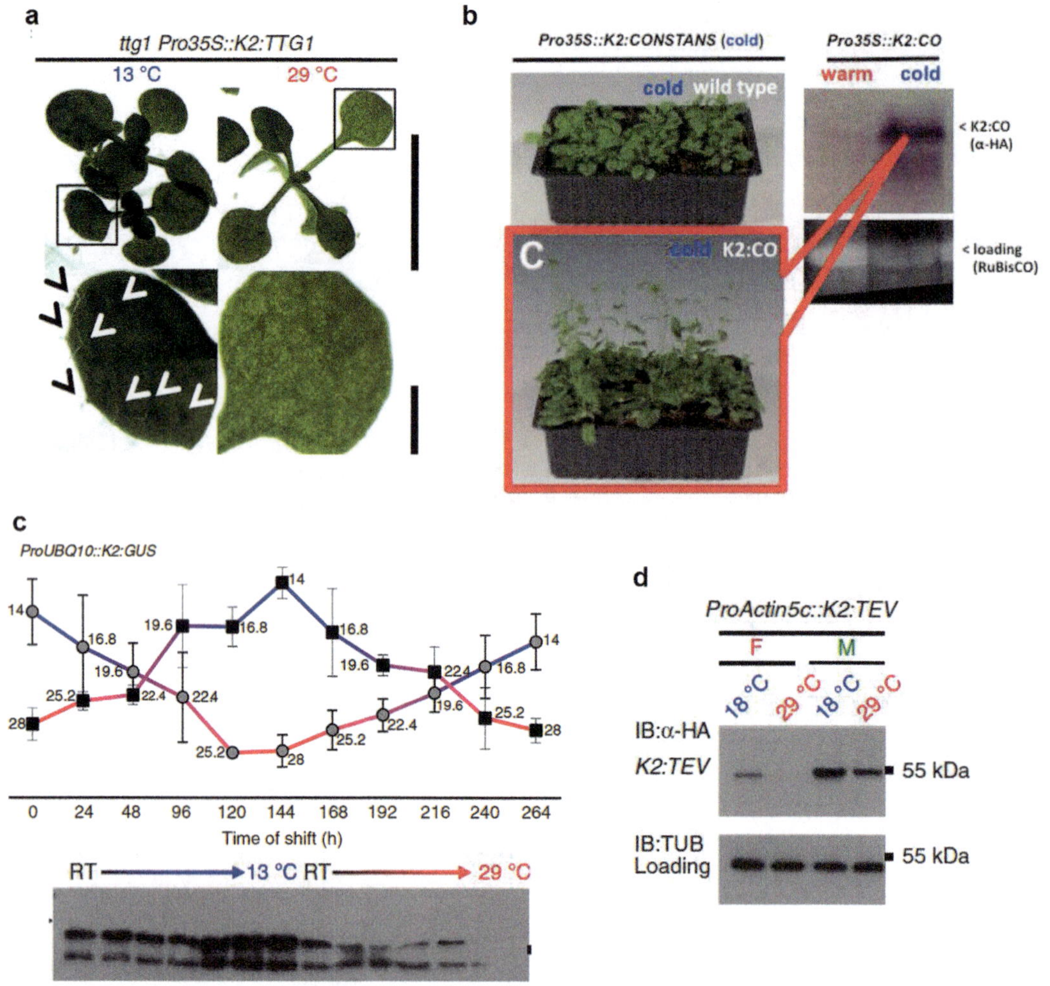

Fig. 3 Examples for use of the lt-degron in *Arabidopsis* and *Drosophila*. (**a**) Conditional protein depletion/accumulation causes developmental phenotypes on demand in vivo and can be used to develop trichome cells in Arabidopsis. (**b**) Switchable flower meristems by conditional accumulation of lt-CONSTANS transcription factor by low temperature. (**c**) In vivo depletion of proteins and inactivation of enzymes as new tool for tuning of activity levels via conditional protein degradation of lt-degron fusion proteins. (**d**) Conditional protein depletion in living *Drosophila* flies. Modified from Faden et al. (2016), *Nature Communications*

be individually determined as they depend on POI, organism or system, and equipment (for examples, *see* Fig. 3) [1].

2 Materials

2.1 Cloning of lt-degron Constructs

1. lt-degron containing plasmid pEN-L1-K2-L2 (Addgene plasmid # 80684; [1]) (*see* **Note 1**).
2. Transformation system (vector) containing promoters of choice, e.g., cell- or tissue-specific or ubiquitously active such

as ProUBQ10 (Addgene plasmid # 79751) or Pro35S (pLEELA; Addgene plasmid # 79752) (*see* **Note 2**).

3. Standard cloning equipment.

2.2 Selection of lt-degron Lines

1. Standard selection media or solutions depending on the transformation vectors used such as Kanamycin or DL-Phosphinotricin (ammonium glufosinate, Basta; Santa Cruz Biotechnology).
2. Anti-HA epitope (hemagglutinin epitope tag) antibody: mouse monoclonal Anti-HA.11 Epitope Tag, Mouse IgG1, Clone: 16B12, MMS-101R, BioLegend or HISS (*see* **Note 3**); in a 1:1000 dilution in standard Tris-buffered saline with Tween 20 (TBST) 5% milk.
3. Optional: anti-DHFR (human dihydrofolate reductase) antibody: mouse monoclonal DHFR A-4 (sc-74593, Santa Cruz Biotechnology); 1:1000 dilution in TBST 5% milk.
4. Optional: Criterion Dodeca Cell (Bio-Rad) (*see* **Note 4**).
5. Optional: methotrexate (Santa Cruz Biotechnology).
6. Optional: Arabidopsis *prt1-1* mutant as control (*see* **Note 5**).

2.3 Application of lt-degron Lines

1. Growth cabinets.
2. Temperature logger (*see* **Note 6**).
3. If applicable, specific needs for selection tailored to targeted POI.

3 Methods

3.1 Cloning of lt-degron Constructs

Standard assembly techniques are used here such as PCR and gateway cloning or construct can be optimized for Modular Cloning such as via Golden Gate Assembly or GoldenBraid (*see* **Note 7**).

3.2 Selection of lt-degron Lines

Selection can be done under restrictive or permissive or both temperatures and in case of stable transformation, the aim is to discard most of the lines where the conditional phenotype does not occur or is difficult to isolate. At the same time, those individuals can be collected with the desired phenotype or phenotypic range to use for further characterization. This phenotype can be investigated on macro- and microscopical as well as on molecular level, including identification of the levels of lt-degron fusion protein and the activity of the POI, if applicable (e.g., when working with a conditional enzyme) (*see* **Note 8**).

1. Start off with choosing the appropriate experimental environment for the selection of lt-degron lines, this can be growth cabinets with highly controlled environments, growth

chambers with controlled environment, or standard growth rooms and greenhouses.

2. If only one growth compartment is available, initially, only one temperature condition will be able for selection; choose between restrictive or permissive temperature depending on the expected read-out of the lt-degron function and the associated feasibility of screening (see above).
3. If possible from the growth and cultivation infrastructure, choose the restrictive or permissive temperatures as "extreme" as possible, that is restrictive as high as 30 °C or permissive as low as 8–10 °C [1].
4. If working with two temperature conditions for selection in parallel, continue as described previously but in two different growth environments.
5. Allow the plants to grow like under standard greenhouse conditions except by keeping them at different temperatures, if this growth regime is not conflicting the function or selection of the chosen POI.
6. If the characteristics of the chosen POI, e.g., when cytotoxic or triggering a developmental phenotype, require specific spatio-temporal induction of the expression of the lt-degron:POI fusion, then this algorithm needs to be adjusted by reducing the time of growth at either restrictive or permissive temperature. If there are no conflicts for growth performance or selectivity are expected, the first round of selection should be done starting after germination to fully grown plants to also observe late phenotypes.
7. Collect, e.g., leaf samples of plants grown at restrictive and/or permissive temperature to compare levels of lt-degron:POI by Western blot or other assays, if other read-outs are expected such as enzymatic activities or fluorescence, *see* Subheading 3.3.2.
8. The restrictive and permissive temperatures should later on be adjusted upon identification of responding lines. Lowering the restrictive and increasing the permissive temperatures will lead to better growth and reduced stress for the specimen which will be beneficial for the experiments and lead to an optimized work and assay flow.

3.3 Application of lt-degron Lines

3.3.1 Developmental Phenotypes on Demand

If a developmental phenotype such as ablation or overproduction of a cell type or target-dependent triggering of a phenotype are envisaged, it could be feasible to plant the identification of functional lines with the bare eye or by using microscopical or histological methods for analysis.

3.3.2 Molecular Phenotypes on Demand

If molecular phenotypes are expected such as enzyme activity or subcellular phenotypes due to structural changes in a cell, it will be difficult at first to identify functional lines with the methods suggested previously in Subheading 3.3.1. These phenotypes simply cannot be easily monitored. We therefore suggest screening on protein levels, i.e., by using large- or high-throughput Western blotting.

1. The easiest way to compare levels of lt-degron:POI is via a side-by-side comparative Western blot using normal or high-throughput gel running and blotting systems, *see* Subheading 2.2.
2. Collect plant material grown, e.g., at two different conditions, here at restrictive versus permissive temperatures, extract total protein and treat as for standard SDS-PAGE and Western blotting.
3. If processing many samples at a time, plan the experiment beforehand carefully to avoid loss of samples, e.g., during loading of multiple gels and make sure to have the capacity to blot all samples either on one semi-dry transfer system or in one wet blotter to exclude artifacts of uneven protein transfer during the blotting procedure. This will inevitably lead to misinterpretation of the results because you will be watching out for strong protein abundance versus low protein abundance in the same experiment. It is advisable to load samples of restrictive and permissive conditions side-by-side to immediately observe potential effects of degradation versus accumulation and identify responding lines.
4. After pre-screening, identified lines need to be verified by subsequent Western blotting and responding standard lines isolated and ideally propagated to a non-segregating homozygous situation.
5. Another possibility to use the lt-degron is in the context of transient expression. If this is preferred, two populations of transformed plants should be risen in parallel under similar general growth conditions under various – at least two – temperatures representing the extreme permissive and restrictive temperatures (*see* **Note 9**).

3.3.3 Controls

Two possible negative controls are conceivable, namely the use of a nondegradable control fusion protein and a mutant background with disturbed N-end rule pathway. For the first option, the codon for the critical neo-N-terminal Phe residue needs to be altered to a Met or Ala. By this, the fusion protein undergoes the same post-translational processing with regard to cleavage of the ubiquitin fusion but cannot be recognized by the E3 ligase PROTEOLYSIS1 (PRT1) [3–7] which is required to initiate lt-degron degradation

[1]. One problem became apparent during the last years when the technique was established in various laboratories. The possible high variability in expression of the lt-degron fusions is directly linked to the copy number in the transformed organism and also to half-life or detectable protein levels and/or active POI. Therefore, in *Arabidopsis*, where N-end rule pathway mutants are available, we strongly recommend to use the *prt1* mutant background introgressed into the lt-degron lines of choice wherever possible. However, in the specific case of the lt-degron, naturally, every single transgenic line or transformation/transfection setup already represents its own control because the responsive temperature is the restrictive one leading to protein degradation and the control condition is the permissive temperature causing protein accumulation.

4 Notes

1. This plasmid is a Gateway-compatible Entry clone containing L1 and L2 recombination sites. It can be used in single and multiple Gateway reactions and as template for PCR fusion constructs.

   The linker at the N-terminus of the lt-degron cassette "K2" may not be altered as functionality might be impaired by exchanging, deleting, or introducing amino acid coding sequences here.
2. These plasmids are Gateway-compatible Destination vectors containing R1 and R2 recombination sites. They can be used in single Gateway reactions to create binary plant expression vectors.
3. This monoclonal antibody is produced from raw ascites fluid derived from the clone 16B12 and available from a multitude of suppliers such as from Covance. There is no need to use the purified version (MMS-101P), it is sufficient to use the "non-purified" MMS-101R.
4. Vertical midi-format electrophoresis cell, includes buffer tank with built-in cooling coil, lid with power cables, 12 gel capacity. Depending on the experience of the experimenter and the comb size, several hundred samples can be run in parallel with this setup.
5. This mutant *Arabidopsis thaliana* line can be used to generate controls that constitutively accumulate the lt-degron:POI fusion protein due to impaired N-end rule-mediated protein degradation. If *prt1* is introgressed in the background of the lt-degron:POI expressing line, the fusion protein cannot be degraded, irrespective of the temperature.

6. Consult, e.g., the website of the German Controlled Environment User Group (www.gceug.de) for further instructions on diverse aspects of controlled plant growth environments from technology to plant growth parameters. This web resource raises awareness for details that need to be considered in instrument and equipment acquisition from measurement devices and growth cabinets to walk-in chambers and whole facilities.
7. Cave! The linker at the N-terminus of the lt-degron cassette "K2" may not be altered as functionality might be impaired by exchanging, deleting, or introducing amino acid coding sequences here.
8. Data logging of growth and selection conditions is very much encouraged. This holds true for both standard greenhouse as well as controlled environment such as in growth chambers or cabinets. According to our experience, only this allows to both identify and document the exact conditions (especially temperatures) the selection and isolation experiments were done at and assist to identify possible interferences by technical problems that might disturb the outcome and influence the selection of lines for the actual follow-up experiments.
9. Extreme permissive and restrictive temperatures should be used for initial screens to quickly eliminate non-functional lines. We made very good experience with temperatures such as 16 °C and 27 °C. But also screens at as low as 8 °C and as high as 30 °C were done in our lab. This all depends strongly on the available infrastructure, the planned time-frame, and the POI.

Acknowledgements

This work was supported by a grant for setting up the junior research group of the *ScienceCampus Halle – Plant-based Bioeconomy* to N.D., a Ph.D. scholarship from the DAAD (Deutscher Akademischer Austauschdienst) to P.R., by DI 1794/3-1 of the German Research Foundation (DFG) to N.D., and by grant LSP-TP2-1 of the Research Focus Program "*Molecular Biosciences as a Motor for a Knowledge-Based Economy*" from the European Regional Development Fund (EFRE) to N.D. Financial support came from the Leibniz Association, the state of Saxony Anhalt, the DFG Graduate Training Center GRK1026 "*Conformational Transitions in Macromolecular Interactions*" at Halle, and the Leibniz Institute of Plant Biochemistry (IPB) at Halle, Germany. N.D. lab is participant of the European Cooperation in Science and Technology (COST) Action BM1307 – "*European network to integrate research on intracellular proteolysis pathways in health and disease (PROTEOSTASIS)*".

References

1. Faden F, Ramezani T, Mielke S, Almudi I, Nairz K, Froehlich MS, Hockendorff J, Brandt W, Hoehenwarter W, Dohmen RJ, Schnittger A, Dissmeyer N (2016) Phenotypes on demand via switchable target protein degradation in multicellular organisms. Nat Commun 7:12202
2. Faden F, Mielke S, Lange D, Dissmeyer N (2014) Generic tools for conditionally altering protein abundance and phenotypes on demand. Biol Chem 395:737–762
3. Bachmair A, Becker F, Schell J (1993) Use of a reporter transgene to generate arabidopsis mutants in ubiquitin-dependent protein degradation. Proc Natl Acad Sci U S A 90:418–421
4. Potuschak T, Stary S, Schlogelhofer P, Becker F, Nejinskaia V, Bachmair A (1998) PRT1 of Arabidopsis thaliana encodes a component of the plant N-end rule pathway. Proc Natl Acad Sci U S A 95:7904–7908
5. Stary S, Yin X, Potuschak T, Schlogelhofer P, Nizhynska V, Bachmair A (2003) PRT1 of Arabidopsis is a ubiquitin protein ligase of the plant N-end rule pathway with specificity for aromatic amino-terminal residues. Plant Physiol 133:1360–1366
6. Dong H, Dumenil J, Lu FH, Na L, Vanhaeren H, Naumann C, Klecker M, Prior R, Smith C, McKenzie N, Saalbach G, Chen L, Xia T, Gonzalez N, Seguela M, Inze D, Dissmeyer N, Li Y, Bevan MW (2017) Ubiquitylation activates a peptidase that promotes cleavage and destabilization of its activating E3 ligases and diverse growth regulatory proteins to limit cell proliferation in Arabidopsis. Genes Dev 31:197–208
7. Mot AC, Prell E, Klecker M, Naumann C, Faden F, Westermann B, Dissmeyer N (2017) Real-time detection of PRT1-mediated ubiquitination via fluorescently labeled substrate probes. New Phytol. doi:10.1111/nph.14497

INDEX

Anja Schmidt (ed.), *Plant Germline Development: Methods and Protocols*, Methods in Molecular Biology, vol. 1669, DOI 10.1007/978-1-4939-7286-9, © Springer Science+Business Media LLC 2017

D

E

F

M

N

O

P

T

U

V

W

Y

Z

Printed by Printforce, the Netherlands